Data Analysis and Graphics
Using R – an Example-based Approach

CAMBRIDGE SERIES IN STATISTICAL AND PROBABILISTIC MATHEMATICS

Editorial Board:

This series of high-quality upper-division textbooks and expository monographs covers all aspects of stochastic applicable mathematics. The topics range from pure and applied statistics to probability theory, operations research, optimization, and mathematical programming. The books contain clear presentations of new developments in the field and also of the state of the art in classical methods. While emphasizing rigorous treatment of theoretical methods, the books also contain applications and discussions of new techniques made possible by advances in computational practice.

Already published
1. *Bootstrap Methods and Their Application*, A.C. Davison and D.V. Hinkley
2. *Markov Chains*, J. Norris
3. *Asymptotic Statistics*, A.W. van der Vaart
4. *Wavelet Methods for Time Series Analysis*, D.B. Percival and A.T. Walden
5. *Bayesian Methods*, T. Leonard and J.S.J. Mu
6. *Empirical Processes in M-Estimation*, S. van de Geer
7. *Numerical Methods of Statistics*, J. Monahan
8. *A User's Guide to Measure-Theoretic Probability*, D. Pollard
9. *The Estimation and Tracking of Frequency*, B.G. Quinn and E.J. Hannan

Data Analysis and Graphics
Using R – an Example-based Approach

John Maindonald

Centre for Bioinformation Science, John Curtin School of Medical Research
and Mathematical Sciences Institute, Australian National University

and

John Braun

Department of Statistical and Actuarial Science University of Western Ontario

CAMBRIDGE
UNIVERSITY PRESS

CAMBRIDGE UNIVERSITY PRESS
Cambridge, New York, Melbourne, Madrid, Cape Town, Singapore, São Paulo

Cambridge University Press
40 West 20th Street, New York, NY 10011-4211, USA

www.cambridge.org
Information on this title: www.cambridge.org/9780521813365

First published 2003
Reprinted 2004, 2005

Printed in the United States of America

A catalogue record for this publication is available from the British Library.

Library of Congress Cataloguing in Publication data

Maindonald, J. H. (John Hilary), 1937–
Data analysis and graphics using R : an example-based approach / John Maindonald and John Braun.
p.   cm. – (Cambridge series in statistical and probabilistic mathematics)
Includes bibliographical references and index.
ISBN 0 521 81336 0
1. Statistical – Data processing.   2. Statistics – Graphic methods – Data processing.   3. R (Computer program
language)   I. Braun, John, 1963–   II. Title.   III. Cambridge series on statistical and probabilistic mathematics.
QA276.4.M245 2003
519.5′0285–dc21   2002031560

ISBN-13   978-0-521-81336-5 hardback
ISBN-10   0-521-81336-0 hardback

It is easy to lie with statistics. It is hard to tell the truth without statistics.

[Andrejs Dunkels]

... technology tends to overwhelm common sense.

[D. A. Freedman]

Contents

Preface

This book is an exposition of statistical methodology that focuses on ideas and concepts, and makes extensive use of graphical presentation. It avoids, as much as possible, the use of mathematical symbolism. It is particularly aimed at scientists who wish to do statistical analyses on their own data, preferably with reference as necessary to professional statistical advice. It is intended to complement more mathematically oriented accounts of statistical methodology. It may be used to give students with a more specialist statistical interest exposure to practical data analysis.

The authors can claim, between them, 40 years of experience in working with researchers from many different backgrounds. Initial drafts of the monograph were constructed from notes that the first author prepared for courses for researchers, first of all at the University of Newcastle (Australia) over 1996–1997, and greatly developed and extended in the course of work in the Statistical Consulting Unit at The Australian National University over 1998–2001. We are grateful to those who have discussed their research with us, brought us their data for analysis, and allowed us to use it in the examples that appear in the present monograph. At least these data will not, as often happens once data have become the basis for a published paper, gather dust in a long-forgotten folder!

We have covered a range of topics that we consider important for many different areas of statistical application. This diversity of sources of examples has benefits, even for those whose interests are in one specific application area. Ideas and applications that are useful in one area often find use elsewhere, even to the extent of stimulating new lines of investigation. We hope that our book will stimulate such cross-fertilization. As is inevitable in a book that has this broad focus, there will be specific areas – perhaps epidemiology, or psychology, or sociology, or ecology – that will regret the omission of some methodologies that they find important.

We use the R system for the computations. The R system implements a dialect of the influential S language that is the basis for the commercial S-PLUS system. It follows S in its close linkage between data analysis and graphics. Its development is the result of a co-operative international effort, bringing together an impressive array of statistical computing expertise. It has quickly gained a wide following, among professionals and non-professionals alike. At the time of writing, R users are restricted, for the most part, to a command line interface. Various forms of graphical user interface will become available in due course.

The R system has an extensive library of packages that offer state-of-the-art-abilities. Many of the analyses that they offer were not, 10 years ago, available in any of the standard packages. What did data analysts do before we had such packages? Basically, they adapted more simplistic (but not necessarily simpler) analyses as best they could. Those whose skills were unequal to the task did unsatisfactory analyses. Those with more adequate skills carried out analyses that, even if not elegant and insightful by current standards, were often adequate. Tools such as are available in R have reduced the need for the adaptations that were formerly necessary. We can often do analyses that better reflect the underlying science. There have been challenging and exciting changes from the methodology that was typically encountered in statistics courses 10 or 15 years ago.

The best any analysis can do is to highlight the information in the data. No amount of statistical or computing technology can be a substitute for good design of data collection, for understanding the context in which data are to be interpreted, or for skill in the use of statistical analysis methodology. Statistical software systems are one of several components of effective data analysis.

The questions that statistical analysis is designed to answer can often be stated simply. This may encourage the layperson to believe that the answers are similarly simple. Often, they are not. Be prepared for unexpected subtleties. Effective statistical analysis requires appropriate skills, beyond those gained from taking one or two undergraduate courses in statistics. There is no good substitute for professional training in modern tools for data analysis, and experience in using those tools with a wide range of data sets. No-one should be embarrassed that they have difficulty with analyses that involve ideas that professional statisticians may take 7 or 8 years of professional training and experience to master.

Influences on the Modern Practice of Statistics

The development of statistics has been motivated by the demands of scientists for a methodology that will extract patterns from their data. The methodology has developed in a synergy with the relevant supporting mathematical theory and, more recently, with computing. This has led to methodologies and supporting theory that are a radical departure from the methodologies of the pre-computer era.

Statistics is a young discipline. Only in the 1920s and 1930s did the modern framework of statistical theory, including ideas of hypothesis testing and estimation, begin to take shape. Different areas of statistical application have taken these ideas up in different ways, some of them starting their own separate streams of statistical tradition. Gigerenzer et al. (1989) examine the history, commenting on the different streams of development that have influenced practice in different research areas.

Separation from the statistical mainstream, and an emphasis on "black box" approaches, have contributed to a widespread exaggerated emphasis on tests of hypotheses, to a neglect of pattern, to the policy of some journal editors of publishing only those studies that show a statistically significant effect, and to an undue focus on the individual study. Anyone who joins the R community can expect to witness, and/or engage in, lively debate that addresses these and related issues. Such debate can help ensure that the demands of scientific rationality do in due course win out over influences from accidents of historical development.

New Tools for Statistical Computing

We have drawn attention to advances in statistical computing methodology. These have led to new powerful tools for exploratory analysis of regression data, for choosing between alternative models, for diagnostic checks, for handling non-linearity, for assessing the predictive power of models, and for graphical presentation. In addition, we have new computing tools that make it straightforward to move data between different systems, to keep a record of calculations, to retrace or adapt earlier calculations, and to edit output and graphics into a form that can be incorporated into published documents.

One can think of an effective statistical analysis package as a workshop (this analogy appears in a simpler form in the JMP Start Statistics Manual (SAS Institute Inc. 1996, p. xiii).). The tools are the statistical and computing abilities that the package provides. The layout of the workshop, the arrangement both of the tools and of the working area, is important. It should be easy to find each tool as it is needed. Tools should float back of their own accord into the right place after use! In other words, we want a workshop where mending the rocking chair is a pleasure!

The workshop analogy is worth pursuing further. Different users have different requirements. A hobbyist workshop will differ from a professional workshop. The hobbyist may have less sophisticated tools, and tools that are easy to use without extensive training or experience. That limits what the hobbyist can do. The professional needs powerful and highly flexible tools, and must be willing to invest time in learning the skills needed to use them.

Good graphical abilities, and good data manipulation abilities, should be a high priority for the hobbyist statistical workshop. Other operations should be reasonably easy to implement when carried out under the instructions of a professional. Professionals also require top rate graphical abilities. The focus is more on flexibility and power, both for graphics and for computation. Ease of use is important, but not at the expense of power and flexibility.

A Note on the R System

The R system implements a dialect of the S language that was developed at AT&T Bell Laboratories by Rick Becker, John Chambers and Allan Wilks. Versions of R are available, at no cost, for 32-bit versions of Microsoft Windows, for Linux and other Unix systems, and for the Macintosh. It is available through the Comprehensive R Archive Network (CRAN). Go to http://cran.r-project.org/, and find the nearest mirror site.

The citation for John Chambers' 1998 Association for Computing Machinery Software award stated that S has "forever altered how people analyze, visualize and manipulate data." The R project enlarges on the ideas and insights that generated the S language. We are grateful to the R Core Development Team, and to the creators of the various R packages, for bringing into being the R system – this marvellous tool for scientific and statistical computing, and for graphical presentation.

Acknowledgements

Many different people have helped us with this project. Winfried Theis (University of Dortmund, Germany) and Detlef Steuer (University of the Federal Armed Forces, Hamburg,

Germany) helped with technical aspects of working with LᴬTEX, with setting up a cvs server to manage the LᴬTEX files, and with helpful comments. Lynne Billard (University of Georgia, USA), Murray Jorgensen (University of Waikato, NZ) and Berwin Turlach (University of Western Australia) gave valuable help in the identification of errors and text that required clarification. Susan Wilson (Australian National University) gave welcome encouragement. Duncan Murdoch (University of Western Ontario) helped set up the *DAAG* package. Thanks also to Cath Lawrence (Australian National University) for her Python program that allowed us to extract the R code, as and when required, from our LᴬTEX files. The failings that remain are, naturally, our responsibility.

There are a large number of people who have helped with the providing of data sets. We give a list, following the list of references for the data near the end of the book. We apologize if there is anyone that we have inadvertently failed to acknowledge. Finally, thanks to David Tranah of Cambridge University Press, for his encouragement and help in bringing the writing of this monograph to fruition.

References

Gigerenzer, G., Swijtink, Z., Porter, T., Daston, L., Beatty, J. & Krüger, L. 1989. *The Empire of Chance.* Cambridge University Press.
SAS Institute Inc. 1996. *JMP Start Statistics.* Duxbury Press, Belmont, CA.

These (and all other) references also appear in the consolidated list of references near the end of the book.

Conventions

Text that is R code, or output from R, is printed in a verbatim text style. For example, in Chapter 1 we will enter data into an R object that we call `austpop`. We will use the `plot()` function to plot these data. The names of R packages, including our own *DAAG* package, are printed in italics.

Starred exercises and sections identify more technical items that can be skipped at a first reading.

Web sites for supplementary information

The DAAG package, the R scripts that we present, and other supplementary information, are available from
  http://cbis.anu.edu/DAAG
  http://www.stats.uwo.ca/DAAG

Solutions to exercises

Solutions to selected exercises are available from the website
  http://www.maths.anu.edu.au/~johnm/r-book.html
  See also www.cambridge.org/0521813360

A Chapter by Chapter Summary

Chapter 1: A Brief Introduction to R

This chapter aims to give enough information on the use of R to get readers started.

Note R's extensive online help facilities. Users who have a basic minimum knowledge of R can often get needed additional information from the help pages as the demand arises. A facility in using the help pages is an important basic skill for R users.

Chapter 2: Style of Data Analysis

Knowing how to explore a set of data upon encountering it for the first time is an important skill. What graphs should one draw?

Different types of graph give different views of the data. Which views are likely to be helpful?

Transformations, especially the logarithmic transformation, may be a necessary preliminary to data analysis.

There is a contrast between exploratory data analysis, where the aim is to allow the data to speak for themselves, and confirmatory analysis (which includes formal estimation and testing), where the form of the analysis should have been largely decided before the data were collected.

Statistical analysis is a form of data summary. It is important to check, as far as this is possible that summarization has captured crucial features of the data. Summary statistics, such as the mean or correlation, should always be accompanied by examination of a relevant graph. For example, the correlation is a useful summary, if at all, only if the relationship between two variables is linear. A scatterplot allows a visual check on linearity.

Chapter 3: Statistical Models

Formal data analyses assume an underlying statistical model, whether or not it is explicitly written down.

Many statistical models have two components: a *signal* (or deterministic) component; and a *noise* (or error) component.

Data from a sample (commonly assumed to be randomly selected) are used to fit the model by estimating the signal component.

The fitted model determines *fitted* or *predicted* values of the signal. The *residuals* (which estimate the noise component) are what remain after subtracting the fitted values from the observed values of the signal.

The normal distribution is widely used as a model for the noise component.

Haphazardly chosen samples should be distinguished from random samples. Inference from haphazardly chosen samples is inevitably hazardous. Self-selected samples are particularly unsatisfactory.

Chapter 4: An Introduction to Formal Inference

Formal analysis of data leads to inferences about the population(s) from which the data were sampled. Statistics that can be computed from given data are used to convey information about otherwise unknown population parameters.

The inferences that are described in this chapter require randomly selected samples from the relevant populations.

A *sampling distribution* describes the theoretical distribution of sample values of a statistic, based on multiple *independent* random samples from the population.

The standard deviation of a sampling distribution has the name *standard error*.

For sufficiently large samples, the normal distribution provides a good approximation to the true sampling distribution of the mean or a difference of means.

A *confidence interval* for a parameter, such as the mean or a difference of means, has the form

$$\text{statistic} \pm t\text{-critical-value} \times \text{standard error}.$$

Such intervals give an assessment of the level of uncertainty when using a sample statistic to estimate a population parameter.

Another viewpoint is that of *hypothesis testing*. Is there sufficient evidence to believe that there is a difference between the means of two different populations?

Checks are essential to determine whether it is plausible that confidence intervals and hypothesis tests are valid. Note however that plausibility is not proof!

Standard chi-squared tests for two-way tables assume that items enter independently into the cells of the table. Even where such a test is not valid, the standardized residuals from the "no association" model can give useful insights.

In the one-way layout, in which there are several independent sets of sample values, one for each of several groups, data structure (e.g. compare treatments with control, or focus on a small number of "interesting" contrasts) helps determine the inferences that are appropriate. In general, it is inappropriate to examine all possible comparisons.

In the one-way layout with quantitative levels, a regression approach is usually appropriate.

Chapter 5: Regression with a Single Predictor

Correlation can be a crude and unduly simplistic summary measure of dependence between two variables. Wherever possible, one should use the richer regression framework to gain deeper insights into relationships between variables.

The line or curve for the regression of a response variable y on a predictor x is different from the line or curve for the regression of x on y. Be aware that the inferred relationship is conditional on the values of the predictor variable.

The model matrix, together with estimated coefficients, allows for calculation of predicted or fitted values and residuals.

Following the calculations, it is good practice to assess the fitted model using standard forms of graphical diagnostics.

Simple alternatives to straight line regression using the data in their raw form are

- transforming x and/or y,
- using polynomial regression,
- fitting a smooth curve.

For size and shape data the allometric model is a good starting point. This model assumes that regression relationships among the logarithms of the size variables are linear.

Chapter 6: Multiple Linear Regression

Scatterplot matrices may provide useful insight, prior to fitting a regression model.

Following the fitting of a regression, one should examine relevant diagnostic plots.

Each regression coefficient estimates the effect of changes in the corresponding explanatory variable when other explanatory variables are held constant.

The use of a different set of explanatory variables may lead to large changes in the coefficients for those variables that are in both models.

Selective influences in the data collection can have a large effect on the fitted regression relationship.

For comparing alternative models, the AIC or equivalent statistic (including Mallows C_p) can be useful. The R^2 statistic has limited usefulness.

If the effect of variable selection is ignored, the estimate of predictive power can be grossly inflated.

When regression models are fitted to observational data, and especially if there are a number of explanatory variables, estimated regression coefficients can give misleading indications of the effects of those individual variables.

The most useful test of predictive power comes from determining the predictive accuracy that can be expected from a new data set.

Cross-validation is a powerful and widely applicable method that can be used for assessing the expected predictive accuracy in a new sample.

Chapter 7: Exploiting the Linear Model Framework

In the study of regression relationships, there are many more possibilities than regression lines! If a line is adequate, use that. But one is not limited to lines!

A common way to handle qualitative factors in linear models is to make the initial level the baseline, with estimates for other levels estimated as offsets from this baseline.

Polynomials of degree n can be handled by introducing into the model matrix, in addition to a column of values of x, columns corresponding to $x^2, x^3, \ldots, x^n$. Typically, $n = 2, 3$ or 4.

Multiple lines are fitted as an interaction between the variable and a factor with as many levels as there are different lines.

Scatterplot smoothing, and smoothing terms in multiple linear models, can also be handled within the linear model framework.

Chapter 8: Logistic Regression and Other Generalized Linear Models

Generalized linear models (GLMs) are an extension of linear models, in which a function of the expectation of the response variable y is expressed as a linear model. A further generalization is that y may have a binomial or Poisson or other non-normal distribution.

Common important GLMs are the logistic model and the Poisson regression model.

Survival analysis may be seen as a further specific extension of the GLM framework.

Chapter 9: Multi-level Models, Time Series and Repeated Measures

In a multi-level model, the random component possesses structure; it is a sum of distinct error terms.

Multi-level models that exhibit suitable balance have traditionally been analyzed within an analysis of variance framework. Unbalanced multi-level designs require the more general multi-level modeling methodology.

Observations taken over time often exhibit time-based dependence. Observations that are close together in time may be more highly correlated than those that are widely separated. The autocorrelation function can be used to assess levels of serial correlation in time series.

Repeated measures models have measurements on the same individuals at multiple points in time and/or space. They typically require the modeling of a correlation structure similar to that employed in analyzing time series.

Chapter 10: Tree-based Classification and Regression

Tree-based models make very weak assumptions about the form of the classification or regression model. They make limited use of the ordering properties of continuous or ordinal explanatory variables. They are unsuitable for use with small data sets.

Tree-based models can be an effective tool for analyzing data that are non-linear and/or involve complex interactions.

The decision trees that tree-based analyses generate may be complex, giving limited insight into model predictions.

Cross-validation, and the use of training and test sets, are essential tools both for choosing the size of the tree and for assessing expected accuracy on a new data set.

Chapter 11: Multivariate Data Exploration and Discrimination

Principal components analysis is an important multivariate exploratory data analysis tool.

Examples are presented of the use of two alternative discrimination methods – logistic regression including multivariate logistic regression, and linear discriminant analysis.

Both principal components analysis, and discriminant analysis, allow the calculation of scores, which are values of the principal components or discriminant functions, calculated observation by observation. The scores may themselves be used as variables in, e.g., a regression analysis.

Chapter 12: The R System – Additional Topics

This final chapter gives pointers to some of the further capabilities of R. It hints at the marvellous power and flexibility that are available to those who extend their skills in the use of R beyond the basic topics that we have treated. The information in this chapter is intended, also, for use as a reference in connection with the computations of earlier chapters.

1

A Brief Introduction to R

This first chapter is intended to introduce readers to the basics of R. It should provide an adequate basis for running the calculations that are described in later chapters.

In later chapters, the R commands that handle the calculations are, mostly, confined to footnotes. Sections are included at the ends of several of the chapters that give further information on the relevant features in R. Most of the R commands will run without change in S-PLUS.

1.1 A Short R Session

1.1.1 R *must be installed!*

An up-to-date version of R may be downloaded from http://cran.r-project.org/ or from the nearest mirror site. Installation instructions are provided at the web site for installing R in Windows, Unix, Linux, and various versions of the Macintosh operating system. Various contributed packages are now a part of the standard R distribution, but a number are not; any of these may be installed as required. Data sets that are mentioned in this book have been collected into a package that we have called *DAAG*. This is available from the web pages http://cbis.anu.edu.au/DAAG and http://www.stats.uwo.ca/DAAG.

1.1.2 *Using the console (or command line) window*

The command line prompt (>) is an invitation to start typing in commands or expressions. R evaluates and prints out the result of any expression that is typed in at the command line in the console window (multiple commands may appear on the one line, with the semicolon (;) as the separator). This allows the use of R as a calculator. For example, type in 2+2 and press the **Enter** key. Here is what appears on the screen:

```
> 2+2
[1] 4
>
```

The first element is labeled [1] even when, as here, there is just one element! The > indicates that R is ready for another command.

In a sense this chapter, and much of the rest of the book, is a discussion of what is possible by typing in statements at the command line. Practice in the evaluation of arithmetic

Table 1.1: *The contents of the file* ACTpop.txt.

Year	ACT
1917	3
1927	8
1937	11
1947	17
1957	38
1967	103
1977	214
1987	265
1997	310

expressions will help develop the needed conceptual and keyboard skills. Here are simple examples:

```
> 2*3*4*5          # * denotes 'multiply'
[1] 120
> sqrt(10)         # the square root of 10
[1] 3.162278
> pi               # R knows about pi
[1] 3.141593
> 2*pi*6378        # Circumference of Earth at Equator (km);
                   # radius is 6378 km
[1] 40074.16
```

Anything that follows a # on the command line is taken as a comment and ignored by R.

There is also a continuation prompt that appears when, following a carriage return, the command is still not complete. By default, the continuation prompt is + (in this book we will omit both the prompt (>) and the continuation prompt (+), whenever command line statements are given separately from output).

1.1.3 Reading data from a file

Our first goal is to read a text file into R using the read.table() function. Table 1.1 displays population in thousands, for Australian Capital Territory (ACT) at various times since 1917. The command that follows assumes that the reader has entered the contents of Table 1.1 into a text file called ACTpop.txt.

When an R session is started, it has a working directory where, by default, it looks for any files that are requested. The following statement will read in the data from a file that is in the working directory (the working directory can be changed during the course of an R session; see Subsection 1.3.2):

```
ACTpop <- read.table("ACTpop.txt", header=TRUE)
```

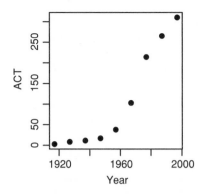

Figure 1.1: ACT population, in thousands, at various times between 1917 and 1997.

This reads in the data, and stores them in the data frame ACTpop. The <- is a left angle bracket (<) followed by a minus sign (-). It means "the values on the right are assigned to the name on the left". Note the use of header=TRUE to ensure that R uses the first line to get header information for the columns.

Type in ACTpop at the command line prompt, and the data will be displayed almost as they appear in Table 1.1 (the only difference is the introduction of row labels in the R output). The object ACTpop is an example of a data frame. Data frames are the usual way for organizing data sets in R. More information about data frames can be found in Section 1.5.

Case is significant for names of R objects or commands. Thus, ACTPOP is different from ACTpop. (For file names on Microsoft Windows systems, the Windows conventions apply, and case does not distinguish file names. On Unix systems letters that have a different case are treated as different.)

We now plot the ACT population between 1917 and 1997 by using

```
plot(ACT ~ Year, data=ACTpop, pch=16)
```

The option pch=16 sets the plotting character to solid black dots. Figure 1.1 shows the graph.

We can make various modifications to this basic plot. We can specify more informative axis labels, change the sizes of the text and of the plotting symbol, add a title, and so on. More information is given in Section 1.8.

1.1.4 Entry of data at the command line

Table 1.2 gives, for each amount by which an elastic band is stretched over the end of a ruler, the distance that the band traveled when released. We can use data.frame() to input these (or other) data directly at the command line. We will assign the name elasticband to the data frame:

```
elasticband <- data.frame(stretch=c(46,54,48,50,44,42,52),
                          distance=c(148,182,173,166,109,141,166))
```

Table 1.2: *Distance (cm)*
versus stretch (mm), for
elastic band data.

Stretch	Distance
46	148
54	182
48	173
50	166
44	109
42	141
52	166

The constructs `c(46,54,48,50,44,42,52)` and `c(148,182,173,166,109,`
`141,166)` join ("concatenate") the separate numbers together into a single vector object.
See later, Subsection 1.4.1. These vector objects then become columns in the data frame.

1.1.5 Online help

Before getting deeply into the use of R, it is well to take time to master the help facilities.
Such an investment of time will pay dividends. R's help files are comprehensive, and are
frequently upgraded. Type in `help(help)` to get information on the help features of the
system that is in use. To get help on, e.g., `plot()`, type in

```
help(plot)
```

Different R implementations offer different choices of modes of access into the help pages
(thus Microsoft Windows systems offer a choice between a form of help that displays the
relevant help file, html help, and compiled html help. The choice between these different
modes of access is made at startup. See `help(Rprofile)` for details).

Two functions that can be highly useful in searching for functions that perform a desired
task are `apropos()` and `help.search()`. We can best explain their use by giving
specific examples. Thus try

```
apropos("sort")          # Try, also, apropos ("sor")
    # This lists all functions whose names include the
    # character string "sort".
help.search("sort")   # Note that the argument is "sort"
    # This lists all functions that have the word 'sort' as
    # an alias or in their title.
```

Note also `example()`. This initiates the running of examples, if available, of the use
of the function specified by the function argument. For example:

```
example(image)
  # for a 2 by 2 layout of the last 4 plots, precede with
  # par(mfrow=c(2,2))
  # to prompt for each new graph, precede with par(ask=T)
```

Much can be learned from experimenting with R functions. It may be helpful to create a simple artificial data set with which to experiment. Another possibility is to work with a subset of the data set to which the function will, finally, be applied. For extensive experimentation it is best to create a new workspace where one can work with copies of any user data sets and functions.

Among the abilities that are documented in the help pages, there will be some that bring pleasant and unexpected surprises. There may be insightful and helpful examples. There are often references to related functions. In most cases there are technical references that give the relevant theory. While the help pages are not intended to be an encyclopedia on statistical methodology, they contain much helpful commentary on the methods whose implementation they document. It can help enormously, before launching into use of an R function, to check the relevant help page!

1.1.6 Quitting R

One exits or quits R by using the q() function:

$$q()$$

There will be a message asking whether to save the workspace image. Clicking **Yes** (the safe option) will save all the objects that remain in the working directory – any that were there at the start of the session and any that were added (and not removed) during the session.

Depending on the implementation, alternatives may be to click on the **File** menu and then on **Exit**, or to click on the × in the top right hand corner of the R window. (Under Linux, clicking on × exits from the program, but without asking whether to save the workshop image.)

Note: In order to quit from the R session we had to type q(). This is because q is a function. Typing q on its own, without the parentheses, displays the text of the function on the screen. Try it!

1.2 The Uses of R

R has extensive capabilities for statistical analysis, that will be used throughout this book. These are embedded in an interactive computing environment that is suited to many different uses. Here we draw attention to abilities, beyond simple one-line calculations, that are not primarily statistical.

R *will give numerical or graphical data summaries*

An important class of R object is the data frame. R uses data frames to store rectangular arrays in which the columns may be vectors of numbers or factors or text strings. Data frames are central to the way that all the more recent R routines process data. For now, think of data frames as rather like a matrix, where the rows are observations and the columns are variables.

As a first example, consider the data frame `cars` that is in the *base* package. This has two columns (variables), with the names `speed` and `dist`. Typing in `summary(cars)` gives summary information on these variables.

```
> data(cars) # Gives access to the data frame cars
> summary(cars)
     speed              dist
 Min.    : 4.0   Min.    :  2.00
 1st Qu.:12.0   1st Qu.: 26.00
 Median :15.0   Median : 36.00
 Mean    :15.4   Mean    : 42.98
 3rd Qu.:19.0   3rd Qu.: 56.00
 Max.    :25.0   Max.    :120.00
```

Thus, we can immediately see that the range of speeds (first column) is from 4 mph to 25 mph, and that the range of distances (second column) is from 2 feet to 120 feet.

R *has extensive abilities for graphical presentation*

The main R graphics function is `plot()`, which we have already encountered. In addition, there are functions for adding points and lines to existing graphs, for placing text at specified positions, for specifying tick marks and tick labels, for labeling axes, and so on. Some details are given in Section 1.8.

R *is an interactive programming language*

Suppose we want to calculate the Fahrenheit temperatures that correspond to Celsius temperatures 25, 26, . . . , 30. Here is a good way to do this in R:

```
> celsius <- 25:30
> fahrenheit <- 9/5*celsius+32
> conversion <- data.frame(Celsius=celsius, Fahrenheit=fahrenheit)
> print(conversion)
  Celsius Fahrenheit
1      25       77.0
2      26       78.8
3      27       80.6
4      28       82.4
5      29       84.2
6      30       86.0
```

1.3 The R Language

R is a functional language that uses many of the same symbols and conventions as the widely used general purpose language C and its successors C++ and Java. There is a language core that uses standard forms of algebraic notation, allowing the calculations that were described in Subsection 1.1.2. Functions – supplied as part of R or one of its packages, or written by the user – allow the limitless extension of this language core.

1.3.1 R *objects*

All R entities, including functions and data structures, exist as objects that can be operated on as data. Type in `ls()` (or `objects()`) to see the names of all objects in the workspace. One can restrict the names to those with a particular pattern, e.g., starting with the letter "*p*". (Type in `help(ls)` and `help(grep)` for more details. The pattern-matching conventions are those used for `grep()`, which is modeled on the Unix grep command. For example, `ls(pattern="p")` lists all object names that include the letter "p". To get all object names that start with the letter "p", specify `ls(pattern="^p")`.) As noted earlier, typing the name of an object causes the printing of its contents.

It is often possible and desirable to operate on objects – vectors, arrays, lists and so on – as a whole. This largely avoids the need for explicit loops, leading to clearer code. Section 1.2 gave an example.

1.3.2 Retaining objects between sessions

Upon quitting an R session, we have recommended saving the workspace image. It is easiest to allow R to save the image, in the current working directory, with the file name `.RData` that R uses as the default. The image will then be loaded automatically when a new session is started in that directory. The objects that were in the workspace at the end of the earlier session will again be available.

Many R users find it convenient to work with multiple workspaces, typically with a different working directory for each different workspace. As a preliminary to loading a new workspace, it will usually be desirable to save the current workspace, and then to clear all objects from it. These operations may be performed from the menu, or alternatively there are the commands `save.image()` for saving the current workspace, `rm(list=ls())` for clearing the workspace, `setwd()` for changing the working directory, and `load()` for loading a new workspace. For details, see the help pages for these functions. See also Subsection 12.9.1.

One should avoid cluttering the workspace with objects that will not again be needed. Before saving the current workspace, type `ls()` to get a complete list of objects. Then remove unwanted objects using

```
rm(<obj1>, <obj2>,...)
```

where the names of objects that are to be removed should appear in place of `<obj1>`, `<obj2>`, ... For example, to remove the objects `celsius` and `fahrenheit` from the workspace image before quitting, type

```
rm(celsius, fahrenheit)
q()
```

In general, we have left it to the reader to determine which objects should be removed once calculations are complete.

1.4 Vectors in R

Vectors may have mode "logical", "numeric", "character" or "list". Examples of vectors are

```
> c(2, 3, 5, 2, 7, 1)
[1] 2 3 5 2 7 1

> 3:10    # The numbers 3, 4,..., 10
[1] 3 4 5 6 7 8 9 10

> c(T, F, F, F, T, T, F)
[1] TRUE FALSE FALSE FALSE TRUE TRUE FALSE

> c("Canberra", "Sydney", "Canberra", "Sydney")
[1] "Canberra" "Sydney"    "Canberra" "Sydney"
```

The first two vectors above are numeric, the third is logical, and the fourth is a character vector. Note the use of the global variables F (=FALSE) and T (=TRUE) as a convenient shorthand when logical values are entered.

1.4.1 Concatenation – joining vector objects

The c in c(2, 3, 5, 2, 7, 1) is an abbreviation for "concatenate". The meaning is: "Collect these numbers together to form a vector." We can concatenate two vectors. In the following, we form numeric vectors x and y, that we then concatenate to form a vector z:

```
> x <- c(2, 3, 5, 2, 7, 1)
> x
[1] 2 3 5 2 7 1
> y <- c(10, 15, 12)
> y
[1] 10 15 12
> z <- c(x, y)
> z
[1] 2 3 5 2 7 1 10 15 12
```

1.4.2 Subsets of vectors

Note three common ways to extract subsets of vectors.

1. Specify the indices of the elements that are to be extracted, e.g.,

```
> x <- c(3, 11, 8, 15, 12)   # Assign to x the values 3,
                             # 11, 8, 15, 12
> x[c(2,4)]                  # Elements in positions 2
[1] 11 15                    # and 4 only
```

2. Use negative subscripts to omit the elements in nominated subscript positions (take care not to mix positive and negative subscripts):

```
> x[-c(2,3)]   # Remove the elements in positions 2 and 3
[1] 3 15 12
```

3. Specify a vector of logical values. The elements that are extracted are those for which the logical value is TRUE. Thus, suppose we want to extract values of x that are greater than 10.

```
> x > 10
[1] FALSE TRUE FALSE TRUE TRUE
> x[x > 10]
[1] 11 15 12
```

1.4.3 Patterned data

Use, for example, 5 : 15 to generate all integers in a range, here between 5 and 15 inclusive.

```
> 5:15
[1]   5   6   7   8   9 10 11 12 13 14 15
```

Conversely, 15 : 5 will generate the sequence in the reverse order.

The function seq() is more general. For example:

```
> seq(from=5, to=22, by=3)   # The first value is 5.   The final
                             # value is <= 22
[1] 5 8 11 14 17 20
```

The function call can be abbreviated to

```
                        seq(5, 22, 3)
```

To repeat the sequence (2, 3, 5) four times over, enter

```
            > rep(c(2,3,5), 4)
            [1] 2 3 5 2 3 5 2 3 5 2 3 5
```

Patterned character vectors are also possible

```
> c(rep("female", 3), rep("male", 2))
[1] "female" "female" "female" "male" "male"
```

1.4.4 Missing values

The missing value symbol is NA. As an example, we may set

```
y <- c(1, NA, 3, 0, NA)
```

Note that any arithmetic operation or relation that involves NA generates an NA. Specifically, be warned that y[y==NA] <- 0 leaves y unchanged. The reason is that all elements of

y==NA evaluate to NA. This does not identify an element of y, and there is no assignment.
To replace all NAs by 0, use the function is.na(), thus

```
> y[is.na(y)] <- 0
> y
[1] 1 0 3 0 0
```

The functions mean(), median(), range(), and a number of other functions, take the
argument na.rm=TRUE; i.e. remove NAs, then proceed with the calculation. By default,
these and related functions will fail when there are NAs. By default, the table() function
ignores NAs.

1.4.5 Factors

A factor is stored internally as a numeric vector with values $1, 2, 3, \ldots, k$. The value k is
the number of levels. The levels are character strings.

Consider a survey that has data on 691 females and 692 males. If the first 691 are females
and the next 692 males, we can create a vector of strings that holds the values thus:

```
gender <- c(rep("female",691), rep("male",692))
```

We can change this vector to a factor, by entering

```
gender <- factor(gender)
```

Internally, the factor gender is stored as 691 1s, followed by 692 2s. It has stored with it
a table that holds the information

1	female
2	male

In most contexts that seem to demand a character string, the 1 is translated into female
and the 2 into male. The values female and male are the levels of the factor. By default,
the levels are chosen to be in sorted order for the data type from which the factor was
formed, so that female precedes male. Hence:

```
> levels(gender)
[1] "female" "male"
```

Note that if gender had been an ordinary character vector, the outcome of the above
levels command would have been NULL. The order of the factor levels determines the
order of appearance of the levels in graphs and tables that use this information. To cause
male to come before female, use

```
gender <- factor(gender, levels=c("male", "female"))
```

This syntax is available both when the factor is first created, and later to change the order
in an existing factor. Take care that the level names are correctly spelled. For example,
specifying "Male" in place of "male" in the levels argument will cause all values
that were "male" to be coded as missing.

One advantage of factors is that the memory required for storage is less than for the corresponding character vector when there are multiple values for each factor level, and the levels are long character strings.

1.5 Data Frames

Data frames are fundamental to the use of the R modeling and graphics functions. A data frame is a generalization of a matrix, in which different columns may have different modes. All elements of any column must, however, have the same mode, i.e. all numeric, or all factor, or all character, or all logical.

Included with our data sets is `Cars93.summary`, created from the `Cars93` data set in the Venables and Ripley *MASS* package, and included in our *DAAG* package. In order to access it, we need first to install it and then to load it into the workspace, thus:

```
> library(DAAG)          # load the DAAG package
> data(Cars93.summary)   # copy Cars93.summary into the workspace ← no longer
> Cars93.summary                                                     necessary
        Min.passengers Max.passengers No.of.cars abbrev
Compact         4              6           16       C
Large           6              6           11       L
Midsize         4              6           22       M
Small           4              5           21       Sm
Sporty          2              4           14       Sp
Van             7              8            9       V
```

Notice that, before we could access `Cars93.summary`, we had first to use `data(Cars93.summary)` to copy it into the workspace. This differs from the S-PLUS behavior, where such data frames in packages that have been loaded are automatically available for access.

The data frame has row labels (accessed using `row.names(Cars93.summary)`) `Compact, Large,....` The column names (accessed using `names(Cars93.summary)`) are `Min.passengers` (i.e. the minimum number of passengers for cars in this category), `Max.passengers`, `No.of.cars`, and `abbrev`. The first three columns are numeric, and the fourth is a factor. Use the function `class()` to check this, e.g. enter `class(Cars93.summary$abbrev)`.

There are several ways to access the columns of a data frame. Any of the following will pick out the fourth column of the data frame `Cars93.summary` and store it in the vector `type` (also allowed is `Cars93.summary[4]`. This gives a data frame with the single column `abbrev`).

```
type <- Cars93.summary$abbrev
type <- Cars93.summary[,4]
type <- Cars93.summary[,"abbrev"]
type <- Cars93.summary[[4]] # Take the object that is stored
                            # in the fourth list element.
```

In each case, one can view the contents of the object `type` by entering `type` at the command
line, thus:

```
> type
[1] C   L   M   Sm Sp V
Levels:   C L M Sm Sp V
```

It is often convenient to use the `attach()` function:

```
> attach(Cars93.summary)
  # R can now access the columns of Cars93.summary directly
> abbrev
[1] C   L   M   Sm Sp V
Levels:   C L M Sm Sp V
> detach("Cars93.summary")
  # Not strictly necessary, but tidiness is a good habit!
  # In R, detach(Cars93.summary) is an acceptable alternative
```

Detaching data frames that are no longer in use reduces the risk of a clash of variable names,
e.g., two different attached data frames that have a column with the name `abbrev`, or an
`abbrev` both in the workspace and in an attached data frame.

In Windows versions, use of `edit()` allows access to a spreadsheet-like display of a
data frame or of a vector. Users can then directly manipulate individual entries or perform
data entry operations as with a spreadsheet. For example,

```
Cars93.summary <- edit(Cars93.summary)
```

To close the spreadsheet, click on the **File** menu and then on **Close**.

1.5.1 Variable names

The `names()` function can be used to determine variable names in a data frame. As an
example, consider the New York air quality data frame that is included with the base R
package. To determine the variables in this data frame, type

```
data(airquality)
names(airquality)
[1] "Ozone"    "Solar.R" "Wind"      "Temp"      "Month"    "Day"
```

The `names()` function serves a second purpose. To change the name of the `abbrev`
variable (the fourth column) in the `Cars93.summary` data frame to `code`, type

```
names(Cars93.summary)[4] <- "code"
```

If we want to change all of the names, we could do something like

```
names(Cars93.summary) <- c("minpass", "maxpass", "number", "code")
```

1.5.2 Applying a function to the columns of a data frame

The `sapply()` function is a useful tool for calculating statistics for each column of a data frame. The first argument to `sapply()` is a data frame. The second argument is the name of a function that is to be applied to each column. Consider the women data frame.

```
> data(women)
> women       # Display the data
   height weight
1      58    115
2      59    117
3      60    120
..............
15     72    164
```

In order to compute averages of each column, type

```
> sapply(women, mean)   # Apply mean() to each of the columns
height weight
  65.0  136.7
```

1.5.3* Data frames and matrices

The numerical values in the data frame women might alternatively be stored in a matrix with the same dimensions, i.e., 15 rows × 2 columns. More generally, any data frame where all columns hold numeric data can alternatively be stored as a matrix. This can speed up some mathematical and other manipulations when the number of elements is large, e.g., of the order of several hundreds of thousands. For further details, see Section 12.7. Note that:

- The `names()` function cannot be used with matrices.
- Above, we used `sapply()` to extract summary information about the columns of the data frame women. If women had been a matrix with the same numerical values in the same layout, the result would have been quite different, and uninteresting – the effect is to apply the function mean to each individual element of the matrix.

1.5.4 Identification of rows that include missing values

Many of the modeling functions will fail unless action is taken to handle missing values. Two functions that are useful for checking on missing values are `complete.cases()` and `na.omit()`. The following code shows how we can identify rows that hold missing values.

```
> data(possum)   # Precede, if necessary, with library(DAAG)
> possum[!complete.cases(possum), ]
      case site Pop sex age hdlngth skullw totlngth taill
BB36    41    2 Vic   f   5    88.4   57.0       83  36.5
BB41    44    2 Vic   m  NA    85.1   51.5       76  35.5
BB45    46    2 Vic   m  NA    91.4   54.4       84  35.0
      footlgth earconch  eye  chest belly
BB36        NA     40.3 15.9   27.0  30.5
BB41      70.3     52.6 14.4   23.0  27.0
BB45      72.8     51.2 14.4   24.5  35.0
```

The function `na.omit()` omits any rows that contain missing values. For example

```
newpossum <- na.omit(possum)      # Has three fewer rows than possum
```

1.6 R Packages

The recommended R distribution includes a number of packages in its library. Note in particular *base*, *eda*, *ts* (time series), and *MASS*. We will make frequent use both of the *MASS* package and of our own *DAAG* package. *DAAG*, and other packages that are not included with the default distribution, can be readily downloaded and installed.

Installed packages, unless loaded automatically, must then be *load*ed prior to use. The *base* package is automatically loaded at the beginning of the session. To load any other installed package, use the `library()` function. For example,

```
library(MASS)                 # Loads the MASS package
```

1.6.1 Data sets that accompany R *packages*

Type in `data()` to get a list of data sets (mostly data frames) in all packages that are in the current search path. To get information on the data sets that are included in the *base* package, specify

```
data(package="base")          # NB. Specify 'package', not 'library'.
```

Replace `"base"` by the name of any other installed package, as required (type in `library()` to get the names of the installed packages).

In order to bring any of these data frames into the workspace, the user must specifically request it. (Ensure that the relevant package is loaded.) For example, to access the data set `airquality` from the *base* package, type in

```
data(airquality)    # Load airquality into the workspace
```

Such objects should be removed (`rm(airquality)`) when they are not for the time being required. They can be loaded again as occasion demands.

1.7* Looping

A simple example of a `for` loop is[1]

```
> for (i in 1:5) print(i)
[1] 1
[1] 2
[1] 3
[1] 4
[1] 5
```

---

[1] Other looping constructs are
```
  repeat <expression>         # Place break somewhere inside
  while (x > 0) <expression>  # Or (x < 0), or etc.
```
Here `<expression>` is an R statement, or a sequence of statements that are enclosed within braces.

Here is a possible way to estimate population growth rates for each of the Australian states and territories:

```
data(austpop)                  # population figures for all
                               # Australian states
growth.rates <- numeric(8)     # numeric(8) creates a numeric
                               # vector with 8 elements, all set
                               # equal to 0
for (j in seq(2,9)) {
    growth.rates[j-1] <- (austpop[9, j]-austpop[1, j])/
    austpop[1, j]}
growth.rates <- data.frame(growth.rates)
row.names(growth.rates) <- names(austpop[c(-1,-10)])
   # We have used row.names() to name the rows of the data frame
```

The result is

```
      growth.rates
NSW            2.30
Vic            2.27
Qld            3.98
SA             2.36
WA             4.88
Tas            1.46
NT            36.40
ACT          102.33
```

Avoiding loops – `sapply()`

The above computation can also be done using the `sapply()` function mentioned in Subsection 1.5.2:

```
> sapply(austpop[,-c(1,10)], function(x){(x[9]-x[1])/x[1]})
   NSW    Vic    Qld     SA     WA    Tas     NT    ACT
  2.30   2.27   3.98   2.36   4.88   1.46  36.40 102.33
```

Note that in contrast to the example in Subsection 1.5.2, we now have an *inline* function, i.e. one that is defined on the fly and does not have or need a name. The effect is to assign the columns of the data frame `austpop[,-c(1,10)]`, in turn, to the function argument x. With x replaced by each column in turn, the function returns `(x[9]-x[1])/x[1]`.

In R there is often a better alternative, perhaps using one of the built-in functions, to writing an explicit loop. Loops can incur severe computational overhead.

1.8 R Graphics

The functions `plot()`, `points()`, `lines()`, `text()`, `mtext()`, `axis()`, `identify()`, etc. form a suite that plot graphs and add features to the graph. To see some of the possibilities that R offers, enter

```
demo(graphics)
```

Press the **Enter** key to move to each new graph.

1.8.1 The function `plot( )` *and allied functions*

The basic command is

```
plot(y ~ x)
```

or

```
plot(x, y)
```

where x and y must be the same length.

Readers may find the following plots interesting (note that `sin()` expects angles to be in radians. Multiply angles that are given in degrees by $\pi/180$ to get radians):

```
plot((0:20)*pi/10, sin((0:20)*pi/10))
plot((1:50)*0.92, sin((1:50)*0.92))
```

Readers might show the second of these graphs to their friends, asking them to identify the pattern! By holding with the left mouse button on the lower border until a double sided arrow appears and dragging upwards, the vertical dimension of the graph sheet can be shortened. If sufficiently shortened, the pattern becomes obvious. The eye has difficulty in detecting patterns of change where the angle of slope is close to the horizontal or close to the vertical.

Then try this:

```
par(mfrow=c(3,1))     # Gives a 3 by 1 layout of plots
plot((1:50)*0.92,   sin((1:50)*0.92))
par(mfrow=c(1,1))
```

Here are two further examples.

```
attach(elasticband)        # R now knows where to find stretch
                           # and distance
plot(stretch, distance)    # Alternative: plot(distance ~ stretch)
detach(elasticband)

attach(austpop)
plot(year, ACT, type="l")  # Join the points ("l" = "line")
detach(austpop)
```

Fine control – parameter settings

When it is necessary to change the default parameter settings, use the `par()` function. We have already used `par(mfrow=c(m, n))` to get an *m* by *n* layout of graphs on a page. Here is another example:

```
par(cex=1.25, mex=1.25)
```

increases the text and plot symbol size 25% above the default. Adding `mex=1.25` makes room in the margin to accommodate the increased text size.

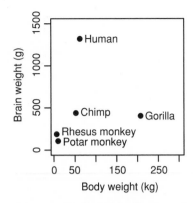

Figure 1.2: Brain weight versus body weight, for the `primates` data frame.

It is good practice to store the existing settings, so that they can be restored later. For this, specify, e.g.,

```
oldpar <- par(cex=1.25, mex=1.25)   # Use par(oldpar) to restore
                                    # earlier settings
```

The size of the axis annotation can be controlled, independently of the setting of `cex`, by specifying a value for `cex.axis`. Similarly, `cex.labels` may be used to control the size of the axis labels.

Type in `help(par)` to get a list of all the parameter settings that are available with `par()`.

Adding points, lines, text and axis annotation

Use the `points()` function to add points to a plot. Use the `lines()` function to add lines to a plot. The `text()` function places text anywhere on the plot. (Actually these functions are identical, differing only in the default setting for the parameter `type`. The default setting for `points()` is `type = "p"`, and that for `lines()` is `type = "l"`. Explicitly setting `type = "p"` causes either function to plot points, `type = "l"` gives lines.) The function `mtext(text, side, line,...)` adds text in the margin of the current plot. The sides are numbered 1 (*x*-axis), 2 (*y*-axis), 3 (top) and 4 (right vertical axis). The `axis()` function gives fine control over axis ticks and labels.

Use of the `text()` function to label points

In Figure 1.2 we have put labels on the points.

We begin with code that will give a crude version of Figure 1.2. The function `row.names()` extracts the row names, which we then use as labels. We then use the function `text()` to add text labels to the points.

```
data(primates)          # The DAAG package must be loaded
attach(primates)        # Needed if primates is not already
                        # attached.
```

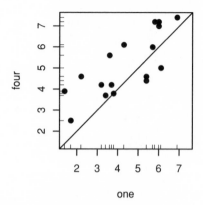

Figure 1.3: Each of 17 panelists compared two milk samples for sweetness. One of the samples had one unit of additive, while the other had four units of additive.

```
plot(Bodywt, Brainwt, xlim=c(0, 300))
  # Specify xlim so that there is room for the labels
text(x=Bodywt, y=Brainwt, labels=row.names(primates), adj=0)
  # adj=0 gives left-justified text
detach(primates)
```

The resulting graph would be adequate for identifying points, but it is not a presentation quality graph. We now note the changes that are needed to get Figure 1.2. In Figure 1.2 we use the `xlab` (*x*-axis) and `ylab` (*y*-axis) parameters to specify meaningful axis titles. We move the labeling to one side of the points by including appropriate horizontal and vertical offsets. We multiply `chw <- par()$cxy[1]` by 0.1 to get an horizontal offset that is one tenth of a character width, and similarly for `chh <- par()$cxy[2]` in a vertical direction. We use `pch=16` to make the plot character a heavy black dot. This helps make the points stand out against the labeling.

Here is the R code for Figure 1.2:

```
attach(primates)
plot(x=Bodywt, y=Brainwt, pch=16, xlab="Body weight (kg)",
     ylab="Brain weight (g)", xlim=c(0,300), ylim=c(0,1500))
chw <- par()$cxy[1]
chh <- par()$cxy[2]
text(x=Bodywt+chw, y=Brainwt+c(-.1, 0, 0,.1, 0)*chh,
     labels=row.names(primates), adj=0)
detach(primates)
```

Where `xlim` and/or `ylim` is not set explicitly, the range of data values determines the limits. In any case, the axis is by default extended by 4% relative to those limits.

Rug plots

The function `rug()` adds vertical bars, showing the distribution of data values, along one or both of the *x*- and *y*-axes of an existing plot. Figure 1.3 has rugs on both the *x*- and *y*-axes. Data were from a tasting session where each of 17 panelists assessed the sweetness of each of two milk samples, one with four units of additive, and the other with one unit of additive. The code that produced Figure 1.3 is

```
data(milk)                       # From the DAAG package
xyrange <- range(milk)
plot(four ~ one, data = milk, xlim = xyrange, ylim =
     xyrange, pch = 16)
rug(milk$one)
rug(milk$four, side = 2)
abline(0, 1)
```

Histograms and density plots

We mention these here for completeness. They will be discussed in Subsection 2.1.1.

The use of color

Try the following:

```
theta <- (1:50)*0.92
plot(theta, sin(theta), col=1:50, pch=16, cex=4)
points(theta, cos(theta), col=51:100, pch=15, cex=4)
palette()                        # Names of the 8 colors in the default
                                 # palette
```

Points are in the eight distinct colors of the default palette, one of which is "white". These are recycled as necessary.

The default palette is a small selection from the built-in colors. The function colors() returns the 657 names of the built-in colors, some of them aliases for the same color. The following repeats the plots above, but now using the first 100 of the 657 built-in colors.

```
theta <- (1:50)*0.92
plot(theta, sin(theta), col=colors()[1:50], pch=16, cex=4)
points(theta, cos(theta), col=colors()[51:100], pch=15, cex=4)
```

1.8.2 Identification and location on the figure region

Two functions are available for this purpose. Draw the graph first, then call one or other of these functions:

- identify() labels points;
- locator() prints out the co-ordinates of points.

In either case, the user positions the cursor at the location for which co-ordinates are required, and clicks the left mouse button. Depending on the platform, the identification or labeling of points may be terminated by pointing outside of the graphics area and clicking, or by clicking with a button other than the first. The process will anyway terminate after some default number n of points, which the user can set. (For identify() the default setting is the number of data points, while for locator() the default is 500.)

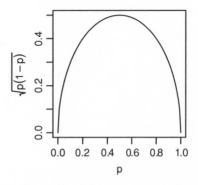

Figure 1.4: The *y*-axis label is a mathematical expression.

As an example, identify two of the plotted points on the `primates` scatterplot:

```
attach(primates)
plot(Bodywt, Brainwt)
identify(Bodywt, Brainwt, n=2) # Now click near 2 plotted points
detach(primates)
```

1.8.3 Plotting mathematical symbols

Both `text()` and `mtext()` allow replacement of the text string by a mathematical expression. In `plot()`, either or both of `xlab` and `ylab` can be an algebraic expression. Figure 1.4 was produced with

```
p <- (0:100)/100
plot(p, sqrt(p*(1-p)), ylab=expression(sqrt(p(1-p))), type="l")
```

Type `help(plotmath)` to get details of available forms of mathematical expression. The final plot from `demo(graphics)` shows some of the possibilities for plotting mathematical symbols. There are brief further details in Section 12.10

1.8.4 Row by column layouts of plots

There are several ways to do this. Here, we will demonstrate two of them.

Multiple plots, each with its own margins

As noted in earlier sections, the parameter `mfrow` can be used to configure the graphics sheet so that subsequent plots appear row by row, one after the other in a rectangular layout, on the one page. For a column by column layout, use `mfcol`. The following example gives a plot that displays four different transformations of the `Animals` data.

```
par(mfrow=c(2,2))          # 2 by 2 layout on the page
library(MASS)              # Animals is in the MASS package
data(Animals)              # Needed if Animals is not already loaded
```

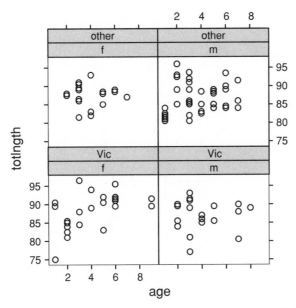

Figure 1.5: Total length of possums versus age, for each combination of population (the Australian state of Victoria or other) and sex (female or male). Further details of these data are in Subsection 2.1.1.

```
attach(Animals)
plot(body, brain)
plot(sqrt(body), sqrt(brain))
plot((body)^0.1, (brain)^0.1)
plot(log(body), log(brain))
detach("Animals")
par(mfrow=c(1,1))          # Restore to 1 figure per page
```

Multiple panels – the lattice *function* xyplot()

The function xyplot() in the *lattice* package gives a rows by columns (*x* by *y*) layout of panels in which the axis labeling appears in the outer margins. Figure 1.5 is an example. Enter

```
> library(lattice)
> data(possum)                 # DAAG must be loaded
> table(possum$Pop, possum$sex)  # Graph reflects layout of this
                               # table
         f  m
  Vic   24 22
  other 19 39
> xyplot(totlngth ~ age | sex*Pop, data=possum)
```

Note that, as we saw in Subsection 1.5.4, there are missing values for age in rows 44 and 46 that xyplot() has silently omitted. The factors that determine the layout of the panels, i.e., sex and Pop in Figure 1.5, are known as *conditioning* variables.

There will be further discussion of the *lattice* package in Subsection 2.1.5. It has functions that offer a similar layout for many different types of plot. To see further examples of the use of `xyplot()`, and of some of the other *lattice* functions, type in

```
example(xyplot)
```

Further points to note about the *lattice* package are:

- The *lattice* package implements trellis style graphics, as used in Cleveland (1993). This is why functions that control stylistic features (color, plot characters, line type, etc.) have *trellis* as part of their name.
- Lattice graphics functions cannot be mixed with the graphics functions discussed earlier in this subsection. It is not possible to use `points()`, `lines()`, `text()`, etc., to add features to a plot that has been created using a *lattice* graphics function. Instead, it is necessary to use functions that are special to *lattice* – `lpoints()`, `llines()`, `ltext()`, etc.
- For inclusion, inside user functions, of statements that will print *lattice* graphs, see the note near the end of Subsection 2.1.5. An explicit print statement is typically required, e.g.

```
print(xyplot(totlngth ~ age | sex*Pop, data=possum))
```

1.8.5 Graphs – additional notes

Graphics devices

On most systems, `x11()` will open a new graphics window. See `help(x11)`. On Macintosh systems that do not have an X11 driver, use `macintosh()`. See `help(Devices)` for a list of devices that can be used for writing to a file or to hard copy. Use `dev.off()` to close the currently active graphics device.

The shape of the graph sheet

It is often desirable to control the shape of the graph page. For example, we might want the individual plots to be rectangular rather than square. The function `x11()` sets up a graphics page on the screen display. It takes arguments `width` (in inches), `height` (in inches) and `pointsize` (in $\frac{1}{72}$ of an inch). The setting of `pointsize` (default = 12) determines character heights.[2]

Plot methods for objects other than vectors

We have seen how to plot a numeric vector y against a numeric vector x. The plot function is a generic function that also has special methods for "plotting" various

[2] It is the relative sizes of these parameters that matter for screen display or for incorporation into Word and similar programs. Once pasted (from the clipboard) or imported into Word, graphs can be enlarged or shrunk by pointing at one corner, holding down the left mouse button, and pulling.

different classes of object. For example, one can give a data frame as the argument to `plot`. Try

```
data(trees)          # Load data frame trees (base package)
plot(trees)          # Has the same effect as pairs(trees)
```

The `pairs()` function will be important when we come to discuss multiple regression. See Subsection 6.1.4, and later examples in that chapter.

Good and bad graphs

There is a difference!

Draw graphs so that they are unlikely to mislead, make sure that they focus the eye on features that are important, and avoid distracting features. In scatterplots, the intention is typically to draw attention to the points. If there are not too many of them, drawing them as heavy black dots or other symbols will focus attention on the points, rather than on a fitted line or curve or on the axes. If they are numerous, dots are likely to overlap. It then makes sense to use open symbols. Where there are many points that overlap, the ink will be denser. If there are many points, it can be helpful to plot points in a shade of gray.[3]

Where the horizontal scale is continuous, patterns of change that are important to identify should have an angle of slope in the approximate range 20° to 70°. (This was the point of the sine curve example in Subsection 1.8.1.)

There are a huge choice and range of colors. Colors, or gray scales, can often be used to useful effect to distinguish groupings in the data. Bear in mind that the eye has difficulty in focusing simultaneously on widely separated colors that appear close together on the same graph.

1.9 Additional Points on the Use of R in This Book

Functions

Functions are integral to the use of the R language. Perhaps the most important topic that we have left out of this chapter is a description of how users can write their own functions. User-written functions are used in exactly the same way as built-in functions. Subsection 12.2.2 describes how users may write their own functions. Examples will appear from time to time through the book.

An incidental advantage from putting code into functions is that the workspace is not then cluttered with objects that are local to the function.

Setting the number of decimal places in output

Often, calculations will, by default, give more decimal places of output than are useful. In the output that we give, we often reduce the number of decimal places below what R gives by default. The `options()` function can be used to make a global change to the number

---

[3] `## Example of plotting with different shades of gray`
`  plot(1:4, 1:4, pch=16, col=c("gray20","gray40","gray60","gray40"), cex=2)`

of significant digits that are printed. For example:

```
> sqrt(10)
[1] 3.162278
options(digits=2)          # Change until further notice, or until
                           # end of session.
> sqrt(10)
[1] 3.2
```

Notice that `options(digits=2)` expresses a wish, which R will not always obey!

Rounding will sometimes introduce small inconsistencies. For example, in Section 4.5, with results rounded to two decimal places,

$$\sqrt{\frac{372}{12}} = 5.57$$

$$\sqrt{2} \times \sqrt{\frac{372}{12}} = 7.88.$$

Note however that $\sqrt{2} \times 5.57 = 7.87$

Other option settings

Type in `help(options)` to get further details. We will meet another important option setting in Chapter 5. (The output that we present uses the setting `options(show.signif.stars=FALSE)`, where the default is TRUE. This affects output in Chapter 5 and later chapters.)

Cosmetic issues

In our R code, we write, e.g., a `<-` b rather than a<-b, and y `~` x rather than y~x. This is intended to help readability, perhaps a small step on the way to literate programming. Such presentation details can make a large difference when others use the code.

Where output is obtained with the simple use of `print()` or `summary()`, we have in general included this as the first statement in the output.

* *Common sources of difficulty*

Here we draw attention, with references to relevant later sections, to common sources of difficulty. We list these items here so that readers have a point of reference when it is needed.

- In the use of `read.table()` for the entry of data that have a rectangular layout, it is important to tune the parameter settings to the input data set. Check Subsection 12.3.1 for common sources of difficulty.
- Character vectors that are included as columns in data frames become, by default, factors. There are implications for the use of `read.table()`. See Subsection 12.3.1 and Section 12.4.
- Factors can often be treated as vectors of text strings, with values given by the factor levels. There are several, potentially annoying, exceptions. See Section 12.4.
- The handling of missing values is a common source of difficulty. See Section 12.5.

- The syntax elasticband[,2] extracts the second column from the data frame elasticband, yielding a numeric vector. Observe however that elasticband[2,] yields a data frame, rather than the numeric vector that the user may require. Specify unlist(elasticband[2,]) to obtain the vector of numeric values in the second row of the data frame. See Subsection 12.6.1. For another instance (use of sapply()) where the difference between a numeric data frame and a numeric matrix is important, see Subsection 12.6.6.
- It is inadvisable to assign new values to a data frame, thus creating a new local data frame with the same name, while it is attached. Use of the name of the data frame accesses the new local copy, while the column names that are in the search path are for the original data frame. There is obvious potential for confusion and erroneous calculations.
- Data objects that individually or in combination occupy a large part of the available computer memory can slow down all memory-intensive computations. See further Subsection 12.9.1 for comment on associated workspace management issues. See also the opening comments in Section 12.7. Note that most of the data objects that are used for our examples are small and thus will not, except where memory is very small, make much individual contribution to demands on memory.

Variable names in data sets

We will refer to a number of different data sets, many of them data frames in our *DAAG* package. When we first introduce the data set, we will give both a general description of the columns of values that we will use, and the names used in the data frame. In later discussion, we will use the name that appears in the data frame whenever the reference is to the particular values that appear in the column.

1.10 Further Reading

An important reference is the R Core Development Web Team's *Introduction to* R. This document, which is regularly updated, is included with the R distributions. It is available from CRAN sites as an independent document. (For a list of sites, go to http://cran.r-project.org.) Books that include an introduction to R include Dalgaard (2002) and Fox (2002).

See also documents, including Maindonald (2001), that are listed under **Contributed Documentation** on the CRAN sites. For a careful detailed account of the R and S languages, see Venables and Ripley (2000). There is a large amount of detailed technical information in Venables and Ripley (2000 and 2002).

Books and papers that set out principles of good graphics include Cleveland (1993 and 1994), Tufte (1997), Wainer (1997), Wilkinson et al. (1999). See also the brief comments in Maindonald (1992).

References for further reading

Cleveland, W.S. 1993. *Visualizing Data.* Hobart Press.
Cleveland, W.S. 1994. *The Elements of Graphing Data*, revised edn. Hobart Press.

Dalgaard, P. 2002. *Introductory Statistics with* R. Springer-Verlag.

Fox, J. 2002. *An* R *and* S-PLUS *Companion to Applied Regression.* Sage Books.

Maindonald J.H. 1992. Statistical design, analysis and presentation issues. *New Zealand Journal of Agricultural Research* 35: 121–141.

Maindonald, J.H. 2001. *Using* R *for Data Analysis and Graphics.* Available as a pdf file at http://wwwmaths.anu.edu.au/~johnm/r/usingR.pdf

R Core Development Team. *An Introduction to* R. This document is available from CRAN sites, updated regularly. For a list, go to http://cran.r-project.org

Tufte, E.R. 1997. *Visual Explanations.* Graphics Press.

Venables, W.N. and Ripley, B.D. 2000. S *Programming.* Springer-Verlag.

Venables, W.N. and Ripley, B.D. 2002. *Modern Applied Statistics with* S, 4th edn., Springer-Verlag, New York. See also "R" Complements to Modern Applied Statistics with S-PLUS, available from http://www.stats.ox.ac.uk/pub/MASS4/.

Wainer, H. 1997. *Visual Revelations.* Springer-Verlag.

Wilkinson, L. and Task Force on Statistical Inference 1999. Statistical methods in psychology journals: guidelines and explanation. *American Psychologist* 54: 594–604.

1.11 Exercises

1. Using the data frame `elasticband` from Subsection 1.1.4, plot `distance` against `stretch`.

2. The following table gives the size of the floor area (ha) and the price ($A000), for 15 houses sold in the Canberra (Australia) suburb of Aranda in 1999.

```
   area sale.price
1   694 192.0
2   905 215.0
3   802 215.0
4  1366 274.0
5   716 112.7
6   963 185.0
7   821 212.0
8   714 220.0
9  1018 276.0
10  887 260.0
11  790 221.5
12  696 255.0
13  771 260.0
14 1006 293.0
15 1191 375.0
```

Type these data into a data frame with column names `area` and `sale.price`.

(a) Plot `sale.price` versus `area`.

(b) Use the `hist()` command to plot a histogram of the sale prices.

(c) Repeat (a) and (b) after taking logarithms of sale prices.

3. The `orings` data frame gives data on the damage that had occurred in US space shuttle launches prior to the disastrous Challenger launch of January 28, 1986. Only the observations in rows 1, 2, 4, 11, 13, and 18 were included in the pre-launch charts used in deciding whether to proceed with the launch.

   Create a new data frame by extracting these rows from `orings`, and plot `total` incidents against `temperature` for this new data frame. Obtain a similar plot for the full data set.

4. Create a data frame called `Manitoba.lakes` that contains the lake's `elevation` (in meters above sea level) and `area` (in square kilometers) as listed below. Assign the names of the lakes using the `row.names()` function. Then obtain a plot of lake area against elevation, identifying each point by the name of the lake. Because of the outlying observation, it is again best to use a logarithmic scale.

	elevation	area
Winnipeg	217	24387
Winnipegosis	254	5374
Manitoba	248	4624
SouthernIndian	254	2247
Cedar	253	1353
Island	227	1223
Gods	178	1151
Cross	207	755
Playgreen	217	657

   One approach is the following:

   ```
   chw <- par()$cxy[1]/3
   chh <- par()$cxy[2]/3
   plot(log(area) ~ elevation, pch=16, xlim=c(170,270),
        ylim=c(6,11))
   text(x=elevation-chw, y=log(area)+chh,
        labels=row.names(Manitoba.lakes), adj=1)
   text(x=elevation+chw, y=log(area)+chh,
        labels=Manitoba.lakes[,2], adj=0)
   title("Manitoba's Largest Lakes")
   ```

5. The following code extracts the lake areas from the `Manitoba.lakes` data frame and attaches the lake names to the entries of the resulting vector.

   ```
   area.lakes <- Manitoba.lakes[[2]]
   names(area.lakes) <- row.names(Manitoba.lakes)
   ```

   Look up the help for the R function `dotchart()`. Use this function to display the data in `area.lakes`.

6. Using the `sum()` function, obtain a lower bound for the area of Manitoba covered by water.

7. The second argument of the `rep()` function can be modified to give different patterns. For example, to get four 2s, then three 3s, then two 5s, enter

   ```
   rep(c(2,3,5), c(4,3,2))
   ```

   (a)   What is the output from the following command?

```
rep(c(2,3,5), 4:2)
```

   (b)   Obtain a vector of four 4s, four 3s and four 2s.

   (c)   The argument `length.out` can be used to create a vector whose length is `length.out`. Use this argument to create a vector of length 50 that consists of the repeated pattern
3   1   1   5   7

8. The ^ symbol denotes exponentiation. Consider the following.

```
1000*((1+0.075)^5 - 1)   # Interest on $1000, compounded
                         # annually at 7.5% p.a. for five years
```

   (a)   Type in the above expression.

   (b)   Modify the expression to determine the amount of interest paid if the rate is 3.5% p.a.

   (c)   What happens if the exponent 5 is replaced by `seq(1, 10)`?

9. Run the following code:

```
gender <- factor(c(rep("female", 91), rep("male", 92)))
table(gender)
gender <- factor(gender, levels=c("male", "female"))
table(gender)

gender <- factor(gender, levels=c("Male", "female"))
                            # Note the mistake
                            # The level was "male", not "Male"
table(gender)
rm(gender)                  # Remove gender
```

Explain the output from the final `table(gender)`.

10. The following code uses the `for()` looping function to plot graphs that compare the relative population growth (here, by the use of a logarithmic scale) for the Australian states and territories.

```
oldpar <- par(mfrow=c(2,4))
for (i in 2:9){
plot(austpop[,1], log(austpop[, i]), xlab="Year",
     ylab=names(austpop)[i], pch=16, ylim=c(0,10))}
par(oldpar)
```

Can this be done without looping? [Hint: The answer is "yes", although this is not an exercise for novices.]

2

Styles of Data Analysis

When a researcher has a new set of data to analyze, what is the best way to begin? What forms of data exploration will draw attention to obvious errors or quirks in the data, or to obvious clues that the data contain? What are the checks that will make it plausible that the data really will support an intended formal analysis? What mix of exploratory analysis and formal analysis is appropriate? Should the analysis be decided in advance, as part of the planning process? To what extent is it legitimate to allow the data to influence what analysis will be performed? What attention should be paid to analyses that other researchers have done with similar data?

In the ensuing discussion, we note the importance of graphical presentation and fore-shadow a view of statistics that emphasizes the role of models. We emphasize the importance of looking for patterns and relationships. Numerical summaries, such as an average, can be very useful, but important features of the data may be missed without a glance at an appropriate graph. The choice of graph to be drawn can make a large difference.

The best modern statistical software makes a strong connection between data analysis and graphics. This close linking of graphs and statistical analysis is the wave of the future. The aim is to combine the computer's ability to crunch numbers and present graphs with the ability of a trained human eye to detect pattern. It is a powerful combination.

We will see in Chapter 3 that an integral part of statistical analysis is the development of a model that accurately describes the data, aids in understanding what the data say, and makes prediction possible. Model assumptions underlie all formal analyses. Without some assumptions, there cannot be a meaningful analysis! As the model assumptions are strengthened, the chances of getting clear results improve. However, there is a price for stronger assumptions; if the assumptions are wrong, then results may be wrong. Graphical techniques have been developed for assessing the appropriateness of most assumptions that must be made in practice.

2.1 Revealing Views of the Data

Exploratory Data Analysis (EDA) is a name for a collection of data display techniques that are intended to let the data speak for themselves, prior to or as part of a formal analysis. Even if a data set has not been collected in a way that makes it suitable for formal statistical analysis, exploratory data analysis techniques can often be used to glean clues from it. However, it is unwise, as too often happens, to rely on this possibility!

An effective EDA display presents data in a way that will make effective use of the human brain's abilities as a pattern recognition device. It looks for what may be apparent from a direct, careful and (as far as possible) assumption-free examination of the data. It has at least four roles:

- EDA examines data, leaving open the possibility that this examination will suggest how data should be analyzed or interpreted. This use of EDA fits well with the view of science as inductive reasoning.
- EDA results may challenge the theoretical understanding that guided the initial collection of the data. EDA then acquires a more revolutionary role. It becomes the catalyst, in the language of Thomas Kuhn, for a paradigm shift.
- EDA allows the data to influence and criticize an intended analysis. EDA facilitates checks on assumptions; subsequent formal analysis can then proceed with confidence. Competent statisticians have always used graphs to check their data; EDA formalizes and extends this practice.
- EDA techniques may reveal additional information, not directly related to the research question. They may, for example, suggest fruitful new lines of research.

There is a risk that data analysts will see patterns that are merely a result of looking too hard. Under torture, the data readily yield false confessions. It remains important to keep a check on the unbridled use of imaginative insight.

EDA is a blend of old and relatively new techniques. Effective practical statisticians have always relied heavily on graphical displays. EDA has systematized ideas on how data should be explored prior to formal analysis. EDA uses whatever help it can get from statistical theory, so that displays are helpful and interpretable.

In the next several subsections, we describe the histogram and density plot, the stem-and-leaf diagram, the boxplot, the scatterplot, the lowess smoother and the trellis style graphics that are available in the *lattice* package. The *lattice* functions greatly extend the available styles and layouts.

2.1.1 Views of a single sample

Histograms and density plots

The histogram is a basic (and over-used) EDA tool. It gives a graphical representation of the frequency distribution of a set of data. The area of each rectangle of a histogram is proportional to the number of observations whose values lie within the width of the rectangle. A mound-shaped histogram may make it plausible that the data follow a normal distribution. In small samples, however, the shape can be highly irregular. In addition, the appearance can depend on the choice of breakpoints; thus, interpretation should always be accompanied with caution. It is often good practice to use more than one set of breakpoints.

The data set `possum` in our *DAAG* package consists of nine morphometric measurements on each of 104 mountain brushtail possums, trapped at seven sites from southern Victoria to central Queensland (data relate to Lindenmayer et al., 1995). We will consider

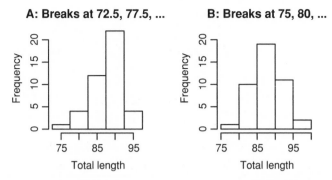

Figure 2.1: The two histograms show the same data, but with a different choice of breakpoints.

measurements for 43 females only using a subset data frame called fossum.[1] Figure 2.1 exhibits two histogram plots of the frequency distribution of the total lengths of the female possums.[2] The only difference in the construction of the two plots is the choice of break-points, but one plot suggests that the distribution is asymmetric (skewed to the left), while the other suggests symmetry.

A histogram is a crude form of a density estimate. A better alternative is, often, a smooth density estimate, as in Figure 2.2. Whereas the width of histogram bars must be chosen somewhat subjectively, density estimates require the choice of a bandwidth parameter that controls the amount of smoothing. In both cases, the software has default choices that can work reasonably well. Figure 2.2 overlays the histograms of Figure 2.1 with a density plot.[3]

Much of the methodology in this book assumes that the data follow a normal distribution (the "bell curve"), which will be discussed in the next chapter. Histograms and density curves can be helpful in drawing attention to particular forms of non-normality, such as that associated with strong skewness in the distribution. It is hard to decide, from visual inspection, when a histogram is adequately close to normal. The density curve is better for this purpose but still not an adequate tool. A more effective way of checking for normality is described in Subsection 3.4.2. Density curves are useful for estimating the population mode, i.e., the value that occurs most frequently.

```
[1] library(DAAG}      # Make sure that the DAAG package is loaded
   data(possum)         # Copy possum into workspace
   fossum <- possum[possum$sex=="f", ]
[2] par(mfrow = c(1, 2))
   attach(fossum)
   hist(totlngth, breaks = 72.5 + (0:5) * 5, ylim = c(0, 22),
        xlab="Total length", main ="A: Breaks at 72.5, 77.5, ...")
   hist(totlngth, breaks = 75 + (0:5) * 5, ylim = c(0, 22),
        xlab="Total length", main="B: Breaks at 75, 80, ...")
   par(mfrow=c(1,1))
[3] dens <- density(totlngth)    # Assumes fossum is still attached
   xlim <- range(dens$x); ylim <- range(dens$y)
   par(mfrow=c(1,2))
   hist(totlngth, breaks = 72.5 + (0:5) * 5, probability = T,
        xlim = xlim, ylim = ylim, xlab="Total length", main=" ")
   lines(dens)
   hist(totlngth, breaks = 75 + (0:5) * 5, probability = T,
        xlim = xlim, ylim = ylim, xlab="Total length", main= " ")
   lines(dens)
   par(mfrow=c(1,1)); detach(fossum)
```

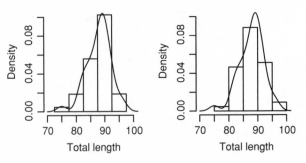

Figure 2.2: Density plots, overlaid on the histograms from Figure 2.1.

Note that `density()` is unsuitable for direct use with data that have sharp lower and/or upper cutoff limits. For example, a failure time distribution may have a mode close to zero, with a sharp cutoff at zero.

The stem-and-leaf display

The stem-and-leaf display is a fine grained alternative to a histogram, for use in displaying a single column of numbers. Here is a simple form of stem-and-leaf plot, for the total lengths of female possums.[4]

```
The decimal point is at the |

  75 | 0
  76 |
  77 |
  78 |
  79 |
  80 |
  81 | 05
  82 | 05
  83 | 00
  84 | 05
  85 | 005
  86 | 055
  87 | 05
  88 | 00055
  89 | 00005555
  90 | 555
  91 | 0055
  92 | 00
  93 | 0
  94 | 0
  95 | 5
  96 | 5
```

The stem is on the left of the vertical bars. The numbers that form the stem (75, 76, . . .) give the whole centimeter part of the length. In this display, the decimal point is exactly at the

---

[4] `stem(fossum$totlngth, scale=4)`

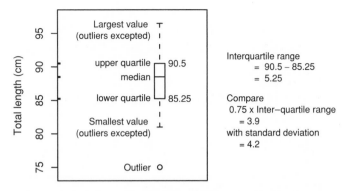

Figure 2.3: Boxplot, with annotation that explains boxplot features.

vertical bar. This will not always be the case. The leaf part, to the right of the vertical bar, is either 0 or 5. This is because measurements are accurate only to the nearest half centimeter. The number of leaves is the frequency for the number of centimeters given in the stem part.

As there are 43 data values, the median or middle value is the 22nd. (There are 21 values that are as small as or smaller than the 22nd largest, and 21 values that are as large or larger.) If one counts leaves, starting at the top and working down, the 22nd leaf corresponds to stem value of 88, with a leaf entry equal to 5. Thus the median (or 50th percentile) is 88.5. The first and third quartiles (the 25th and 75th percentiles) can be recovered in a similar way. The lower quartile (= 85.25) lies half-way between the 11th and 12th observations, while the upper quartile (= 90.5) lies half-way between the 32nd and 33rd observations.[5]

Boxplots

Like the histogram, the boxplot is a coarse summary of a set of data. It gives a graphical summary that a trained eye can comprehend at a glance, focusing on specific important features of a set of data. Figure 2.3 shows a boxplot of total lengths of females in the `possum` data set.[6]

2.1.2 Patterns in grouped data

Example: eggs of cuckoos

Boxplots allow convenient side by side comparisons of different groups, as in the cuckoo egg data that we now present (data are from Latter (1902). Tippett (1931) presents them in a summarized form). Cuckoos lay eggs in the nests of other birds. The eggs are then unwittingly adopted and hatched by the host birds. In Figure 2.4 the egg lengths are grouped by the species of the host bird.[7]

---

[5] `quantile(fossum$totlngth, prob=c(0.25, 0.5, 0.75))` # Get this information directly
    # NB also that median() gives the 50% quantile
[6] `boxplot(fossum$totlngth)` # Simplified boxplot
[7] `data(cuckoos)`   # Assumes DAAG package is loaded.
    `library(lattice)`
    `stripplot(species ~ length, xlab="Length of egg", aspect =0.5, data=cuckoos)`
    # See Section 4.5 for a description of stripplot()
    ## To get the species names as shown, precede the above with:
    `levnam <- strsplit(levels(cuckoos$species), "\\.")`
    # See Subsection 12.2.1 for a description of strsplit()
    `levels(cuckoos$species) <- sapply(levnam, paste, collapse=" ")`

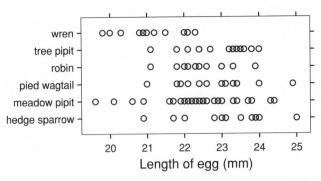

Figure 2.4: Strip plot of cuckoo egg lengths, with a different "strip" for each different host species.

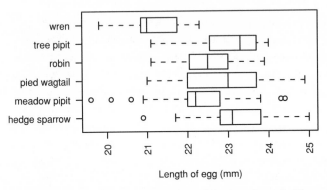

Figure 2.5: Side by side boxplot presentation of the data shown in Figure 2.4.

Compare Figure 2.4 with the boxplot form of presentation in Figure 2.5.[8]

Eggs planted in wrens' nests appear clearly smaller than eggs planted in other birds' nests. Apart from several outlying egg lengths in the meadow pipit nests, the length variability within each host species' nest is fairly uniform.

2.1.3 Patterns in bivariate data – the scatterplot

The examination of pattern and relationship is a central theme of science. Pattern and relationship are core themes of statistics. The scatterplot is a simple but important tool for the examination of pairwise relationships. Some graphs are helpful. Some are not. We will illustrate with specific examples.

The fitting of a smooth trend curve

The data that are plotted in Figure 2.6 are from a study that examined how the electrical resistance of a slab of kiwifruit changed with the apparent juice content. It can be useful to make a comparison with a curve provided by a data-smoothing routine that is not restricted

---

[8] `## Simplified version of the figure`
```
boxplot(length ~ species, data=cuckoos, xlab="Length of egg",
        horizontal=TRUE, las=2)
library(lattice) # Alternative using bwplot() from lattice package
bwplot(species ~ length, xlab="Length of egg", data=cuckoos)
```

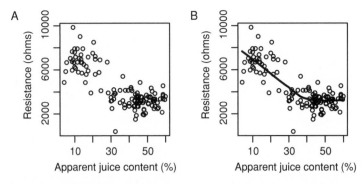

Figure 2.6: Electrical resistance versus apparent juice content. The right panel is the same as the left panel but with a smooth curve that has been fitted to the data.

to using a particular mathematical form of curve, as in the right panel of Figure 2.6 which is an estimate of the relationship between electrical resistance and apparent juice content. The curve was fitted using the lowess method. Note that R has both a `lowess()` function and (in the *modreg* package) the more general `loess()` function. The `lowess()` function does smoothing in one dimension only, while `loess()` will handle multi-dimensional smoothing. For technical details of lowess and loess, see Cleveland (1981), and references given in that paper. The fitted smooth curve shows a form of response that is clearly inconsistent with a straight line.[9] It suggests that there is an approximate linear relationship for juice content up to somewhat over 35%; at that point the curve gradually merges into a horizontal line, and there is no further change in resistance after the juice content reaches around 45%.

A smooth trend curve that has been superimposed on a scatterplot can be a useful aid to interpretation. When the data appear to scatter about a simple mathematical curve, the curve-fitting methods that we discuss in other chapters can be used to obtain a "best fit" or regression line or curve to pass through the points.

What is the appropriate scale?

Figures 2.7A and 2.7B plot brain weight (g) against body weight (kg), for a number of different animals.[10]

Figure 2.7A is almost useless. The axes should be transformed so that the data are spread out more evenly. Here, we can do this by choosing a logarithmic scale. Multiplication by the same factor (e.g., for the tick marks in Figure 2.7B, by a factor of 10) always gives the same distance along the scale. If we marked points 1, 5, 25, 125, ... along one or other axis, they would also lie an equal distance apart.

---

[9]
```
## code for the right panel of the figure
data(fruitohms) # From DAAG package
plot(ohms ~ juice, xlab="Apparent juice content (%)",
     ylab="Resistance (ohms)", data=fruitohms)
lines(lowess(fruitohms$juice, fruitohms$ohms), lwd=2)
```
[10]
```
## The following omits the labeling information
library(MASS)
data(Animals)
x11(width=7.5, height=4)
oldpar <- par(mfrow = c(1,2))   # Two panels
plot(brain ~ body, data=Animals)
plot(log(brain) ~ log(body), data=Animals)
par(oldpar)
```

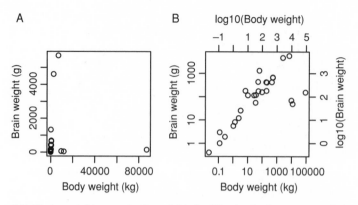

Figure 2.7: Brain weight versus body weight, for 27 animals that vary greatly in size.

A logarithmic scale is appropriate for scaling quantities that change in a multiplicative manner. For example, if cells in a growing organism divide and produce new cells at a constant rate, then the total number of cells changes in a multiplicative manner, resulting in so-called exponential growth. Random changes in the relative growth rate will produce adult organisms whose size (e.g., height) is, on the logarithmic scale, approximately normally distributed. The reason for this is that the growth rate on a natural logarithmic scale ($\log_e$) equals the relative growth rate. Derivation of this result is a straightforward use of the differential calculus.

The logarithmic transformation is so commonly needed that we have felt it necessary to introduce it at this point. Biologists, economists and others should become comfortable working with it. As we have indicated, there are many circumstances in which it makes good sense to work on a logarithmic scale, i.e., to use a logarithmic transformation. There is a brief discussion of other transformations in Chapter 5.

2.1.4* Multiple variables and times

Overlaying plots of several time series (sequences of measurements taken at regular intervals) might seem appropriate for making direct comparisons. However, this approach will only work if the scales are similar for the different series.

As an example, consider the number of workers (in thousands) in the Canadian labor force broken down by region (BC, Alberta, Prairies, Ontario, Quebec, Atlantic) for the 24-month period from January, 1995 to December, 1996 (a time when Canada was emerging from a deep economic recession). Data are in the data frame `jobs`. Columns 1–6 of the data frame have the respective numbers for the six different regions. Here are the ranges of values in its columns.[11]

```
      BC Alberta Prairies Ontario Quebec Atlantic      Date
[1,] 1737    1366      973    5212   3167      941 95.00000
[2,] 1840    1436      999    5360   3257      968 96.91667
```

---

[11] `data(jobs)`   `# Assumes that DAAG package is loaded.`
     `sapply(jobs, range)`

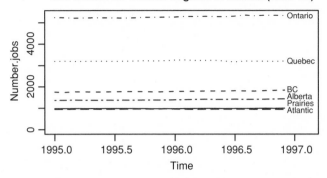

Figure 2.8: Numbers in the labor force (thousands) for various regions of Canada.

In order to see where the economy was taking off most rapidly, it is tempting to plot all of the series on the same graph (as in Figure 2.8).[12] Because the labor forces in the various regions do not have similar sizes, it is impossible to discern any differences among the regions from this plot. Even plotting on the logarithmic scale does not remedy this problem, if the same limits are used for each plot.

Figure 2.9 uses the *lattice* package, which we will discuss in more detail in Subsection 2.1.5. The six different panels use different *slices* of the same logarithmic scale.[13] As a consequence, equal distances correspond to equal relative changes. It is now clear that Alberta and BC experienced the most rapid job growth during the period, and that there was little or no job growth in Quebec and the Atlantic region.

The axis labels show the actual numbers of jobs, in powers of 2. Notice also that, although the scale is logarithmic, equal changes in the numbers of jobs still correspond, roughly, to equal distances on each of the vertical scales. This is because the relative change between the lower and upper vertical axis limits is in each case small, less than 10%.

2.1.5 Lattice (trellis style) graphics

Lattice plots allow the use of the layout on the page to reflect meaningful aspects of data structure. The R *lattice* graphics package, which is one of the recommended packages, implements *trellis* style graphics as in the S-PLUS *trellis* package. The *lattice* package

---

```
12  ## The following code gives a presentation similar to the figure
    matplot(jobs[,7], jobs[,-7], type="l", xlim=c(95,97.1))
      # Arguments 1 & 2 to matplot() can be vector or matrix or data frame
      # For more information on matplot(), type help(matplot)
    text(rep(jobs[24,7], 6), jobs[24,1:6], names(jobs)[1:6], adj=0)
13  library(lattice)
    Jobs <- stack(jobs, select = 1:6)
      # stack() concatenates selected data frame columns into a single
      # column, & adds a factor that has the column names as levels
    Jobs$Year <- rep(jobs[,7], 6)
    names(Jobs) <- c("Number", "Province", "Year")
    xyplot(Number ~ Year | Province, data = Jobs,
      scales = list(y = list(log = 2, relation="sliced")),
      type = "l", layout=c(2,3), par.strip.text=list(cex = 0.8))
      # The parameter relation="sliced" causes the length of the
      # relevant scale(s) ('slice(s)') to be the same for all panels
```

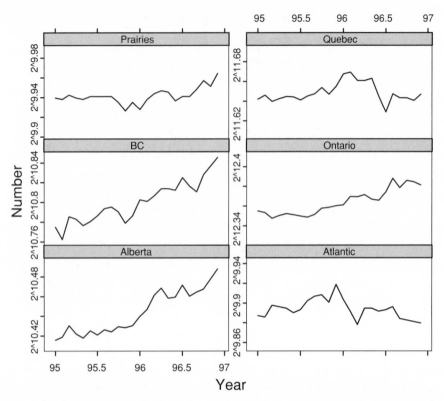

Figure 2.9: This shows the same data as in Figure 2.8, but now with separate logarithmic scales on which the same percentage increase, e.g., by 1%, corresponds to the same distance on the scale, for all plots.

sits on top of the *grid* package. Provided the *grid* package is installed, it will be loaded automatically when *lattice* is loaded.

Examples that present panels of scatterplots – using xyplot()

The basic function for use in drawing panels of scatterplots is xyplot(). We will use the data frame tinting (in our *DAAG* package) to demonstrate the use of xyplot(). Data are from an experiment that aimed to model the effects of the tinting of car windows on visual performance (data relate to Burns et al., 1999). The authors were mainly interested in effects on side window vision, and hence in visual recognition tasks that would be performed when looking through side windows.

In the tinting data frame, the columns are

- Variables csoa (critical stimulus onset asynchrony, i.e., the time in milliseconds required to recognize an alphanumeric target), it (inspection time, i.e., the time required for a simple discrimination task) and age (age to the nearest year).
- The ordered factor tint (levels no, lo, hi).
- Factors target (locon, i.e., low contrast, hicon, i.e., high contrast), sex (f = female, m = male) and agegp (Younger = a younger participant, in the 20s; Older = an older participant, in the 70s).

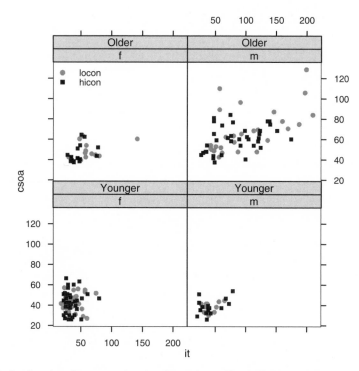

Figure 2.10: Lattice plot of `csoa` against `it`, for each combination of `sex` and `agegp`. In addition, different colors (gray and black) show different levels of `target`.

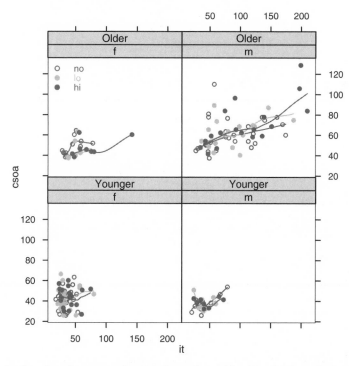

Figure 2.11: Lattice plot of `csoa` against `it`, for each combination of `sex` and `agegp`. Different colors (printed in grayscale) now show different levels of `tint`. Notice the addition of smooth curves.

Each of 28 individuals was tested at each level of tint, for each of the two levels of target.

A first step might be to plot csoa against it for each combination of sex and agegp. The code is

```
data(tinting)            # Assumes that DAAG package is loaded.
xyplot(csoa ~ it | sex * agegp, data=tinting)
# Simple use of xyplot()
```

We can however improve on this very simple graph, by using different symbols (in a black and white plot) or different colors, depending on whether the target is low contrast or high contrast. Also we can ask for a key. The code becomes

```
xyplot(csoa ~ it | sex*agegp, data=tinting,
       panel=panel.superpose, groups=target, auto.key=TRUE)
```

There are more possibilities for refinement yet. Figure 2.10 has used parameter settings that specify the choice of colors (here gray or black), plotting symbols and the placement of the key.[14]

There are long response times for some of the elderly males that occur, as might have been expected, with the low contrast target. The relationship between csoa and it seems much the same for both levels of contrast.

Fitting a trend curve would help make the relationship clearer. Figure 2.11 (printed in grayscale) includes smooth trend curves, now with different colors for different levels of tinting. Observe that the longest times are for the high level of tinting.[15]

Inclusion of lattice graphics functions in user functions

The function xyplot() does not itself print the graph. Instead, it returns an object of class trellis which, if the statement is typed on the command line, is then "printed" by the function print.trellis(). Thus, typing

```
xyplot(csoa ~ it | sex * agegp, data=tinting)
```

on the command line is equivalent to

```
print(xyplot(csoa ~ it | sex * agegp, data=tinting))
```

[14]
```
colr <- c("gray", "black")
plotchar <- c(16, 15)
xyplot(csoa ~ it|sex*agegp, data=tinting,
       groups=target, col=colr, pch=plotchar,
       key=list (x=0.14, y=0.84,
                 points=list(pch=plotchar, col=colr),
                 text=list(levels(tinting$target))))
```
[15]
```
colr <- c("skyblue1", "skyblue4") [c(2,1,2)]
plotchar <- c(1,16,16)  # open, filled, filled
xyplot(csoa~it|sex*agegp, groups=tint, data=tinting,
       col=colr, pch=plotchar, type=c("p", "smooth"), span=0.8,
       key=list(x=.14, y=.84,
                points=list(col=colr, pch=plotchar),
                text=list(levels(tinting$tint), col=colr)))
# The parameter "span" controls the extent of smoothing.
```

In a function, unless the lattice command appears as the final statement of the function, the `print` statement must be explicit, i.e.

```
print(xyplot(csoa ~ it | sex * agegp, data=tinting))
```

or its equivalent, e.g.

```
tinting.trellis <- xyplot(csoa ~ it | sex * agegp,
                          data=tinting)
print(tinting.trellis)
```

An incomplete list of lattice functions

There are a number of alternatives to `xyplot()`, including the following:

```
bwplot(factor ~ numeric, . .)          # Box and whisker plot
qqmath(factor ~ numeric, . .)          # normal & other
                                       # probability plots
dotplot(factor ~ numeric, . .)         # 1-dim. Display
stripplot(factor ~ numeric, . .)       # 1-dim. Display
barchart(character ~ numeric, . .)
histogram( ~ numeric, . .)
densityplot( ~ numeric, . .)           # Smoothed version of
                                       # histogram
splom( ~ data frame, . .)              # Scatterplot matrix,
                                       # cf pairs()
parallel( ~ data frame, . .)           # Parallel coordinate plots
```

To get a succession of examples of the use of `bwplot()`, type in

```
example(bwplot)
```

Similarly for `qqmath`, `dotplot`, etc. The examples are in each instance taken from the relevant help page.

2.1.6 What to look for in plots

Outliers

We are interested in noting points that appear to be isolated from the main body of the data. Such points (whether errors or genuine values) are liable to distort any model that we fit. What we see as an outlier depends, inevitably, on the view we have of the data. On a fairly simple level, what we see may depend on whether or not, and on how, we transform the data. Boxplots are useful for highlighting outliers in one dimension. Outliers can also be detected using the normal probability plot that we will encounter in Subsection 3.4.2.

Scatterplots may highlight outliers in two dimensions. Some outliers will be apparent only in three or more dimensions. The presence of outliers can indicate departure from model assumptions.

Asymmetry of the distribution

Most asymmetric distributions can be characterized as either positively skewed or negatively skewed. Positive skewness is the commonest form of asymmetry. There is a long tail to the right, values near the minimum are bunched up together, and the largest values are widely dispersed. Provided that all values are greater than zero, a logarithmic transformation typically makes such a distribution more symmetric. A distribution that is skew cannot be normal. Severe skewness is typically a more serious problem for the validity of results than other types of non-normality.

If values of a variable that takes positive values range by a factor of more than 10:1 then, depending on the application area context, positive skewness is to be expected. A logarithmic transformation should be considered.

Changes in variability

Boxplots and histograms readily convey an impression of the amount of variability or scatter in the data. Side by side boxplots, such as in Figure 2.5, or strip charts such as in Figure 2.4 are particularly useful for comparing variability across different samples or treatment groups. Many statistical models depend on the assumption that variability is constant across treatment groups.

When variability increases as data values increase, the logarithmic transformation will often help. If the variability is constant on a logarithmic scale, then the relative variation on the original scale is constant. (Measures of variability will be discussed in Subsection 2.2.2.)

Clustering

Clusters in scatterplots may suggest structure in the data that may or may not have been expected. When we proceed to a formal analysis, this structure must be taken into account. Do the clusters correspond to different values of some relevant variable? Outliers are a special form of clustering.

Non-linearity

We should not fit a linear model to data where relationships are demonstrably non-linear. Often it is possible to transform variables so that terms enter into the model in a manner that is closer to linear. If not, the possibilities are endless, and we will canvass only a small number of them. See especially Chapter 7.

If there is a theory that suggests the form of model that we need to use, then we should use this as a starting point. Available theory may, however, incorporate various approximations, and the data may tell a story that does not altogether match the available theory. The data, unless they are flawed, have the final say!

2.2 Data Summary

Means, standard deviations, medians and correlations are all forms of data summary. Summaries, in graphical or tabular form, are essential for the presentation of data. One or another form of data summary may be a useful preliminary to the analysis of data, allowing a simplification of the analysis.

With any form of summary, it is important to ask whether there is information in the data that has been lost or obscured. This has special pertinence for the use of the correlation coefficient.

2.2.1 Mean and median

We give a simple example where the analysis can be simplified by first taking means over subgroups of the data. In the data frame `kiwishade`, there are four results for each plot, one for each of four vines. Shading treatments were applied to whole plots, i.e., to groups of four vines. The plots were divided into north-facing, west-facing, and east-facing blocks, with each treatment occurring once in each block. (Because the trial was conducted in the Southern hemisphere, there is no south-facing block!) For details of the layout, look ahead to Figure 9.3.

We do not lose any information, for purposes of comparing treatments, if we base the analysis on the plot means. (There is however loss of information for other forms of scrutiny of the data, as we note below.) Here are the first few rows of the data frame:

```
    yield block shade         plot
1 101.11 north   none north.none
2 108.02 north   none north.none
3 106.67 north   none north.none
4 100.30 north   none north.none
5  92.64  west   none  west.none
```

We now use the `aggregate()` function to form the data frame that holds the means for each combination of block and plot:[16]

```
    block   shade meanyield
1   east    none   99.0250
2  north    none  104.0250
3   west    none   97.5575
4   east Aug2Dec  105.5550
```

What have we lost in performing this aggregation? Use of the aggregated data for analysis commits us to working with plot means. If there were occasional highly aberrant values, use of medians might be preferable. The data should have a say in determining the form of summary that is used.

When data sets are large, there may be special advantages in working with a suitable form of data summary. Thus consider a data set that has information on the worldwide distribution of dengue (Hales et al., 2002). Dengue is a mosquito-borne disease that is a

---

[16]
```
data(kiwishade)       # Assumes that DAAG package is loaded.
attach(kiwishade)
kiwimeans <- aggregate(yield, by=list(block, shade), mean)
names(kiwimeans) <- c("block","shade","meanyield")
detach(kiwishade)
kiwimeans[1:4, ]
```

risk in hot and humid regions. Dengue status – information on whether dengue had been reported during 1965–1973 – is available for 2000 administrative regions, while climate information is available on a much finer scale, on a grid of about 80 000 pixels at 0.5° latitude and longitude resolution. Should the analysis work with a data set that consists of 2000 administrative regions, or with the much larger data set that has one row for each of the 80 000 pixels? There are three issues here that warrant attention:

- Dengue status is a summary figure that is given by administrative region. An analysis that uses the separate data for the 80 000 pixels will, in effect, predict dengue status for climate variable values that are in some sense averages for the administrative region. Explicit averaging, prior to the analysis, gives the user control over the form of averaging that will be used. If, for example, values for some pixels are extreme relative to other pixels in the administrative region, medians may be more appropriate than means. In some regions, the range of climatic variation may be extreme. The mean will give the same weight to sparsely populated cold mountainous locations as to highly populated hot and humid locations on nearby plains.
- Correlation between observations that are close together geographically, though still substantial, will be less of an issue for the data set in which each row is an administrative region. Points that repeat essentially identical information are a problem both for the interpretation of plots and, often, for the analysis. Regions that are geographically close will often have similar climates and the same dengue status.
- Analysis is more straightforward with data sets that are of modest size. It is easier to do the various checks that are desirable. The points that appear on plots are more nearly independent. Plots are less likely to appear as a dense mass of black ink.

For all these reasons it is preferable to base the main analysis on some form of average of climate data by administrative region. There are many possible ways to calculate an average, of which the mean and the median are the most common.

2.2.2 Standard deviation and inter-quartile range

An important measure of variation in a population is the population standard deviation (often written σ). This is a population parameter that is almost always unknown. The variance, which is the square of the standard deviation, is widely used in formal inference.

The population standard deviation can be estimated from a random sample, using the sample standard deviation

$$s = \sqrt{\frac{\sum(x - \bar{x})^2}{n - 1}}.$$

In words, given n data values, take the difference of each data value from the mean, square, add the squared differences together, divide by $n - 1$, and take the square root. (In R, the standard deviation is calculated using the function sd(), and the variance is calculated using var().)

For s to be an accurate estimate of σ, the sample must be large. The standard deviation is similar in concept to the inter-quartile range H, which we saw in Subsection 2.1.1 is the

Table 2.1: *Standard deviations for cuckoo eggs data.*

Hedge sparrow	Meadow pipit	Pied wagtail	Robin	Tree pipit	Wren
1.049	0.920	1.072	0.682	0.880	0.754

difference between the first and third quartiles. For data that are approximately normally distributed, we have the relationship

$$s \approx 0.75H.$$

If the data are approximately normally distributed, one standard deviation either side of the mean takes in roughly 68% of the data, whereas the region between the lower and upper quartiles takes in 50% of the data.

Cuckoo eggs example

Consider again the data on cuckoo eggs that we discussed in Subsection 2.1.2. The group standard deviations are listed in Table 2.1.[17]
    The variability in egg length is smallest when the robin or the wren is the host.

Degrees of freedom

If in calculating s we had divided by n rather than $n - 1$ before taking the square root, we would have computed the average of the squared differences of the observations from their sample average. The denominator $n - 1$ is the number of degrees of freedom remaining after estimating the mean. With one data point, the sum of squares about the mean is zero, the degrees of freedom are zero, and no estimate of the variance is possible. The degrees of freedom are the number of data values, additional to the first data value.
    The number of degrees of freedom is reduced by 1 for each parameter estimated. The standard deviation calculation described above sums squared differences of data values from their sample mean, divides by the degrees of freedom, and takes the square root. The standard deviation is in the same units as the original measurements.

The pooled standard deviation

Consider two independent samples of sizes n_1 and n_2, respectively, randomly selected from populations that have the same amount of variation but for which the means may differ. Thus, two means must be estimated. The number of degrees of freedom remaining for estimating the (common) standard deviation is $n_1 + n_2 - 2$. We compute the so-called pooled standard deviation by summing squares of differences of each data value from their respective sample mean, dividing by the degrees of freedom $n_1 + n_2 - 2$, and taking the square root:

$$s_p = \sqrt{\frac{\sum(x - \bar{x})^2 + \sum(y - \bar{y})^2}{n_1 + n_2 - 2}}.$$

[17] ```
Code to calculate group standard deviations
data(cuckoos)
sapply(split(cuckoos$length, cuckoos$species), sd)
# Subsection 12.6.7 has information on split()
```
```

Use of this pooled estimate of the standard deviation is appropriate if variations in the two populations are plausibly similar. The pooled standard deviation is estimated with more degrees of freedom, and therefore, more accurately, than either of the separate standard deviations.

Elastic bands example

Consider data from an experiment in which 21 elastic bands were randomly divided into two groups, one of 10 and one of 11. Bands in the first group were immediately tested for the amount that they stretched under a weight of 1.35 kg. The other group were dunked in hot water at 65°C for four minutes, then left at air temperature for ten minutes, and then tested for the amount that they stretched under the same 1.35 kg weight as before. The results were:

  Ambient: 254 252 239 240 250 256 267 249 259 269 (Mean = 253.5)

  Heated: 233 252 237 246 255 244 248 242 217 257 254 (Mean = 244.1)

The pooled standard deviation estimate is $s = 10.91$, with 19 ($= 10 + 11 - 2$) degrees of freedom. Since the separate standard deviations ($s_1 = 9.92$; $s_2 = 11.73$) are quite similar, the pooled standard deviation estimate is a sensible summary of the variation in the data set.

2.2.3 Correlation

The usual Pearson or product–moment correlation is a summary measure of linear relationship. Calculation of a correlation should always be accompanied by examination of a relevant scatterplot. The scatterplot provides a useful visual check that the relationship is linear. Often the addition of a smooth trend line helps the assessment. If the relationship is not linear, but is monotonic, it may be appropriate to use a Spearman rank correlation. Use, e.g., `cor.test(Animals$body, Animals$brain)` to calculate the Pearson product–moment correlation, and `cor.test(Animals$body, Animals$brain, method="spearman")` to calculate the Spearman correlation. The correlation is included with other output.

Figure 2.12 gives a few graphs to consider. For which does it make sense to calculate:

1. a Pearson correlation coefficient?
2. a Spearman rank correlation?

The figure that appears in the upper left in each panel is the Pearson correlation. For the lower right panel, the Pearson correlation is 0.882, while the Spearman correlation, which better captures the strength of the relationship, is 0.958. Here a linear fit clearly is inadequate. The magnitude of the correlation r, or of the squared correlation r^2, does not of itself indicate whether the fit is adequate.

Here are ways in which the use of correlation may mislead:

*   Values may not be independent. If we wish to make anything of the magnitude of the coefficient, then it is necessary to assume that sample pairs (x, y) have been taken at random from a bivariate normal distribution.
*   There may be some kind of subgroup structure in the data. If, for example, values of x and/or y are quite different for males and females, then the correlation may only

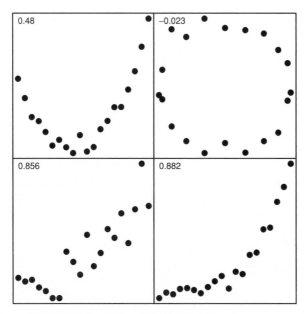

Figure 2.12: Different relationships between y and x. In the lower right panel, the Pearson correlation is 0.882, while the Spearman rank correlation is 0.958.

reflect a difference between the sexes. Or if random samples are taken from each of a number of villages, then it is unclear whether the correlation reflects a correlation between village averages or a correlation between individuals within villages, or a bit of each. The interpretation is confused because the two correlations may not be the same, and may even go in different directions. See Cox and Wermuth (1996).

• Any correlation between a constituent and a total amount is likely to be, in part at least, a mathematical artifact. Thus, consider a study of an anti-hypertensive drug that hopes to determine whether the change $y - x$ is larger for those with higher initial blood pressure. If x and y have similar variances then, unfortunately, $y - x$ will have a negative correlation with x, whatever the influence of the initial blood pressure.

Note that while a correlation coefficient may sometimes be a useful single number summary of the relationship between x and y, regression methods offer a much richer framework for the examination of such relationships.

2.3 Statistical Analysis Strategies

We have emphasized the importance of careful initial scrutiny of the data. Techniques of a broadly EDA type have, in addition, a role in scrutinizing results from formal analysis, in checking for possible model inadequacies and perhaps in suggesting remedies. In later chapters, we will discuss the use of diagnostic statistics and graphs in examination both of the model used and of output from the analysis. These are an "after the event" form of EDA. In the course of an analysis, the data analyst may move backwards and forwards between exploratory analysis and more formal analyses.

2.3.1 Helpful and unhelpful questions

Analyses ask questions of data. The questions may be simple: "Does increasing the amount of an additive in milk make it seem sweeter? If so, by how much does its sweetness increase?"

Or they may ask questions about relationships, as in the electrical resistance example plotted in Figure 2.6: "What is the relationship between electrical resistance and apparent juice content? How accurately can we predict resistance?" This is far more informative than doing repeated trials, some with a juice content of 30% and some with a juice content of 50%. With the data we have, it would be bad practice to do a formal statistical test to compare, for example, juice content of less than 30% with juice content of more than 50%. Even worse would be a study comparing resistance at 40% juice content with resistance at 50% juice content; the result would be a complete failure to detect the relatively rich relationship that exists between the apparent juice content and resistance.

With these data, we can do much better than that. We can study the relationship and ask whether a line or other simple curve is an adequate description of it. If we fit a line to part of the data, we might ask where the pattern changes. We can also ask how we could change the experiment so that we can obtain a more accurate curve.

2.3.2 Planning the formal analysis

Where existing data are available, a cautious investigator will use them to determine the form of analysis that is appropriate for the main body of data. If the data to be collected are closely comparable to data that have been analyzed previously, the analyst will likely know what to expect. It is then possible and desirable to plan the analysis in advance. This reduces the chance of biasing the results of the analysis in a direction that is closest to the analyst's preference! Even so, graphical checks of the data should precede formal analysis. There may be obvious mistakes. The data may have surprises for the analyst.

If available at the beginning of the study, the information from the analysis of earlier data may, additionally, be invaluable in the design of data collection for the new study. When prior data are not available, a pilot study involving a small number of experimental runs can sometimes be used to provide this kind of information.

Where it is not altogether clear what to expect, careful preliminary examination is even more necessary. In particular, the analyst should look for

- outliers,
- clusters in the data,
- unexpected patterns within groups,
- between group differences in the scatter of the data,
- whether there are unanticipated time trends associated, e.g., with order of data collection.

In all studies, it is necessary to check for obvious data errors or inconsistencies. In addition, there should be checks that the data support the intended form of analysis.

2.3.3 Changes to the intended plan of analysis

Planning the formal analysis is one aspect of planning a research study. Such advance planning should allow for the possibility of limited changes as a result of the exploratory analysis. As noted above, the best situation is where there are existing data that can be used for a practice run of the intended analysis.

What departures from the original plan are acceptable, and what are not? If the exploratory analysis makes it clear that the data should be transformed in order to approximate normality more closely, then use the transformation. It is sometimes useful to do both analyses (with the untransformed as well as with the transformed data) and compare them.

On the other hand, if there are potentially a large number of comparisons that could be made, the comparisons that will be considered should be specified in advance. Prior data, perhaps from a pilot study, can assist in this choice. Any investigation of other comparisons may be undertaken as an exploratory investigation, a preliminary to the next study.

Data-based selection of one or two comparisons from a much larger number is not appropriate, since huge biases may be introduced. Alternatively there must be allowance for such selection in the assessment of model accuracy. The issues here are non-trivial, and we defer further discussion until later.

2.4 Recap

Exploratory data analysis aims to allow the data to speak for themselves, often prior to or as part of a formal analysis. It gives multiple views of the data that may provide useful insights. Histograms, density plots, stem-and-leaf displays and boxplots are useful for examining the distributions of individual variables. Scatterplots are useful for looking at relationships two at a time. If there are several variables, the scatterplot matrix provides a compact visual summary of all two-way relationships.

Before analysis, look especially for

- outliers,
- skewness (e.g., a long tail) in the distribution of data values,
- clustering,
- non-linear bivariate relationships,
- indications of heterogeneous variability (i.e., differences in variability across samples),
- whether transformations seem necessary.

After analysis, check residuals for all these same features. Where the analysis examines relationships involving several variables, adequate checks may be possible only after analysis.

As a **preliminary to formal data analysis**, exploratory data analysis provides a useful set of checks. It may be hard to detect some types of departures from the assumptions that may be required for later formal analysis. Failure of the independence assumption is hard to detect, unless the likely form of dependence is known and the sample is large. Be aware of any structure in the data that may be associated with lack of independence.

2.5 Further Reading

The books and papers on graphical presentation that were noted in Chapter 1 are equally relevant to this chapter. Cleveland (1993) is especially pertinent to the present chapter. Chatfield (2002) has a helpful and interesting discussion, drawing on consulting experience, of approaches to practical data analysis.

On statistical presentation issues, and deficiencies in the published literature, see Andersen (1990), Chanter (1981), Gardner et al. (1983), Maindonald (1992), Maindonald and Cox (1984) and Wilkinson et al. (1999). The Wilkinson et al. paper has helpful comments on the planning of data analysis, the role of exploratory data analysis, and so on. Nelder (1999) is challenging and controversial.

Two helpful web pages are
http://www.math.yorku.ca/SCS/friendly.html#graph
and http://www.rdg.ac.uk/ssc/dfid/booklets.html

References for further reading

Andersen, B. 1990. *Methodological Errors in Medical Research: an Incomplete Catalogue.* Blackwell Scientific.

Chanter, D.O. 1981. The use and misuse of regression methods in crop modelling. In: *Mathematics and Plant Physiology*, eds. D.A. Rose and D.A. Charles-Edwards. Academic Press.

Chatfield, C. 2002. Confessions of a statistician. *The Statistician* 51: 1–20.

Cleveland, W.S. 1993. *Visualizing Data.* Hobart Press.

Gardner, M.J., Altman, D.G., Jones, D.R. and Machin, D. 1983. Is the statistical assessment of papers submitted to the *British Medical Journal* effective? *British Medical Journal* 286: 1485–1488.

Maindonald J.H. 1992. Statistical design, analysis and presentation issues. *New Zealand Journal of Agricultural Research* 35: 121–141.

Maindonald, J.H. and Cox, N.R. 1984. Use of statistical evidence in some recent issues of DSIR agricultural journals. *New Zealand Journal of Agricultural Research* 27: 597–610.

Nelder, J.A. 1999. From statistics to statistical science. *Journal of the Royal Statistical Society, Series D*, 48: 257–267.

Wilkinson, L. and Task Force on Statistical Inference 1999. Statistical methods in psychology journals: guidelines and explanation. *American Psychologist* 54: 594–604.

2.6 Exercises

1. For the possum data set, use hist (possum$age) to draw a histogram of possum ages. Where are the breaks, i.e., the class boundaries between successive rectangles? Repeat the exercise, this time specifying

    ```
    hist(possum$age, breaks=c(0,1.5,3,4.5,6,7.5,9))
    ```

    Use table(cut(possum$age, breaks=c(0,1.5,3,4.5,6,7.5,9))) to obtain the table of counts. In which interval are possums with age=3 included; in (1.5,3] or in

(3,4.5]? List the values of age that are included in each successive interval. Explain why setting `breaks=c(0,1.5,3,4.5,6,7.5,9)` leads to a histogram that is misleading.

2. Now try `plot(density(possum$age, na.rm=TRUE))`. Which is the most useful representation of the distribution of data – one or other histogram, or the density plot? What are the
benefits and disadvantages in each case?

3. Examine the help for the function `mean()`, and use it to learn about the trimmed mean. For the
total lengths of female possums, calculate the mean, the median, and the 10% trimmed mean.
How does the 10% trimmed mean differ from the mean for these data? Under what circumstances
will the trimmed mean differ substantially from the mean?

4. Assuming that the variability in egg length for the cuckoo eggs data is the same for all host birds,
obtain an estimate of the pooled standard deviation as a way of summarizing this variability.
[Hint: Remember to divide the appropriate sums of squares by the number of degrees of freedom
remaining after estimating the six different means.]

5. Plot a histogram of the `earconch` measurements for the `possum` data. The distribution should
appear *bimodal* (two peaks). This is a simple indication of clustering, possibly due to sex differences. Obtain side by side boxplots of the male and female `earconch` measurements. How
do these measurement distributions differ? Can you predict what the corresponding histograms
would look like? Plot them to check your answer.

6. Download and load the package *Devore5*, available from the CRAN sites. Then gain access to
data on tomato yields by typing

```
library(Devore5)
data(ex10.22)
tomatoes <- ex10.22
```

This data frame gives tomato yields at four levels of salinity, as measured by electrical conductivity
(EC, in nmho/cm).

(a) Obtain a scatterplot of `yield` against EC.
(b) Obtain side by side boxplots of `yield` for each level of EC.
(c) The third column of the data frame is a factor representing the four different levels of
    EC. Comment upon whether the yield data are more effectively analyzed using EC as a
    quantitative or qualitative factor.

3

Statistical Models

An engineer may build a scale model of a proposed bridge or a building. Medical researchers may speak of using some aspect of mouse physiology as a model for human physiology. The hope is that experiments in the mouse will give a good idea of what to expect in humans. The model captures important features of the object that it represents, enough features to be useful for the purpose in hand. A scale model of a building may be helpful for checking the routing of the plumbing but may give little indication of the acoustics of seminar rooms that are included in the building. Clarke (1968) comments that: "Models and hypotheses succeed in simplifying complex situations by ignoring information outside their frame and by accurate generalization within it."

Our interest is in mathematical and statistical models. Mathematical or deterministic models are used to describe regular law-like behavior. In fundamental research in the physical sciences, deterministic models are often adequate. Statistical variability may be so small that it can, for many purposes, be ignored. Applications of the natural sciences may be a different matter. In studying how buildings respond to a demolition charge, there will be variation from one occasion to another, even for identical buildings and identically placed charges. There will be variation in which parts of the building break first, in what parts remain intact, and in the trajectories of fragments.

Statistical models rely on probabilistic forms of description that have wide application over all areas of science. They often consist of a deterministic component as well as a random component that attempts to account for variation that is not accounted for by a law-like property.

Models should, wherever possible, be scientifically meaningful, but not at the cost of doing violence to the data. The scientific context includes the analyses, if any, that other researchers have undertaken with related or similar data. It can be important to note and use such analyses critically. While they may give useful leads, there can be serious inadequacies in published analyses. Further discussion on this point can be found in articles and books given in the list of references at the end of Chapter 2.

As we saw in Chapter 2, consideration of a model stays somewhat in the background in initial efforts at exploratory data analysis. The choice of model is of crucial importance in formal analysis. The choice may be influenced by previous experience with comparable data, by subject area knowledge, and by exploratory analysis of the data that are to be analyzed.

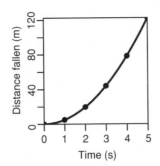

Figure 3.1: Distance fallen versus time, for a stone that starts at rest and falls freely under gravity.

3.1 Regularities

We take a variety of regularities for granted in our everyday lives. We expect that the sun will rise in the morning and set in the evening, that fire will burn us, and so on. It is our experience of the world, rather than logical deductive argument, that has identified these regularities. Scientific investigation, especially in the natural sciences, has greatly extended and systematized our awareness of regularities. Mathematical models have been crucial for describing and quantifying these regularities.

3.1.1 Mathematical models

Mathematical models may be deterministic, or they may incorporate a random component. We begin with a deterministic mathematical model. Figure 3.1 is a graphical representation of the formula for the distance that a falling object, starting at rest above the earth's surface, travels under gravity in some stated time.[1] The formula is

$$d = \frac{1}{2}gt^2$$

where t is the time in seconds, g (≈ 9.8 m/s^2) is the acceleration due to gravity, and d is the distance in meters. Thus, a freely falling object will fall 4.9 meters in the first second, 19.6 meters in the first two seconds, and so on.

There are important aspects of the fall about which the formula tells us nothing. It gives no indication of the likely damage if the stone were to strike one's foot! For the limited purpose of giving information about distance fallen it is, though, a pretty good formula. A good model captures those aspects of a phenomenon that are relevant for the purpose at hand, allowing us to get a handle on complex phenomena. The reward for simplifying by ignoring what seems irrelevant for present purposes is that the model is tractable – we can use it to make predictions.

The formula is not totally accurate – it neglects the effects of air resistance. Even if it were totally accurate, measurement inaccuracy would introduce small differences between

---

[1]
```
x <- 1:5
y <- 0.5 * 9.8 * x^2
plot(y ~ x, xlab = "Time (s)", ylab = "Distance fallen (m)", pch = 16)
lines(spline(x, y)) # Pass a smooth curve through the points
rm(x, y)
```

Table 3.1: *Depression* (`depression`), *and Depression/Weight Ratio, for different weights* (`weight`) *of lawn roller.*

	weight (t)	depression (mm)	depression/weight
1	1.9	2	1.1
2	3.1	1	0.3
3	3.3	5	1.5
4	4.8	5	1.0
5	5.3	20	3.8
6	6.1	20	3.3
7	6.4	23	3.6
8	7.6	10	1.3
9	9.8	30	3.1
10	12.4	25	2.0

observations and predictions from the formula. We could, if we wished, model these inaccuracies, thus making the model a statistical model.

Regular law-like behavior is harder to discern in the biological sciences, and even more difficult in the social sciences. The equation describing the trajectory of a falling object is a striking contrast with our incomplete understanding of the "forces" that give rise to antisocial behavior. No deterministic law exists to explain such phenomenona, though statistical models that incorporate both regular and random components may have a useful role.

3.1.2 Models that include a random component

Statistical models typically include at least two components. One component describes deterministic law-like behavior. In engineering terms, that is the *signal*. The other component is random, often thought of as *noise*, i.e., subject to statistical variation. Typically, we assume that the elements of the noise component are uncorrelated.

Figure 3.2, which plots the data that are shown in Table 3.1, is an example. Different weights of roller were rolled over different parts of a lawn, and the depression noted (data are from Stewart et al., 1988).

We might expect depression to be proportional to roller weight. That is the signal part. The variation exhibited in the values for Depression/Weight makes it clear that this is not the whole story. The points in Figure 3.2 do not lie on a line.[2] We therefore model deviations from the line as random *noise*:

$$\text{depression} = b \times \text{weight} + \text{noise}.$$

Here b is a constant, which we do not know but can try to estimate. The noise is different for each different part of the lawn. If there were no noise, all the points would lie exactly on a line, and we would be able to determine the slope of the line exactly.

---

[2] `data(roller)`   `# The DAAG package must be loaded`
  `plot(depression ~ weight, data = roller,`
      `xlab = "Weight of roller (t)", ylab = "Depression(mm)", pch = 16)`
  `abline(0, 2.25)`   `# A slope of 2.25 looks about right`

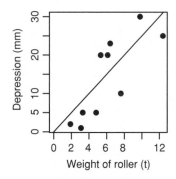

Figure 3.2: Depression in lawn, versus roller weight. The line, through the origin, was drawn by eye.

3.1.3 Smooth and rough

In discussing statistical models, it may be helpful to speak of the "smooth" and the "rough". Many models have the form

$$\text{observed value} = \text{model prediction} + \text{statistical error},$$

or, in terminology perhaps more familiar to engineers,

$$\text{observed value} = \text{signal} + \text{noise}.$$

Mathematically, one may write

$$Y = \mu + \varepsilon$$

(often μ is a function of explanatory variables). The model prediction (μ) is the "smooth" component of the model. The statistical error (ε) is the "rough" component of the model. Using the mathematical idea of *expected value*, it is usual to define $\mu = E(Y)$, where the $E(Y)$ denotes "expected value of Y". The expected value generalizes the mean.

In regression analysis, a line or curve is fitted through the plotted points. The fitted model is the "smooth" component, used to predict the signal or *response*. For example, after fitting a line through the lawn roller data, we can predict the depression that would result for a given weight. For assessing the accuracy of the model fit, the residuals are needed. These are the differences between observed values of `depression` (the "response"), and predictions of depression at the respective values of `weight` (the "predictor").

The accuracy of the prediction depends upon the amount of noise. As noted earlier, if there is no noise, there will be no error when using the line to make a prediction. A lot of noise renders accurate prediction difficult. Residuals are what is "left over" after fitting the linear model. The residuals are the "rough" component of the model, and estimate the noise.

Figure 3.3A exhibits the residuals for the lawn roller data after fitting a straight line.

An alternative to fitting a line is to fit a smooth curve. Sometimes, this helps indicate whether a line really is appropriate. In Figure 3.3B, the smooth curve has been obtained using `lowess()`. Note that there is just one point that seems to be causing the line, and the fitted curve, to bend down. In any case, there is no statistical justification for fitting a curve rather than a line, as can be verified with a formal analysis. There is serious "over-fitting" in Figure 3.3B.

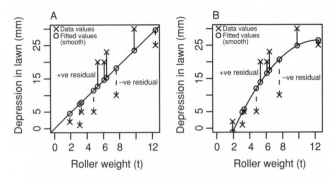

Figure 3.3: In A a line has been fitted, while in B the loess method was used to fit a smoothed curve. Residuals (the "rough") appear as vertical lines. Positive residuals are black lines, while negative residuals are dashed. Figures 3.3A and 3.3B were created using our function g3.3(), which is available from our web pages. Interested readers can check the code.

3.1.4 The construction and use of models

Any model must, in the first place, be useful. It must yield inferences that, for its intended use, are acceptably accurate. Often, the intended use is prediction. Alternatively, or additionally, there may be an interest in model parameters. Thus for the lawn roller data of Table 3.1, one focus of interest is the rate of increase of depression with increasing roller weight, i.e., the slope of the line.

Accurate description of the data, over the range of data values, is a desirable requisite. Another is scientific meaningfulness, which is a reason for making the line in Figure 3.3 pass through the origin.

Model structure should reflect data structure. We think of the pattern of change of depression with roller weight as a deterministic or *fixed* effect. Actual measurements incorporate, in addition, a random effect that reflects variation from one part of the lawn to another, differences in the handling of the roller, and measurement error. We require a model that accurately reflects both fixed and random sources of variation. Elementary statistics courses typically emphasize fixed effects, with a single random source of variation superimposed.

In practical contexts, multiple random sources of variation are the rule rather than the exception. If there had been multiple lawns, the effects would undoubtedly have differed from one lawn to another. Thus we would have to reckon with between lawn variation, superimposed on the within lawn variation on which our data give information. Data from multiple lawns are essential, for anything more than qualitative generalization to other lawns.

3.1.5 Model formulae

R's modeling functions use model formulae to describe the role of variables and factors in models. A large part of the data analyst's task is to find the model formula that will be effective for the task in hand. For example, the following model statement, with model formula depression ~ weight, fits a line to the data of Table 3.1: (if the data frame `roller` is not already in the workspace, it will be necessary to load it from our *DAAG* package):

```
lm(depression ~ weight, data=roller)    # Prints out summary
                                        # information
```

Note however that this fits a model that has an intercept term. For a model that omits the intercept term, specify

```
lm(depression ~ -1 + weight, data=roller)
```

Model formulae give an economical way to specify the way that variables and factors enter into a model. Once understood, they aid economy of thought and description.

3.2 Distributions: Models for the Random Component

A distribution is a model for a population of data values. Mathematically, the notion of distribution allows us to deal with variability. In comparing two groups, a model for the variability is required in each group. In regression, a model is required for variation of the responses about a line or curve.

Discrete distributions are conceptually simpler than continuous distributions. We discuss important discrete distributions immediately below, before going on to discuss the normal and other continuous distributions.

3.2.1 Discrete distributions

Bernoulli distribution

Successive tosses of a fair coin come up tails (which we count as zero) with probability 0.5, and heads (which we count as one) with probability 0.5, independently between tosses. More generally, we may have a probability $1 - \pi$ for tails and π for heads. We say that the number of heads has a Bernoulli distribution with parameter π. Alternatively, we could count the number of tails and note that it has a Bernoulli distribution with parameter $1 - \pi$.

Binomial distribution

The total number of heads in n tosses of a fair coin follows a binomial distribution (the Bernoulli distribution is the special case for which $n = 1$) with size n and $\pi = .5$. The numbers of female children in two-child families can be modeled as binomial with $n = 2$ and $\pi \approx .5$. We can use the function `dbinom()` to determine probabilities of having 0, 1 or 2 female children:[7]

```
   0    1    2
0.25 0.50 0.25
```

---

[7] `dbinom(0:2, size=2, prob=0.5)    # Simple version`
   `## To get the labeling (0, 1, 2) as in the text, specify:`
   `probs <- dbinom(0:2, size=2, prob=0.5)`
   `names(probs) <- 0:2`
   `probs`

On average, 25% of two-child families will have no female child, 50% will have one female child, and 25% will have two female children. For another example, we obtain the distribution of female children in four-child families:[8]

```
     0        1        2        3        4
0.0625   0.2500   0.3750   0.2500   0.0625
```

If we want to calculate the probability that a four-child family has no more than 2 females, we must add up the probabilities of 0, 1, and 2 females (.0625 + .2500 + .3750 = .6875). The function `pbinom()` can be used to determine such cumulative probabilities.[9]

As a final example, suppose a sample of 50 manufactured items is taken from an assembly line that produces 20% defective items, on average.[10] The probability of observing fewer than 5 defectives in the sample is .0185.

The function `qbinom()` goes in the other direction, i.e., from cumulative probabilities to number of events. Thus, in the four-child family example, `qbinom(p = 0.65, size = 4, prob = 0.5)` gives the minimum number of females such that the cumulative probability is greater than or equal to 0.65. We will make no further use of `qbinom()`.

Poisson distribution

The Poisson distribution is often used to model counts of defects or events that occur relatively rarely. The R functions follow the same pattern as for the functions above, i.e., they are `dpois()`, `ppois()` and `qpois()`.

As an example, consider a population of raisin buns for which there are an average of 3 raisins per bun. Any individual raisin has a small probability of finding its way into any individual bun. We have the following probabilities for 0, 1, 2, 3, or 4 raisins per bun:[11]

```
     0        1        2        3        4
0.0498   0.1494   0.2240   0.2240   0.1680
```

The cumulative probabilities are[12]

```
     0        1        2        3        4
0.0498   0.1991   0.4232   0.6472   0.8153
```

Thus, for example, the probability of observing 2 or fewer raisins in a bun is .4232.

3.2.2 Continuous distributions

Normal distribution

The normal distribution, which has the bell-shaped density curve pictured in Figure 3.4, is often used as a model for continuous measurement data (sometimes a transformation of the data is required in order for the normal model to be useful). The height of the curve is a function of the distance, measured in number of standard deviations, from the mean.

---

[8] dbinom(0:4, size=4, prob=0.5)
[9] pbinom(q=2, size=4, prob=0.5)
[10] pbinom(q=4, size=50, prob=0.2)
[11] dpois(x = 0:4, lambda = 3)   # lambda = mean number of raisins per bun
[12] ppois(q = 0:4, lambda = 3)

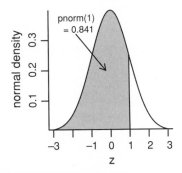

Figure 3.4: A plot of the normal density. The horizontal axis is labeled in standard deviations distance from the mean. The area of the shaded region is the probability that a normal random variable has a value less than one standard deviation above the mean.

The area under the density curve is 1. The density curve plotted in Figure 3.4 corresponds to a normal distribution with a mean of 0, located at the peak or mode of the density. By adding a fixed value μ to a population of such normal variates, we can change the mean to μ, leaving the standard deviation unchanged.

Functions for calculations with continuous distributions work a little differently from functions for calculation with discrete distributions. Whereas `dbinom()` gives probabilities, `dnorm()` gives values of the probability density function.[13] The function `pnorm()` calculates the cumulative probability, i.e., the area under the curve up to the specified ordinate or x-value. For example, there is a probability of .841 that a normal deviate is less than 1.[14] This corresponds to the area of the shaded region in Figure 3.4.[15] The function `qnorm()` calculates the deviate that corresponds to a given cumulative probability, i.e., the area under the curve up to the specified ordinate. The q stands for *quantile*. Another term, that has in mind the division of the area into 100 equal parts, is *percentile*. For example, the 90th percentile is 1.645.[16]

Other continuous distributions

There are many other statistical models for continuous observations. The simplest model is the uniform distribution, for which an observation is equally likely to take any value in

---

[13]
```
## The following plots the normal density, in the range -3 to 3
z <- pretty(c(-3,3), 30) # Divide the interval up into about 30 numbers
ht <- dnorm(z)            # By default: mean=0, variance=1
plot(z, ht, type="l", xlab="Normal deviate", ylab="Ordinate", yaxs="i")
axis(1, at=1, tck=dnorm(1)/par()$usr[4], labels=FALSE)
# Around 84.1% of the total area is to the left of the vertical line.
```
[14] `pnorm(1.0)   # assuming mean=0 and standard deviation=1`
[15]
```
## Additional examples
pnorm(0)           # .5     (exactly half the area is to the left of the mean)
pnorm(-1.96)       # .025
pnorm(1.96)        # .975
pnorm(1.96, mean=2) # .484  (a normal distribution with mean 2 and SD 1)
pnorm(1.96, sd=2)  # .836  (sd = standard deviation)
```
[16]
```
qnorm(.9)          # 90th percentile for normal distn with mean=0 and SD=1
## Additional examples are
qnorm(0.841)       # 1.0
qnorm(0.5)         # 0
qnorm(0.975)       # 1.96
qnorm(c(.1,.2,.3)) # -1.282 -0.842 -0.524  (10th, 20th and 30th percentiles)
qnorm(.1, mean=100, sd=10) # 87.2 (10th percentile, mean=100, SD=10)
```

a given interval. In more precise technical language, the probability density of values is constant on a fixed interval.

Another model is the exponential distribution that gives high probability density to positive values lying near 0; the probability density decays exponentially as the values increase. This simple model has been used to model times between arrivals of customers to a queue. The exponential distribution is a special case of the chi-squared distribution. The latter distribution arises, for example, when dealing with contingency tables. Details on computing probabilities for these distributions can be found in the exercises.

3.3 The Uses of Random Numbers

3.3.1 Simulation

R has functions that generate random numbers from some specified distributions. To take a simple example, we can simulate a random sequence of 10 0s and 1s from a population with some specified proportion of 1s, e.g., 50% (i.e., a Bernoulli distribution):[17]

```
1 0 0 0 1 1 1 0 1 0
```

The random sample is different on each occasion, depending on the setting of a starting number that is called the *seed*. Occasionally, it is desirable to set the seed for the random number generator so that the selection of sample elements is the same on successive executions of a calculation.

Use of the function `set.seed()` makes it possible to use the same random number seed in two or more successive calls to a function that uses the random number generator. Ordinarily, this is undesirable. However, users will sometimes, for purposes of checking a calculation, wish to repeat calculations with the same sequence of random numbers as was generated in an earlier call. (To obtain the sample above, specify `set.seed(23826)` before issuing the `rbinom` command.)

We can simulate values from several other distributions. We will offer examples of simulated binomial, Poisson, and normal samples.

To generate the numbers of daughters in a simulated sample of 25 four-child families, assuming that males and females are equally likely, we use the `rbinom()` function:[18]

```
3 1 2 4 1 2 0 3 2 1 2 3 2 4 2 1 1 1 2 2 3 2 0 2 2
```

Now consider the raisin buns, with an overall average of 3 raisins per bun. Here are numbers of raisins for a simulated random sample of 20 buns:[19]

```
1 3 2 3 6 1 5 7 1 6 2 3 6 2 2 3 5 4 2 3
```

The function `rnorm()` calculates random deviates from the normal distribution. For example, the following are 10 random values from a normal distribution with mean 0 and standard deviation 1:[20]

```
0.187   0.053   1.689  -1.106  -0.646   0.213  -0.474  -0.913
0.400   0.039
```

---

[17] `rbinom(10, size=1, p=.5)`   # 10 Bernoulli trials, prob=0.5
[18] `rbinom(25, size=4, prob=0.5)` # For our sequence, precede with set.seed(9388)
[19] `rpois(20, 3)`   # For our sequence, precede with set.seed(4982)
[20] `options(digits=2)`   # controls the number of displayed digits
    `rnorm(10)`   # 10 random values from the normal distribution
                                    # For our sequence, precede with set.seed(3663)

Calculations for other distributions follow the same pattern. For example, uniform random numbers are generated using `runif()` and exponential random numbers are generated using `rexp()`.[21] The reader is encouraged to try the related exercises at the end of this chapter.

3.3.2 Sampling from populations

Here we will discuss probability-based methods for choosing random samples. These methods have important practical uses. They contrast with more informal methods for choosing samples.

Consider first a sample from a *finite* population. Suppose, for example, that names on an electoral roll are numbered, actually or notionally, from 1 to 9384. We can obtain a random sample of 15 individuals thus:[22]

```
[1] 9178 2408 8724  173  106 4664 3787 6381 5098 3228 8321
165 7332 9036  540
```

This gives us the numerical labels for the 15 individuals that we should include in our sample. The task is then to go out and find them! The option `replace=FALSE` gives a *without replacement* sample, i.e., it ensures that no one is included more than once.

Alternatively, we may have a sample from an *infinite* population. The weights of 5 plants that were given a nutritional supplement are a random sample from the population that consists of all weights that we might have observed. Suppose we have 10 plants (labeled from 1 to 10, inclusive) that we wish to randomly assign to two equal groups, control and treatment. One such random assignment is as follows:[23]

```
$Control
[1] 6 8 3 7 9

$Treatment
[1]  5  4  2  1 10
```

We then assign plants 6, 8, 3, 7, and 9 to the control group. By choosing the plants in such a manner, we avoid biases that could arise, for example, due to choosing healthier looking plants for the treatment group.

Cluster sampling

Cluster sampling is one of many different probability-based variants on simple random sampling. In surveys of human populations cluster-based sampling, e.g., samples of households or of localities, with multiple individuals from each chosen household or locality, is likely to introduce a cluster-based form of dependence. The analysis must then take account of

---

[21] `runif(n = 20, min=0, max=1) # 20 numbers, uniform distribution on (0, 1)`
   `rexp(n=10, rate=3)        # 10 numbers, exponential distribution, mean 1/3.`
[22] `sample(1:9384, 15, replace=FALSE)`
   `   # For our sequence, precede with set.seed(3676)`
[23] `split(sample(seq(1:10)), rep(c("Control","Treatment"),5))`
   `   # sample()  gives a random re-arrangement (permutation) of 1, 2, ..., 10`
   `   # For our sequence, precede with set.seed(366)`

this clustering. Standard inferential methods require adaptation to take account of the fact that it is the clusters that are independent, not the individuals within the clusters.

Resampling methods

We will later encounter methods that take repeated random samples from what are already random samples, in order to make inferences about the population from which the original sample came. These methods are called, not surprisingly, *resampling* methods.

3.4 Model Assumptions

Common model assumptions are normality, independence of the elements of the *error* or *noise* term, and homogeneity of variance. There are some assumptions whose failure is unlikely to compromise the validity of analyses. We say that the method used is *robust* against those assumptions. Other assumptions matter a lot. How do we know which is which? Much of the art of applied statistics comes from knowing which assumptions are important, and need careful checking. There are few hard and fast rules.

3.4.1 Random sampling assumptions – independence

Typically, the data analyst has a sample of values that will be used as a window into a wider population. The inferential methods that we will discuss require the assumption that the sample has been generated by a specific random mechanism. Almost all the standard elementary methods assume that all population values are chosen with equal probability, independently of the other sample values. Use of these methods can be extended in various ways to handle modifications of the simple independent random sampling scheme. For example, we can modify the methodology to handle analyses of data from a random sample of clusters of individuals.

Often samples are chosen haphazardly, e.g., an experimenter may pick a few plants from several different parts of a plot. Or a survey interviewer may, in a poor quality survey, seek responses from individuals who can be found in a shopping center. Self-selected samples can be particularly unsatisfactory, e.g., those readers of a monthly magazine who are sufficiently motivated to respond to a questionnaire that is included with the magazine.

In practice, analysts may make the random sampling assumption when the selection mechanism does not guarantee randomness. Inferences from data that are chosen haphazardly are inevitably less secure than where we have random samples. Random selection avoids the conscious or unconscious biases that result when survey or other samplers make their own selection, or take whatever items seem suitable.

Failure of the independence assumption is a common reason for wrong statistical inferences. Failure is, at the same time, hard to detect. Ideally, data should be gathered in such a way that the independence assumption is guaranteed. This is why randomization is so important in designed experiments, and why random sampling, whether of individuals or of clusters, is so important in designed sample surveys.

With observational data, it is necessary to think carefully about the mechanisms that may have generated the data, and consider whether these are likely to have generated dependencies. Common sources of dependence are clustering in the data, and temporal or spatial dependence. Values within a cluster, or close together in time and/or space, may be more similar than those in different clusters. Temporal and spatial dependence arise because values that are close together in time or space are relatively more similar.

Tests for independence are at best an occasionally useful guide. They are of little use unless we have some idea how the assumption may have failed, and the sample is large! It is in general better to try to identify the nature of the dependence, and use a form of analysis that allows for it.

Ideally, we want the model to reflect the design of the data collection, i.e., the experimental or sampling design. Often however, the mechanisms that generated the data are not totally clear, and the model is at best a plausible guess. Models that do not obviously reflect mechanisms that generated the data can sometimes be useful for prediction. They can also, if their deficiencies are not understood or if they are used inappropriately, be misleading. Careful checking that the model is serving its intended purpose, and caution, are necessary.

3.4.2 Checks for normality

Many data analysis methods rest on the assumption that the data are normally distributed. Real data are unlikely to be exactly normally distributed. For practical purposes, two questions are important:

- How much departure from normality can we tolerate?
- How can we decide if it is plausible that the data are from a normal distribution?

The first question is difficult. Broadly, it is necessary to check for gross departures from normality, and particularly skewness. Small departures are of no consequence. For modest sized samples, only gross departures will be detectable.

Examination of histograms

What checks will detect gross departures? The approach taken in Chapter 2 was to draw a histogram or density plot. While histograms and density plots have their place, they are less useful than we might think. Figure 3.5 shows five histograms, obtained by taking five independent random samples of 50 values from a normal distribution.[24] None of these histograms show a close resemblance to a theoretical normal distribution.

---

[24] ## The following gives a rough equivalent of the figure:
```
set.seed (21)  # Use to reproduce the data in the figure
par(mfrow=c(2,3))
x <- pretty(c(6.5,13.5), 40)
for(i in 1:5){
    y <- rnorm(50, mean=10, sd=1)
    hist(y, prob=TRUE, xlim=c(6.5,13.5), ylim=c(0,0.5), main="")
    lines(x, dnorm(x,10,1))
    }
rm(x, y)
```

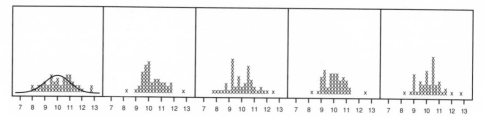

Figure 3.5: Each panel shows a simulated distribution of 50 values from a normal distribution with mean = 10 and sd = 1. The underlying theoretical normal curve is overlaid on the left panel.

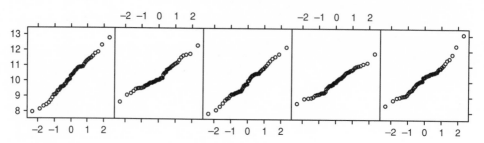

Figure 3.6: Normal probability plots for the same random normal data as were displayed in Figure 3.5.

The normal probability plot

A better tool for assessing normality is the normal probability plot. First the data values are sorted. These are then plotted against the ordered values that might be expected if the data really were from a normal distribution. If the data are from a normal distribution, the plot should approximate a straight line. Figure 3.6[25] shows normal probability plots for the same five sets of 50 normally distributed values as we displayed in Figure 3.5.

Figures such as Figure 3.6 help train the eye to know what to expect in samples of size 50. The process can be repeated several times. Such plots give a standard against which to compare the normal probability plot for the sample. They allow the data analyst to calibrate the eye, to obtain an idea of the sorts of departures from linearity that are to be expected in samples of random normal data.

The sample plot, set alongside plots for random normal data

Consider data from an experiment that tested the effect on stretchiness of leaving elastic bands in hot water for four minutes. Eighteen bands were first tested for amount of stretch under a load that stretched the bands by a small amount (the actual load was 425 g, thought small enough not to interfere with the elastic qualities of the bands). This information was used to arrange bands into nine pairs, such that the two members of a pair had similar initial stretch. One member of each pair, chosen at random, was placed in hot water (60–65 °C) for

---

[25] `set.seed(21)  # Use the same setting as for the previous figure`
```
library(lattice)
qqmath(~rnorm(50*5)|rep(1:5,rep(50,5)), layout=c(5,1), aspect=1)
## Alternatively, use the function qreference() from our DAAG package
qreference(m=50, seed=21, nrep=5, nrows=1) # 50 values in each panel
```

Table 3.2: *Eighteen elastic bands were divided into nine pairs, with bands of similar stretchiness placed in the same pair. One member of each pair was placed in hot water (60–65 °C) for four minutes, while the other was left at ambient temperature. After a wait of about ten minutes, the amounts of stretch, under a 1.35 kg weight, were recorded.*

	Pair #								
	1	2	3	4	5	6	7	8	9
Heated (mm)	244	255	253	254	251	269	248	252	292
Ambient	225	247	249	253	245	259	242	255	286
Difference	19	8	4	1	6	10	6	−3	6

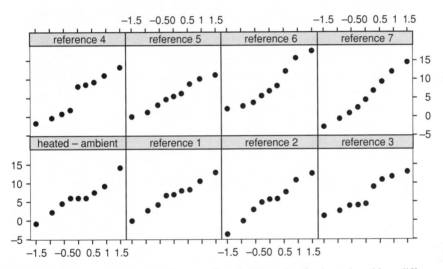

Figure 3.7: The lower left panel is the normal probability plot for heated–ambient differences. Remaining panels show plots for samples of nine numbers from a normal distribution.

four minutes. The other member of the pair remained at ambient temperature. All bands were then measured for amount of stretch under a load of 1.35 kg weight. Table 3.2 shows the results.

We are interested in the distribution of the differences. In the next chapter, these will be the basis for various statistical calculations. We present the normal probability plot for these data in the lower left panel of Figure 3.7.[26] The other seven plots are for samples (all of size 9) of simulated random normal values.

---

[26]
```
data(pair65)           # From our DAAG package
test <- pair65$heated - pair65$ambient; n <- length(test)
y <- c(test, rep(rnorm(n), 7))
fac <- c(rep("heated-ambient",n), paste("reference", rep(1:7, rep(n,7))))
qqmath(~ y|fac, aspect=1, layout=c(4,2))
## Alternative, using our function qreference()
qreference(pair65$heated - pair65$ambient, nrep=8)
```

The lower left panel shows the normal probability plot for the data that we have. Remaining plots give a standard against which to compare the plot for the experimental data. There is no obvious feature that distinguishes the plot in the lower left panel from the seven reference plots.

Formal statistical testing for normality?

There are formal statistical tests for normality. A difficulty with such tests is that normality is difficult to rule out in small samples, while in large samples the tests will almost inevitably identify departures from normality that are too small to have any practical consequence for standard forms of statistical analysis.

Additionally, in large samples, the effects of averaging may lead to normality of the statistic under consideration, even when the underlying population is clearly not normally distributed. Here, we obtain help from an important theoretical result, called the "Central Limit Theorem". This theorem states that the distribution of the sample mean approximates the normal distribution with arbitrary accuracy, provided a large enough sample is taken (there are regularity conditions that must be satisfied, but these can usually be taken for granted). There are similar results for a number of other sample statistics. A consequence is that, depending on the analysis that is to be performed, normality may not be an important issue for analyses where samples are large; tests for normality will detect non-normality in contexts where we have the least reason to be concerned about it.

3.4.3 Checking other model assumptions

In Chapter 2, we discussed a number of exploratory techniques that can aid in checking whether the standard deviation is the same for all observations in a data set. Following analysis, a plot of residuals against fitted values may give useful indications. For example, residuals may tend to fan out as fitted values increase, giving a "funnel" effect, a fairly sure sign that the standard deviation is increasing. Alternatively, or additionally, there may be evidence of outliers – one or more unusually large residuals. The major concern may however be to identify points, whether or not outliers, that have such high *influence* that they distort model estimates.

3.4.4 Are non-parametric methods the answer?

While sometimes useful, classical non-parametric tests that purport to be assumption-free are not the answer to every problem of failure of assumptions. These tests do rest on assumptions, and we still have to be assured that these assumptions are realistic. If used in a way that ignores structure in the data that we should be modeling, we risk missing insights that parametric methods may provide. Building too little structure into a model can be just as bad as building in too much structure.

There is a trade-off between the strength of model assumptions and our ability to find effects. We have already seen that some of the newer methodologies such as lowess smoothing are a welcome addition to the statistical toolbox. However, if we assume a linear relationship,

Table 3.3: *An example that illustrates the dangers of adding over contingency tables.*

	Engineering			Sociology			Total	
	Male	Female		Male	Female		Male	Female
Admit	30	10	Admit	15	30	Admit	45	40
Deny	30	10	Deny	5	10	Deny	35	20

we may be able to find it, where we will find nothing if we look for a general form of smooth curve or a completely arbitrary relationship. This is why simple non-parametric approaches are often unsatisfactory – they assume too little. Often they assume much less than we know to be true. Johnson (1995) has useful comments on the role of non-parametric tests. In part the objection is to a view of non-parametric modeling that is too limited.

3.4.5 Why models matter – adding across contingency tables

Table 3.3 is a contrived example that shows admission patterns in two separate university faculties. Looking at the table of totals on the right, we gain the clear impression that females are more likely to be admitted (40 out of 60 = 67%) than males (45 out of 80 = 56%). Indeed, the proportion of female applicants who are admitted is, overall, larger than the proportion of male applicants admitted.

The puzzle is that when we look at the individual faculties, females and males are accepted in exactly the same proportions, lower for Engineering (50%) than for Sociology (75%). A greater proportion of males apply in Engineering, while a greater proportion of females apply in Sociology. Thus the overall rate for males is biased towards the rate for Engineering, while the overall rate for females is biased towards the rate for Sociology.

What model is in mind? For simplicity, we will suppose that the university in question has just two faculties – Engineering and Sociology! Do we have in mind choosing a female at random, then comparing her chances of admission with the chances for a male that has been chosen at random? The randomly chosen male, who is more likely to be an engineer, will have a poorer chance than a randomly chosen female – 56% as against 67%.

Or, is our interest in the chances of a particular student, be they engineer or sociologist? If the latter, then we first note the faculty to which the student will apply. The admission rate is 50% for engineering applicants, as against 75% for sociology applicants, irrespective of sex.

This is an example of Simpson's paradox. In order to combine the odds ratios for the admission rates, we should use the Mantel–Haenszel method. There are further details in Chapter 4. (The odds of admission in each faculty are the same for the two sexes. Hence, the male:female odds ratio, with faculty as the classifying factor, is 1:1 in both faculties.)

The data set **UCBAdmissions** that is in the R *base* package has aggregate data on applicants to graduate school at Berkeley for the six largest departments in 1973, classified by admission and sex. An apparent association between admission chances and gender stemmed from the tendency of females to apply to departments with higher rejection rates. Type in

```
help(UCBAdmissions)
```

to get further details. See also Bickel et al. (1975). We will discuss these data further in Section 8.3.

3.5 Recap

Statistical models have both *signal* components and *noise* components. In simpler cases, which include most of the cases we consider,

$$\text{observation} = \text{signal} + \text{noise}.$$

After fitting a model, we have

$$\text{observation} = \text{fitted value} + \text{residual}$$

which we can think of as

$$\text{observation} = \text{smooth component} + \text{rough component}.$$

The hope is that the fitted value will recapture most of the signal, and that the residual will contain mostly noise. Unfortunately, as the relative contribution of the noise increases,

- it becomes harder to distinguish between signal and noise,
- it becomes harder to decide between competing models.

Model assumptions, such as normality, independence and constancy of the standard deviation, should be checked, to the extent that this is possible.

3.6 Further Reading

Finding the right statistical model is an important part of statistical problem solving. Chatfield (2002, 2003) has helpful comments. Clarke (1968) has a useful discussion of the use of models in archaeology. See also the very different points of view of Breiman and Cox (as discussant) in Breiman (2001). Our stance is much closer to Cox than to Breiman. See also our brief comments on Bayesian modeling in Section 4.11.

Johnson (1995) comments critically on the limitations of widely used non-parametric methods. See Hall (2001) for an overview of non-parametrics from a modern perspective.

References for further reading

Bickel, P.J., Hammel, E.A., and O'Connell, J.W. 1975. Sex bias in graduate admissions: data from Berkeley. *Science* 187: 398–403.

Breiman, L. 2001. Statistical modeling: the two cultures. *Statistical Science* 16: 199–215.

Chatfield, C. 2002. Confessions of a statistician. *The Statistician* 51: 1–20.

Chatfield, C. 2003. *Problem Solving. A Statistician's Guide, 2nd edn.* Chapman and Hall/CRC.

Clarke, D. 1968. *Analytical Archaeology.* Methuen.

Hall, P. 2001. Biometrika centenary: non-parametrics. *Biometrika* 88: 143–165.

Johnson, D.H. 1995. Statistical sirens: the allure of non-parametrics. *Ecology* 76: 1998–2000.

3.7 Exercises

1. An experimenter intends to arrange experimental plots in four blocks. In each block there are seven plots, one for each of seven treatments. Use the function `sample()` to find four random permutations of the numbers 1 to 7 that will be used, one set in each block, to make the assignments of treatments to plots.

2. Use `<- rnorm(100)` to generate a random sample of 100 numbers from a normal distribution. Calculate the mean and standard deviation of y. Now put the calculation in a loop and repeat 25 times. Store the 25 means in a vector named `av`. Calculate the standard deviation of the values in `av`.

3. Create a function that does the calculations of exercise 2.

4. To simulate samples from normal populations having different means and standard deviations, the `mean` and `sd` arguments can be used in `rnorm()`. Simulate a random sample of size 20 from a normal population having a mean of 100 and a standard deviation of 10.

5. Use the parameter `mfrow` to `par()` to set up the layout for a 3 by 4 array of plots. In the top 4 panels, show normal probability plots for 4 separate "random" samples of size 10, all from a normal distribution. In the middle 4 panels, display plots for samples of size 100. In the bottom 4 panels, display plots for samples of size 1000. Comment on how the appearance of the plots changes as the sample size changes.

6. The function `runif()` generates a sample from a uniform distribution, by default on the interval 0 to 1. Try `x <- runif(10)`, and print out the resulting numbers. Then repeat exercise 5 above, but taking samples from a uniform distribution rather than from a normal distribution. What shape do the plots follow?

7. The function `pexp(x, rate=r)` can be used to compute the probability that an exponential variable is less than x. Suppose the time between accidents at an intersection can be modeled by an exponential distribution with a rate of .05 per day. Find the probability that the next accident will occur during the next three weeks.

8. Use the function `rexp()` to simulate 100 exponential random numbers with rate .2. Obtain a density plot for the observations. Find the sample mean of the observations. Compare with the the population mean (the mean for an exponential population is 1/rate).

9. *This exercise investigates simulation from other distributions. The statement `x <- rchisq(10, 1)` generates 10 random values from a chi-squared distribution with one degree of freedom. The statement `x <- rt(10, 1)` generates 10 random values from a t distribution with one degree of freedom. Make normal probability plots for samples of various sizes from each

of these distributions. How large a sample is necessary, in each instance, to obtain a consistent shape?

10. The following data represent the total number of aberrant crypt foci (abnormal growths in the colon) observed in seven rats that had been administered a single dose of the carcinogen azoxymethane and sacrificed after six weeks (thanks to Ranjana Bird, Faculty of Human Ecology, University of Manitoba for the use of these data):

```
87 53 72 90 78 85 83
```

Enter these data and compute their sample mean and variance. Is the Poisson model appropriate for these data? To investigate how the sample variance and sample mean differ under the Poisson assumption, repeat the following simulation experiment several times:

```
x <- rpois(7, 78.3)
mean(x); var(x)
```

4

An Introduction to Formal Inference

A random sample is a set of values drawn independently from a larger population. A (uniform) random sample has the characteristic that all members of the population have an equal chance of being drawn. In the previous chapter, we discussed the implications of drawing repeated random samples from a normally distributed population, where the probability for a value is proportional to the normal density. In this chapter, we will expand upon that discussion by introducing the use of standard error to assess estimation accuracy. Confidence intervals and tests of hypotheses formalize the approach. We will comment on weaknesses in the hypothesis testing framework.

4.1 Standard Errors

In this section we will encounter the abbreviations:

- SEM: The standard error of a mean.
- SED: The standard error of a difference between two means.

4.1.1 Population parameters and sample statistics

Quantities that are functions of values in the total population are known as population *parameters*. The (usually unknown) population mean, often denoted by the symbol μ, is an example of a parameter. We commonly use the sample mean $\bar{x}$ (which is a sample *statistic*) to estimate the population mean. How accurate, for this purpose, is it? The standard error is a good starting point for assessing accuracy.

Other examples of statistics are the median, the quartiles, the standard deviation, the variance, the slope of a regression line, and the correlation coefficient. Each may be used as an estimate of the corresponding population parameter. A starting point for assessing accuracy of an estimate is the standard error that can be calculated for it.

4.1.2 Assessing accuracy – the standard error

The standard error is defined as the standard deviation of the sampling distribution of the statistic. An important special case is the standard error of the mean (SEM). The sampling distribution of the mean is the distribution of means from repeated independent random samples. Taking such repeated random samples is a theoretical idea, rather than (in most

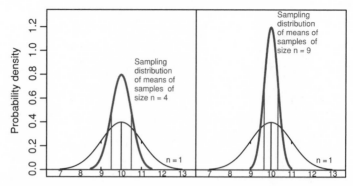

Figure 4.1: Sampling distribution of the mean ($n = 4$ and $n = 9$), compared with the distribution of individual sample values, for a normal population with mean $= 10$ and standard deviation $= 1$.

situations) a practical possibility. It is also an idea that we can illustrate using repeated computer-generated random samples from one or other theoretical distribution.

If data values are independent, then a suitable estimate for the SEM is

$$s_{\bar{x}} = \frac{s}{\sqrt{n}}.$$

This deceivingly simple formula, relating the SEM to the standard deviation, hides quite complex mathematical ideas. The formula relies crucially on the assumption of independence. If the data are not independent, then this formula does not apply. Chapter 9 gives further details of the effects of some forms of dependence.

The SEM indicates the extent to which the sample mean might be expected to vary from one sample to another. Figure 4.1 compares the theoretical sampling distribution of the standard error of the mean, for samples of sizes 4 and 9, with the distribution of individual sample values. Note how the distribution becomes narrower as the sample size increases; this reflects the decrease in the SEM with increasing sample size: the sample mean estimates the population mean with an accuracy that improves as the sample size increases.

There is a further important result, known as the *Central Limit Theorem*. As the sample size increases, the normal distribution is an increasingly accurate approximation to the sampling distribution of the mean. If the population distribution is close to symmetric, the approximation may be adequate even for sample sizes as small as 3 or 4.

4.1.3 Standard errors for differences of means

We often wish to compare means of different samples. Where there are two independent samples of size n_1 and n_2, the comparison is usually in the form of a difference

$$\bar{x}_1 - \bar{x}_2$$

where $\bar{x}_1$ and $\bar{x}_2$ denote the respective sample means. If the corresponding standard errors are denoted by SEM$_1$ and SEM$_2$, then the standard error of the difference (SED) can be computed using the formula

$$SED = \sqrt{SEM_1^2 + SEM_2^2}.$$

If all SEMs are the same, then for all comparisons,

$$SED = \sqrt{2} \times SEM.$$

It is often reasonable to assume equality of the standard deviations in the populations from which the samples are drawn. Then

$$SEM_1 = \frac{s}{\sqrt{n_1}}, \quad SEM_2 = \frac{s}{\sqrt{n_2}}$$

and the formula can be written as

$$SED = s\sqrt{\frac{1}{n_1} + \frac{1}{n_2}}$$

where s is the pooled standard deviation estimate described in Subsection 2.2.2.

As an example, consider the unpaired elastic band experiment data of Subsection 2.2.2. The pooled standard deviation estimate is 10.91. Hence, the SED is $10.91 \times \sqrt{\frac{1}{10} + \frac{1}{11}} =$ 4.77.

4.1.4* The standard error of the median

For data that come from a normal distribution, there is a similarly simple formula for the standard error of the median. It is

$$SE_{median} = \sqrt{\frac{\pi}{2}}\frac{s}{\sqrt{n}} \approx 1.25\frac{s}{\sqrt{n}}.$$

This indicates that the standard error of the median is about 25% greater than the standard error of the mean. Thus, for data from a normal distribution, the population mean can be estimated more precisely than can the population median. (A different formula for the standard error of the median, one that depends on the distribution, must be used when the data are not normally distributed.)

Consider again the `cuckoos` data. The median and standard error for the median of the egg lengths in the wrens' nests are 21.0 and .244, respectively.[1]

4.1.5* Resampling to estimate standard errors: bootstrapping

The formula for the SEM has the same form, irrespective of the distribution. By contrast, the formula that we gave for the standard error of the median applies only when data are normally distributed. The bootstrap estimate of the standard error of the median avoids this requirement, and avoids also the need to look for some alternative distribution that may be a better fit to the data. A comparison between the bootstrap estimate and the normal

---

[1]
```
data(cuckoos)                      # From our DAAG package
wren <- split(cuckoos$length, cuckoos$species)$wren
median(wren)
n <- length(wren)
sqrt(pi/2)*sd(wren)/sqrt(n)   # this SE computation assumes normality
```

theory estimate allows an assessment of the seriousness of any bias that may result from non-normality. We proceed to calculate the bootstrap estimate of the standard error for the median length for the eggs that were in wrens' nests. (The *boot* package is needed for all bootstrap examples.) We will use the result as a check on our earlier computation.

The idea is as follows. In estimating the standard error of the median, we are seeking the standard deviation of medians that could be obtained for all possible samples of egg lengths in wrens' nests. Of course, we have access to one sample only, but if our sample has been collected properly it should give us a good approximation to the entire population. The bootstrap idea is to compute with the sample in much the same way as, if they were available, we would compute with repeated samples from the population.

Instead of taking additional samples from the whole egg length population, we take *resamples* from the egg length sample that we have in hand. This sampling is done with replacement to ensure that we obtain different "new" samples. The resample size is taken to be the same as the original sample size. We estimate the standard deviation of the median by computing sample medians for each of the resamples and taking the standard deviation of all of these medians. Even though the resamples are not genuine new samples, this estimate for the standard error of the median has good statistical properties.

Here is the output from R for the egg lengths from the wrens' nests:[2]

```
ORDINARY NONPARAMETRIC BOOTSTRAP


Call:
boot(data = wren, statistic = median.fun, R = 999)


Bootstrap Statistics :
    original  bias    std. error
t1*      21  0.061       0.225
```

The original estimate of the median was 21. The bootstap estimate of the standard error is 0.225, based on 999 resamples. Compare this with the slightly larger standard error estimate of 0.244 given by the normal theory formula in Section 4.1.4. The bootstrap estimate of the standard error will of course differ somewhat between different runs of the calculation. Also given is an estimate of the bias, i.e., of the tendency to under- or over-estimate the median.

4.2 Calculations Involving Standard Errors: the *t* Distribution

Readers who are comfortable with Figures 4.2 to 4.4 may wish to skip this section. The calculations follow the same pattern as those we considered for the normal distribution in Subsection 3.2.2, but now with a distribution whose standard deviation is the SEM. Typically, the SEM is not known precisely, but has to be estimated from the data, with a

---

[2] `library(boot)`
```
  ## First define the function median.fun, with two required arguments:
  ##         data specifies the data vector,
  ##         indices selects vector elements for a particular resample
  median.fun <- function(data, indices){median(data[indices])}
  ## This now becomes the statistic argument to the boot() function:
  wren.boot <- boot(data = wren, statistic = median.fun, R = 999)
   # R = number of resamples
  wren.boot
```

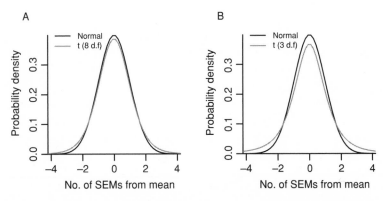

Figure 4.2: In A we have overlaid the density for a normal distribution with the density for a *t*-distribution with 8 d.f. In B we have overlaid the density for a *t*-distribution with 3 d.f.

precision that depends on the amount of data that are available. We assume that the normal distribution is an adequate approximation to the sampling distribution of the mean, but must account for the uncertainty in our estimate of the SEM.

For the data of Table 3.2 (which are also in the data frame `pair65` in the *DAAG* package) the mean change in amount of stretch before and after heating is 6.33, with a standard deviation of 6.10. The standard error (SEM) of this change is thus $6.10/\sqrt{9} = 2.03$. The mean is 3.11 times the magnitude of the standard error. We may report: "The mean change is 6.33 [SEM 2.03], based on $n = 9$ values."

As the estimate of the mean was based on 9 data values, 8 $(= 9 - 1)$ d.f. remain for estimating the standard error. One data value would not allow an estimate of the standard deviation, and hence of the SEM. Every additional value above one gives us one extra piece of information, and hence increases the d.f. by one. We adjust for the uncertainty in the estimate of the standard error by working, not with a normal distribution, but with a *t*-distribution with the appropriate number of degrees of freedom.

Figures 4.2A and 4.2B show the density curve for a normal distribution overlaid with those for *t*-distributions with 8 and 3 d.f. respectively. The density curve for a *t*-distribution with 3 d.f. is much more clearly different. The main difference, in each case, is in the tails.

The *t*-distribution is less concentrated around the mean than is the normal distribution and more spread out in the tails, with the difference greatest when the number of degrees of freedom is small.

Figure 4.3 illustrates the calculation of the area under the curve (= probability), between one SEM below the mean and one SEM above the mean, (A) for the normal distribution and (B) for the *t*-distribution. Details of the calculations are (with the setting `options(digits=3)`)

```
## Normal distribution, as in (A)
> pnorm(1)              # P[Z <= 1], where Z is distance from
                        # the mean, measured in SEMs.
[1] 0.841
> pnorm(-1)             # P[Z <= -1]
[1] 0.159
```

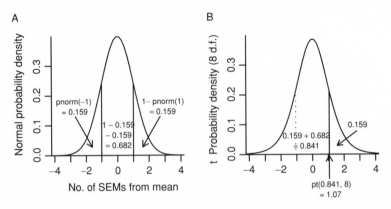

Figure 4.3: The distribution A is for a normal density function, taken as the sampling distribution of the mean. The graphs demonstrate the calculation of the probability of lying within one SEM of μ, (A) for a normal distribution, and (B) for a t-distribution with 8 d.f.

```
> pnorm(1) - pnorm(-1) # Between 1 SEM below & 1 SEM above
[1] 0.683
## t-distribution with 8 d.f., as in (B)
> pt(1, 8)
[1] 0.827
> pt(-1, 8)
[1] 0.173
> pt(1, 8) - pt(-1, 8)
[1] 0.653
```

We can alternatively turn the calculation around, and calculate limits that enclose some specified probability, typically 95% or 99% or 99.9%, rather than 68% or 65% as in the calculations above. Consider first the calculation of limits that enclose 95% of the probability. If the central region encloses 95% of the probability, then the lower and upper tails will each account for 2.5%. The cumulative probability, up to the desired upper limit, is 97.5% = 2.5% + 95%. We find

```
> qnorm(0.975)          # normal distribution
[1] 1.96
> qt(0.975, 8)          # t-distribution with 8 d.f.
[1] 2.31
```

Table 4.1 summarizes information on the multipliers, for a normal distribution, and for a t-distribution with 8 d.f., for several different choices of area under the curve. We leave it as an exercise for the reader to add further columns to this table, corresponding to different numbers of d.f. Changing from a normal distribution to a t-distribution with 8 d.f. led to a small change, from 1.0 to 1.07, for enclosing the central 68.3% of the area. It makes a substantial difference, giving an increase from 1.96 to 2.31, if we wish to enclose 95% of the area.

Figure 4.4 illustrates the calculation of endpoints that enclose 95% of the probability, for a normal distribution, and for a t-distribution with 8 d.f.

Table 4.1: *A comparison of normal distribution endpoints (multipliers for the SEM) with the corresponding t-distribution endpoints on 8 d.f.*

Probability enclosed between limits	Cumulative probability	Number of SEMs	
		Normal distribution	*t*-Distribution (8 df)
68.3%	84.1% .	1.0	1.07
95%	97.5%	1.96	2.31
99%	99.5%	2.58	3.36
99.9%	99.95%	3.29	5.04

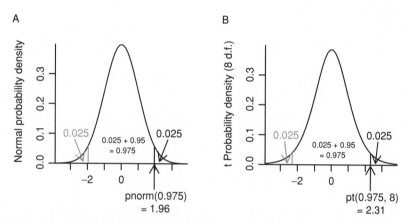

Figure 4.4: Calculation of the endpoints of the region that encloses 95% of the probability: (A) for a normal distribution, and (B) for a *t*-distribution with 8 d.f.

How good is the normal theory approximation?

For random samples from a distribution that is close to symmetric, the approximation is often adequate, even for samples as small as 3 or 4. In practice, we may know little about the population from which we are sampling. Even if the main part of the population distribution is symmetric, occasional aberrant values are to be expected. Such aberrant values do, perhaps fortunately, work in a conservative direction – they make it more difficult to detect genuine differences. The take-home message is that, especially in small samples, the probabilities and quantiles can be quite imprecise. They are rough guides, intended to assist researchers in making a judgement.

4.3 Confidence Intervals and Hypothesis Tests

Often we want an interval that most often, when samples are taken in the way that our sample has been taken, will include the population mean. There are two common choices for the long run proportion of similar samples that should contain the population mean – 95% and 99%.

We can use confidence intervals as the basis for tests of hypotheses. If the confidence interval for the population mean does not contain zero, we will reject the hypothesis that the population mean is zero.

4.3.1 One- and two-sample intervals and tests for means

Confidence intervals of 95% or 99%

We will use the paired elastic band data of Table 3.2 to illustrate the calculations. We noted above, in Section 4.2, that the mean change was 6.33 [SEM 2.03], based on $n = 9$ pairs of values. From Table 4.1, we note that, in 95% of samples of size 9 (8 d.f.), the sample mean will lie within $2.31 \times$ SEM of the population mean. Thus a 95% confidence interval for the mean has the endpoints[3]

$$(6.33 - 2.03 \times 2.31, 6.33 + 2.03 \times 2.31) = (1.64, 11.02).$$

From Table 4.1, the multiplier for a 99% confidence interval is 3.36, much larger than the multiplier of 2.58 for the case where the standard deviation is known.[4] A 99% confidence interval is

$$(6.33 - 2.03 \times 3.36, 6.33 + 2.03 \times 3.36) = (-0.49, 13.15).$$

Tests of hypotheses

We could ask whether the population mean difference for the paired elastic bands really differs from zero. Since the 95% confidence interval does not contain zero, we can legitimately write:

Based on the sample mean of 6.33, with SEM = 2.03, the population mean is greater than zero ($p < 0.05$).

The 99% confidence interval for the elastic band differences did contain zero. Therefore we cannot replace ($p < 0.05$) by ($p < 0.01$) in the statement above. The smallest p that we could substitute into the above statement thus lies between 0.01 and 0.05. This smallest p is called the p-value. A first step, for this example, is to use the pt() function to calculate the probability that the t-statistic is less than $-$mean/SEM $= -6.10/2.03$:[5]

```
[1] 0.00713
```

---

[3] ## R code that calculates the endpoints
```
attach(pair65)        # From our DAAG package
pair65.diff <- heated-ambient
pair65.n <- length(pair65.diff)
pair65.se <- sd(pair65.diff)/sqrt(pair65.n)
mean(pair65.diff) + qt(c(.025,.975),8)*pair65.se
detach(pair65)
```
[4] `qt(0.995, 8)`
[5] `1-pt(6.33/2.03, 8)      # Equals pt(-6.33/2.03, 8)`

We then double this to determine the sum of the probabilities in the two tails, leading to $p = 0.014$. We may wish to write:

Based on the sample mean of 6.33, the population mean is greater than zero ($p = 0.014$).

Formally, taking the population mean to be μ, we say that the *null* hypothesis is

H_0: $\mu = 0$

while the alternative hypothesis is $\mu \neq 0$.

The formal methodology of hypothesis testing may seem contorted. A small *p*-value makes the null hypothesis appear implausible. It is not a probability statement about the null hypothesis itself, or for that matter about its alternative. All it offers is an assessment of implications that flow from accepting the null hypothesis. A straw man is set up, the statement that $\mu = 0$. The typical goal is to knock down this straw man. By its very nature, hypothesis testing lends itself to various abuses.

What is a small p-value?

At what point is a *p*-value small enough to be convincing? Conventionally, $p = 0.05$ ($= 5\%$) is used as the cutoff. But 0.05 is too large, if results from the experiment are to be made the basis for a recommendation for changes to farming practice or to medical treatment. It may be too small when the interest is in deciding which effects merit further experimental or other investigation.

A summary of one- and two-sample calculations

Confidence intervals for a mean difference, or for a difference of means, have the form

$$\text{difference} \pm t\text{-critical value} \times \text{standard error of difference}.$$

The *t*-statistic has the form

$$t = \frac{\text{difference}}{\text{standard error of difference}}.$$

Given t, the *p*-value for a (two-sided) test is defined as

$$P(T > t) + P(T < -t)$$

where T has a *t*-distribution with the appropriate number of degrees of freedom. A small *p*-value corresponds to a large value of $|t|$, regarded as evidence that the true difference is nonzero and leading to the rejection of the *null hypothesis*.

Table 4.2 lists confidence intervals and tests in the one- and two-sample cases.[6] The single sample example is for the paired elastic band data that we discussed at the beginning

---

[6] ## t-test and confidence interval calculations

```
heated <- c(254, 252, 239, 240, 250, 256, 267, 249, 259, 269)
ambient <- c(233, 252, 237, 246, 255, 244, 248, 242, 217, 257, 254)
t.test(heated, ambient, var.equal=TRUE)
rm(heated, ambient)
```

Table 4.2: *Formulae for confidence intervals and tests of hypothesis based on the t-distribution.*

	Confidence interval	Test statistic	Df
One-sample t	$\bar{d} \pm t_{\text{crit}}\text{SE}[\bar{d}]$	$t = \dfrac{\bar{d}}{\text{SE}[\bar{d}]}$	$n-1$
e.g.	$6.33 \pm 2.306 \times \frac{6.10}{\sqrt{9}}$	$t = \frac{6.33}{6.10/\sqrt{9}}$	8
Two-sample t	$\bar{x}_2 - \bar{x}_1 \pm t_{\text{crit}}\text{SE}[\bar{x}_2 - \bar{x}_1]$	$t = \dfrac{\bar{x}_2 - \bar{x}_1}{\text{SE}[\bar{x}_2 - \bar{x}_1]}$	$n_1 + n_2 - 2$
e.g.	$253.5 - 244.1 \pm 2.09 \times 10.91\sqrt{\frac{1}{10} + \frac{1}{11}}$ $= 253.5 - 244.1\ 2.09 \times 4.77 = (-0.6, 19.4)$	$t = \dfrac{253.5 - 244.1}{10.91 \times \sqrt{\frac{1}{10} + \frac{1}{11}}}$	19

Here, t_{crit} is the 97.5th percentile of a t-statistic with 8 df (1-sample example) or 19 df (2-sample example). (The 97.5th percentile is the same as the two-sided 5% critical value.)

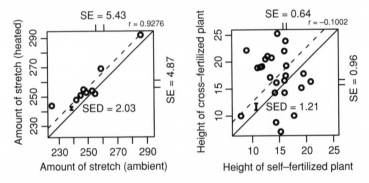

Figure 4.5: Second versus first member, for each pair. The first panel is for the ambient/heated elastic band data from Section 4.2, while the second is for Darwin's plants.

of this section. The example that we use for the two-sample calculations is discussed in Subsection 2.2.2.

When is pairing helpful?

Figure 4.5 shows, for two different sets of paired data, a plot of the second member of the pair against the first.[7] The first panel is for the paired elastic band data of Section 4.2, while the second panel (for the data set `mignonette`) is from the biologist Charles Darwin's experiments that compared the heights of crossed plants with the heights of self-fertilized plants (data, for the wild mignonette *Reseda lutea*, are from p. 118 of Darwin, 1877).

```
7 par(mfrow=c(1,2))
  data(pair65); data(mignonette)
  attach(mignonette); attach(mignonette)            # DAAG package
  plot(heated ~ ambient); abline(0, 1)              # left panel
  abline(mean(heated-ambient), 1, lty=2)
  plot(cross ~ self, pch=rep(c(4,1), c(3,12))); abline(0, 1)  # right panel
  abline(mean(cross-self), 1, lty=2)
  detach(pair65); detach(mignonette); par(mfrow = c(1,1))
```

Plants were paired within the pots in which they were grown, with one plant on one side and one on the other.

For the paired elastic band data there is a clear correlation, and the standard error of the difference is much less than the root mean square of the two separate standard errors. For Darwin's data there is little evidence of correlation. The standard error of differences of pairs is about equal to the root mean square of the two separate standard errors. For the elastic band data, the pairing was helpful; it led to a low SED. The pairing was not helpful for Darwin's data (note that Darwin (cited above) gives other data sets where the pairing was helpful, in the sense of allowing a more accurate comparison).

If the data are paired, then the two-sample t-test corresponds to the wrong model! It is appropriate to use the one-sample approach, whether or not there is evidence of correlation between members of the same pair.

What if the standard deviations are unequal?

If the assumption of equal variances or standard deviations fails we have heterogeneity of variance. There has been extensive attention to implications, for the two-sample t-test, of unequal variances in the two groups. The Welch procedure gives an adequate practical answer, unless degrees of freedom are very small. The Welch statistic is the difference in means divided by the standard error of difference, i.e.,

$$t = \frac{\bar{x}_2 - \bar{x}_1}{\text{SED}},$$

where

$$\text{SED} = \sqrt{\frac{s_2^2}{n_2} + \frac{s_1^2}{n_1}}.$$

If the two variances are unequal this does not have a t-distribution. However, criti-cal values are quite well approximated by a t-distribution with degrees of freedom given by a suitable approximation. The most commonly used approximation is that of Welch (1949), leading to the name *Welch test*. For details, see Miller (1986). The function t.test() has the Welch test as its default; it assumes equal variances only if the user specifies this.

Note that if $n_1 = n_2$ then the statistic is the same as for the t-test that is based on the pooled estimate of variance. However, the degrees of freedom are reduced.

Different ways to report results

There are many situations where means and standard errors are all that is needed. Where treatment differences are large, more than about six times the SEM for any individual treat-ment or four times the SED for comparing two means, there may be no point in presenting the results of significance tests. It may be better to quote only means and standard errors.

For the paired elastic band data of Table 3.2, the mean difference in amount of stretch before and after heating is 6.33, with a standard deviation of 6.10. The standard error of this difference (SED) is thus $6.10/\sqrt{9} = 2.03$. The mean is 3.11 times the magnitude of the standard error. Especially in engineering and physical science contexts where the aim is to accompany a report of the mean with a statement of its precision, it would be enough to report: "The mean change is 6.33 [SED 2.03]."

Confidence intervals and hypothesis testing give this result a more interpretive twist. It is useful to set these various alternatives side by side:[8]

1. The mean change is 6.33 [SED 2.03].
2. The t-statistic is $t = 6.333/2.034 = 3.11$, on 8 ($= 9 - 1$) degrees of freedom. In other words, the difference is 3.11 times the standard error.
3. A 95% confidence interval for the change is

$$(6.33 - 2.306 \times 2.034, 6.33 + 2.306 \times 2.034),$$

   i.e. (1.64, 11.02).
   [The multiplier, equal to 2.306, is the 5% two-sided critical value for a t-statistic on 8 ($= 9 - 1$) d.f.]
4. We reject the null hypothesis that the true mean difference is 0 ($p = 0.014$) – see Subsection 4.3.1 for definitions.
   [The two-sided p-value for $t = 3.11$ on 8 d.f. is 0.014.]

Alternative 1 is straightforward. The t-statistic (alternative 2) expresses the change as a multiple of its standard error. Often, and especially if the difference is more than four or five times the SED, it is all that is needed. The conventional wisdom is that the change is worthy of note if the p-value is less than 0.05 or, equivalently, if the 95% confidence interval does not contain 0. For this, the t-statistic must be somewhat greater than 1.96, i.e., for all practical purposes >2.0. For small degrees of freedom, the t-statistic must be substantially greater than 2.0.

Readers who have difficulty with alternatives 3 and 4 may find it helpful to note that these restate and interpret the information in alternatives 1 and 2. Those who have problems with confidence intervals and (especially) tests of hypotheses are in good company. There is an increasing body of opinion that wishes to relegate them to a relatively minor role in statistical analysis or eliminate them altogether.

If standard errors are not enough and formal inferential information is required, confidence intervals may be preferable to formal tests of hypotheses.

4.3.2 Confidence intervals and tests for proportions

We assume that individuals are drawn independently and at random from a binomial population where individuals are in one of two categories – male as opposed to female, a favorable treatment outcome as opposed to an unfavorable outcome, survival as opposed to non-survival, defective as opposed to non-defective, Democrat as opposed to Republican,

---

[8]
```
diffs <- pair65$heated - pair65$ambient
n <- length(diffs)
av <- mean(diffs); sd <- sqrt(var(diffs)); se <- sd/sqrt(n)
print(c(av, se))   # Item 1
print(av/se)       # Item 2
t.test(diffs)      # Items 3 and 4
rm(diffs, n, av, sd, se)
```

Table 4.3: *Approximate 95%*
confidence interval, assuming
$0.35 \leq \pi \leq 0.65$.

n	Approximate 95% confidence interval
25	$p \pm 20\%$
100	$p \pm 10\%$
400	$p \pm 5\%$
1000	$p \pm 3.1\%$

etc. Let π be the population proportion. In a sample of size n, the proportion in the category of interest is denoted by p. Then,

$$SE[p] = \sqrt{\pi(1-\pi)/n}.$$

An upper bound for $SE[p]$ is $1/(2\sqrt{n})$. If π is between about 0.35 and 0.65, the inaccuracy in taking $SE[p]$ as $1/(2\sqrt{n})$ is small.

This approximation leads to the confidence intervals shown in Table 4.3. Note again that the approximation is poor if π is much outside the range 0.35 to 0.65.

An alternative is to use the estimator

$$SE[p] = \sqrt{\frac{p(1-p)}{n}}.$$

An approximate 95% confidence bound for the proportion π is then

$$p \pm 1.96\sqrt{\frac{p(1-p)}{n}}.$$

4.3.3 Confidence intervals for the correlation

The correlation measure that we discuss here is the Pearson or product–moment correlation, which measures linear association.

The standard error of the correlation coefficient is typically not a useful statistic. The distribution of the sample correlation, under the usual assumptions (e.g., bivariate normality), is too skew. The function `cor.test()` in the *ctest* package may be used to test the null hypothesis that the sample has been drawn from a population in which the correlation ρ is zero. This requires the assumption that the conditional distribution of y, given x, is normal, independently for different ys and with mean given by a linear function of x.

Classical methods for comparing the magnitudes of correlations, or for calculation of a confidence interval for the correlation, rely on the assumption that the joint distribution of (x, y) is bivariate normal. In addition to the assumption for the test that $\rho = 0$, we need to know that x is normally distributed, independently between (x, y) pairs. In practice, it may be enough to check that both x and y have normal distributions. This methodology is not, at the time of writing, implemented in R.

Table 4.4: *Contingency table derived from*
data that relates to the Lalonde (1986) paper.

	High School Graduate Certificate	
	Yes	No
NSW74 male trainees	54	131
PSID3 males	63	65

The function boot.ci() in the R *boot* package offers an alternative methodology, with less reliance on normality assumptions. Independence between (x, y) pairs is as important as ever. This approach will be demonstrated in Subsection 4.8.3.

4.4 Contingency Tables

Table 4.4 is from US data that were used in the evaluation of labor training programs, aimed at individuals who had experienced economic and social difficulties.[9] (See Lalonde 1986 and Dehejia and Wahba 1999). The NSW74 trainee group had participated in a labor training program, while the PSID3 group had not. (The NSW74 group is a subset of the intervention (training) group from the US federally funded National Supported Work Demonstration program in the mid-1970s. They are the subset for which information is available on 1974 earnings. The PSID data is from a Panel Study of Income Dynamics that was conducted at the same time. The PSID3 subset was designed to mimic more closely the characteristics of the NSW training and control groups.)

A chi-squared test for no association has[10]

```
X-squared = 13.0, df = 1, p-value = 0.00032
```

Clearly, high school dropouts are more strongly represented in the NSW74 data.

The mechanics of the calculation

The null hypothesis is that the proportion of the total in each cell is, to within random error, the result of multiplying a row proportion by a column proportion. The independence assumption, i.e. the assumption of independent allocation to the cells of the table, is crucial.

Assume there are I rows and J columns. The expected value in cell (i, j) is calculated as

$$E_{ij} = \text{(proportion for row } i) \times \text{(proportion for column } j) \times \text{total.}$$

We can then obtain a score for each cell of the table by computing the square of the difference between the expected value and the observed value, and dividing the result by the expected value. Summing up all scores gives the chi-squared statistic. Under the null hypothesis

---

[9] data(nsw74psid3)   # From our DAAG package
   table(nsw74psid3$trt, nsw74psid3$nodeg)
      # This leads to a table in which, relative to Table 4.4, the rows
      # are interchanged. The PSID3 males are coded 0, while the NSW74
      # male trainees are coded 1.
[10] chisq.test(table(nsw74psid3$trt, nsw74psid3$nodeg), correct=FALSE)

Table 4.5: *The calculated* expected values *for the contingency table in Table 4.4.*

| | High School Graduate | | | |
	Yes	No	Total	Row proportion
NSW74 trainees	54 (69.15)	131 (115.85)	185	$185/313 = 0.591$
PSID3	63 (47.85)	65 (80.15)	128	$128/313 = 0.409$
Total	117	196	313	
Column proportion	$117/313 = 0.374$	$196/313 = 0.626$		

Table 4.6: *Contingency table compiled from Hobson (1988, Table 12.1, p. 248).*

| | Object moves | |
Dreamer moves	Yes	No
Yes	5	17
No	3	85

this statistic has an approximate chi-squared distribution with $(I - 1)(J - 1)$ degrees of freedom.

In Table 4.5, the values in parentheses are the expected values E_{ij}.

The expected values are found by multiplying the column totals by the row proportions. (Alternatively, the row totals can be multiplied by the column proportions.) Thus $117 \times 0.591 = 69.15$, $196 \times 0.591 = 115.85$, etc.

An example where a chi-squared test may not be valid

In Table 4.6 we summarize information that Hobson (1988) derived from drawings of dreams, made by an unknown person that he calls "The Engine Man". Among other information Hobson notes, for each of 110 drawings of dreams made, whether the dreamer moves, and whether an object moves. Dreamer movement may occur if an object moves, but is relatively rare if there is no object movement. (Note that Hobson does not give the form of summary that we present in Table 4.6.)

It may also seem natural to do a chi-squared test for no association.[11] This gives $X^2 = 7.1$ (1 d.f.), $p = 0.008$.

A reasonable rule, for the use of the chi-squared approximation, may be that all expected values should be at least 2 (Miller 1986), so that for these data we should be all right. We can check the result by doing a Fisher exact test. This is available in a number of different statistical packages, including R. Astonishingly, the Fisher exact test[12] gives exactly the same result as the chi-squared test, i.e. $p = 0.008$.

[11] `engineman <- matrix(c(5,3,17,85), 2,2)`
`chisq.test(engineman)`
[12] `fisher.test(engineman)`

Table 4.7: *Cross-classification of
species occurring in South Australia/
Victoria and in Tasmania.*

Common/rare classification	Habitat type		
	D	W	WD
CC	37	190	94
CR	23	59	23
RC	10	141	28
RR	15	58	16

More seriously, there is a time sequence to the dreams. Thus, there could well be a clustering in the data, i.e. runs of dreams of the same type. Hobson gives the numbers of the dreams in sequence. Assuming these represent the sequence in time, this would allow a check of the strength of any evidence for runs. Hobson's table has information that our tabular summary (Table 4.6) has not captured.

4.4.1 Rare and endangered plant species

The calculations for a test for no association in a two-way table can sometimes give useful insight, even where a formal test of statistical significance would be invalid. The example that now follows illustrates this point. Data are from species lists for various regions of Australia. Species were classified CC, CR, RC and RR, with C denoting common and R denoting rare. The first code letter relates to South Australia and Victoria, and the second to Tasmania. They were further classified by habitat according to the Victorian register, where D = dry only, W = wet only, and WD = wet or dry.[13]

We use a chi-squared calculation to check whether the classification into the different habitats is similar for the different rows.[14] We find

```
X2 = 35.0 df = 6 p-value = 0.0000043
```

This looks highly significant. This low *p*-value should attract a level of scepticism. We do not have a random sample from some meaningful larger population. Suppose that there is clustering, so that species come in closely related pairs, with both members of the pair always falling into the same cell of the table. This will inflate the chi-squared statistic by a factor of 2 (the net effect of inflating the numerator by 2^2, and the denominator by 2). There probably is some such clustering, though not of the type that we have suggested by way of this simplistic example. Such clustering will inflate the chi-squared statistic by an amount that the available information does not allow us to estimate.

---

[13] ## Enter the data thus:
```
   rareplants <- matrix(c(37,190,94,23,59,23,10,141,28,15,58,16), ncol=3,
       byrow=TRUE, dimnames=list(c("CC","CR","RC","RR"), c("D","W","WD")))
```
[14] chisq.test(rareplants, correct=FALSE)
```
    # For simplicity, we have suppressed the continuity correction.
```

The standard Pearson chi-squared tests rely on multinomial sampling assumptions, with counts entering independently into the cells. Where it is possible to form replicate tables, the assumption should be tested.

The nature of departures from a consistent overall row pattern

The investigator then needs to examine the nature of variation with the row classification. For this, it is helpful to look at the residuals.[15] The null hypothesis implies that the expected relative numbers in different columns are the same in every row. The chi-squared residuals show where there may be departures from this pattern. In large tables these will, under the null hypothesis, behave like random normal deviates with mean zero and variance one. The values that should be noted are those that are somewhat greater than 2.0. Notice that the CC species are, relative to the overall average, over-represented in the WD classification, the CR species are over-represented in the D classification, while the RC species are under-represented in D and WD and over-represented in W:

```
Chi-squared residuals, i.e., (obs-exp)/exp^0.5
         D        W        WD
CC    -0.369   -1.1960    2.263
CR     2.828   -1.0666   -0.275
RC    -2.547    2.3675   -2.099
RR     1.242    0.0722   -1.023
```

For reference, here is the table of expected values.

```
           Expected values
          D       W       WD
CC      39.3   207.2    74.5
CR      12.9    67.8    24.4
RC      21.9   115.6    41.5
RR      10.9    57.5    20.6
```

4.4.2 Additional notes

Interpretation issues

Having found an association in a contingency table, what does it mean? The interpretation will differ depending on the context. The incidence of gastric cancer is relatively high in Japan and China. Do screening programs help? Here are two ways in which the problem has been studied:

- In a long term follow-up study, patients who have undergone surgery for gastric cancer may be classified into two groups – a "screened" group whose cancer was detected by mass screening, and an "unscreened" group who presented at a clinic or hospital with gastric cancer. The death rates over the subsequent five- or ten-year period are then compared. For example, the five-year mortality may be around 58% in the unscreened

---

[15] ## Calculate standardized residuals
```
x2 <- chisq.test(rareplants)
round( (x2$observed-x2$expected)/sqrt(x2$expected), 3)
```

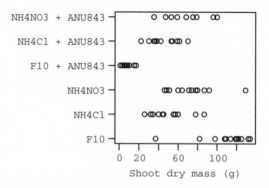

Figure 4.6: One-way strip plot, with different strips corresponding to different treatment regimes, for rice shoot mass data.

group, compared with 72% in the screened group, out of approximately 300 patients in each group.

- In a prospective cohort study, two populations – a screened population and an unscreened population – may be compared. The death rates in the two populations over a ten-year period may then be compared. For example, the annual death rate may be of the order of 60 per 100 000 for the unscreened group, compared with 30 per 100 000 for the screened group, in populations of several thousand individuals.

In the long term follow-up study, the process that led to the detection of cancer was different between the screened and unscreened groups. The screening may lead to surgery for some cancers that would otherwise lie dormant long enough that they would never attract clinical attention. The method of detection is a confounding factor. It is necessary, as in the prospective cohort study, to compare all patients in a screened group with all patients in an unscreened group. Even so, it is necessary, in a study where assignment of participants is not random, to be sure that the two populations are comparable.

Modeling approaches

Modeling approaches typically work with data that record information on each case separately. Data where there is a binary (yes/no) outcome, and where logistic regression may be appropriate, are an important special case. Chapter 8 gives further details.

4.5 One-Way Unstructured Comparisons

We will use the shoot dry mass data that are presented in Figure 4.6 as a basis for discussion.[16] The data are from an experiment that compared wild type (wt) and genetically modified rice plants (ANU843), each with three different chemical treatments. The total number of observations is 72.

There is one "strip" for each factor level. The strips appear in the order of factor levels, starting from the lowest of the strips. The order that we set is deliberate – notice that F10,

[16]
```
library(lattice)
  data(rice)            # From our DAAG package
  lev <- c("F10", "NH4NO3", "F10 +ANU843", "NH4Cl +ANU843",
         "NH4NO3 +ANU843")
  rice$trt <- factor(rice$trt, levels=lev)
  stripplot(trt ~ ShootDryMass, data=rice, xlab="Shoot dry mass (g)", aspect=0.5)
```

NH4Cl and NH4NO3 appear first without ANU843, i.e., they are results for "wild type" plants. The final three levels repeat these same treatments, but for ANU843. For the moment, we ignore this two-way structure, and carry out a one-way analysis (Perrine et al. (2001) has an analysis of these data).

Figure 4.6 is one of a number of graphical presentation possibilities for a one-way layout. Others are (1) a side by side comparison of the histograms – but there are too few values for that; (2) density plots – again there are too few values; and (3) a comparison of the boxplots – this works quite well with 12 values for each treatment.

The strip plot displays "within group" variability, as well as giving an indication of variability among the group means. The one-way analysis of variance formally tests whether the variation among the means is greater than what might occur simply because of the natural variation within each group. An *F*-statistic that is much larger than 1 points to the conclusion that the means are different. The *p*-value is designed to assist this judgement.

The analysis of variance table is[17]

```
Analysis of Variance Table

Response: ShootDryMass
          Df Sum Sq Mean Sq F value    Pr(>F)
trt        5  68326   13665  36.719 < 2.2e-16
Residuals 66  24562     372
```

Here, we note that the mean squared error (effectively, a pooled variance estimate for the 5 samples) is 372. Note that 5 degrees of freedom are associated with estimating 5 group means, 1 degree of freedom is associated with estimating the overall mean, leaving 66 d.f. for estimating the mean squared error.

For this table,

$$\text{SEM} = \sqrt{\frac{372}{12}} = 5.57.$$

(Divide by 12 because each mean is the average of 12 values. We use the variance estimate $\widehat{\sigma}^2 = 372$, on 66 d.f.)

$$\text{SED} = 7.87 \, (= \sqrt{2} \times \text{SEM}).$$

The very small *p*-value for the *F*-statistic is a strong indication that there are indeed differences among the treatment means. Interest then focuses on teasing out the nature of those differences. A first step is to print out the coefficients (S-PLUS users will need, in order to reproduce this output, to change options()$contrasts from its default). These are[18]

	Estimate	Std. Error	t value	Pr(>|t|)
(Intercept)	108.3	5.57	19.45	0.00e+00
trtNH4Cl	-58.1	7.88	-7.38	3.47e-10
trtNH4NO3	-35.0	7.88	-4.44	3.45e-05
trtF10 +ANU843	-101.0	7.88	-12.82	0.00e+00
trtNH4Cl +ANU843	-61.8	7.88	-7.84	5.11e-11
trtNH4NO3 +ANU843	-36.8	7.88	-4.68	1.49e-05

---

[17] rice.aov <- aov(ShootDryMass ~ trt, data=rice)
   anova(rice.aov)
[18] summary.lm(rice.aov)$coef

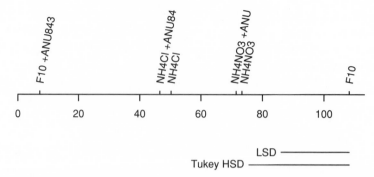

Figure 4.7: Graphical presentation of results from the one-way analysis of the rice shoot dry mass data. Means that differ by more than the LSD (least significant difference) are different, at the 5% level, in a *t*-test that compares the two means. Tukey's Honest Significant Difference (HSD) takes into account the number of means that are compared. See the text for details.

The initial level, which is `F10`, has the role of a reference or baseline level. Notice that it appeared on the lowest level of Figure 4.6. The "(Intercept)" line gives the estimate for `F10`. Other treatment estimates are differences from the estimates for `F10`. Thus the treatment estimates are

108.3	108.3−58.1	108.3−35.0	108.3−101.0	108.3−61.8	108.3−36.8
= 108.3	= 50.2	= 73.3	= 7.3	= 46.5	= 71.5

The standard errors (`Std. Error`) are, after the first row, all standard errors for differences between `F10` and later treatments. Notice that these standard errors all have the same value, i.e. 7.88. Because all treatments occur equally often, the standard deviations for all pairs of treatment differences equals 7.88. In order to check this, use the `relevel()` function to set the reference level to be another level than `F10`, and re-run the analysis.[19] Readers who want further details of the handling of analysis of variance calculations may wish to look ahead to the discussion in Section 7.1, which the present discussion anticipates.

The results are in a form that facilitates tests of significance for comparing treatments with the initial reference treatment. It is straightforward to derive a confidence interval for the difference between any treatment and the reference. We leave the details of this for discussion in Chapter 7.

4.5.1 Displaying means for the one-way layout

For genuinely one-way data, we want a form of presentation that reflects the one-way structure. Figure 4.7 shows a suitable form of presentation.[20]

Results come in pairs – one result for wild type plants and one for the `ANU843` variety. Notice that for `F10` there is a huge difference, while for the other two chemicals there is no detectable difference between wild type and `ANU843`. This emphasizes the two-way structure of the data. In general, use of a one-way analysis for data that have a two-way

[19] `## Make NH4Cl the reference (or baseline) level`
`   rice$trt <- relevel(rice$trt, ref="NH4Cl")`
[20] `## Use the function oneway.plot() from our DAAG package`
`   oneway.plot(aov(ShootDryMass ~ trt, data=rice))`

structure is undesirable. Important features may be missed. Here, detecting the interaction of variety with chemical relied on recognizing this structure.

Is the analysis valid?

Figure 4.6 suggested that variance was much lower for the F10 +ANU843 combination than for the other treatments. The variance seems lower when the mass is lower. Because only one treatment seems different from the rest, the analysis of variance table would not change much if we omitted it. The main concern is that the standard error of difference will be too large for comparisons involving F10 +ANU843. It will be too small for other comparisons.

4.5.2 Multiple comparisons

In Figure 4.7 we give two "yardsticks" against which to compare differences between means. We did the F-test because we wanted protection from finding spurious differences – a result of looking at all possible comparisons rather than just one. Tukey's Honest Significant Difference (HSD) provides an alternative form of protection from finding such spurious differences. It takes us into the arena of multiple comparisons. It is the appropriate yardstick against which to compare treatment differences if, not having a preliminary F-test, we want a 5% probability of finding a difference between the largest and smallest means when there was no difference in the populations from which they were drawn.

In practice, it makes sense to do a preliminary analysis of variance F-test. If that suggests differences between the means, then it is of interest to use both yardsticks in comparing them. The Least Significant Difference gives an anti-conservative yardstick, i.e., one that is somewhat biased towards finding differences. Tukey's HSD gives a stricter conservative yardstick, i.e., one that is somewhat biased against finding differences. Ignoring changes in degrees of freedom and possible changes in the standard error as the number of means to be compared increases, the HSD will increase as the number of means that are to be compared increases.[21]

*Microarray data – severe multiplicity

Multiple tests are a serious issue in the analysis of microrray data, where an individual slide may yield information on some thousands of genes. For example, Callow et al. (2000) compare gene expression in liver tissue from eight mice in which a gene had been knocked out, with liver tissue from eight normal mice. The interest is in identifying other genes whose expression has been simultaneously inhibited. The comparison is repeated for each of 5462 spots, most of which correspond to different genes. A good way to get an indication of the genes that are differentially expressed, i.e., expressed in one group of mice but not in the other, is to do a normal probability plot of the 5462 t-statistics. Even better may be to plot the ordered t-statistics against the quantiles of t-distribution with perhaps

```
[21] ## The SED is 7.88 with 66 degrees of freedom.   There are 6 means.
    qt(p=.975, df=66) * 7.88                        # Equals 15.7
    qtukey(p=.95, nmeans=6, df=66) * 7.88 / sqrt(2)  # Equals 23.1
     # NB: We call qtukey() with p=0.95, not with p=.975 as for the pairwise
     # t-test.  The test works with |difference|, and is inherently one-sided.
     # We divide by sqrt(2) because the Tukey statistic is expressed as a
     # multiple of the SEM = SED/sqrt(2) = 7.88/sqrt(2)
```

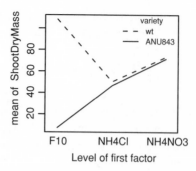

Figure 4.8: Interaction plot, showing how the effect of levels of the first factors changes with the level of the second factor.

14 ($= 16 - 2$) d.f. The spots that show clear evidence of differential expression lie below the trend line to the left or above the trend line to the right of the plot. For further details, see Dudoit et al. (2002).

4.5.3 Data with a two-way structure

The rice data, analysed earlier in this section with a one-way analysis of variance, really have a two-way structure. A first factor relates to whether F10 or NH4Cl or NH4NO3 is applied. A second factor relates to whether the plant is wild type (wt) or ANU843. Figure 4.8 is designed to show this structure.[22]

This shows a large difference between ANU843 and wild type (wt) for the F10 treatment. For the other treatments, there is no detectable difference. A two-way analysis would show a large interaction. We leave this as an exercise for the reader. Note, finally, that the treatments were arranged in two blocks. In general, this has implications for the analysis. In this instance it turns out that there is no detectable block effect, which is why we could ignore it. Chapter 9 has an example of the analysis of a randomized block design.

4.5.4 Presentation issues

The discussion so far has treated all comparisons as of equal interest. Often they are not. There are several possibilities:

- Interest may be in comparing treatments with a control, with comparisons between treatments of lesser interest.
- Interest may be in comparing treatments with one another, with any controls used as a check that the order of magnitude of the treatment effect is pretty much what was expected.
- There may be several groups of treatments, with the chief interest in comparisons between the different groups.

---

[22]
```
trtchar <- as.character(rice$trt)
fac1 <- factor(sapply(strsplit(trtchar," \\+"), function(x)x[1]))
variety <- sapply(strsplit(trtchar, " \\+"),
                  function(x)c("wt","ANU843")[length(x)])
variety <- factor(variety, levels=c("wt", "ANU843"))
attach(rice)
interaction.plot(fac1, variety, ShootDryMass)
detach(rice)
```

Table 4.8: *Distance traveled* (`distance.traveled`) *by model car, as a function of starting point* (`starting.point`) *up a 20° ramp.*

starting.point	distance.traveled		
3	31.38	30.38	33.63
6	26.63	25.75	27.13
9	18.75	22.50	21.63
12	13.88	11.75	14.88

Any of these situations should lead to specifying in advance the specific treatment comparisons that are of interest.

Often, however, scientists prefer to regard all treatments as of equal interest. Results may be presented in a graph that displays, for each factor level, the mean and its associated standard error. Alternatives to displaying bars that show the standard error may be to show a 95% confidence interval for the mean, or to show the standard deviation. Displaying or quoting the standard deviation may be appropriate when the interest is, not in comparing level means, but in obtaining an idea of the extent to which the different levels are clearly separated.

In any case:

- For graphical presentation, use a layout that reflects the data structure, i.e., a one-way layout for a one-way data structure, and a two-way layout for a two-way data structure.
- Explain clearly how error bars should be interpreted – ± SE limits, ± 95% confidence interval, ± SED limits, or whatever. Or if the intention is to indicate the variation in observed values, the SD (standard deviation) may be more appropriate.
- Where there is more than one source of variation, explain what source(s) of "error" is/are represented. It is pointless and potentially misleading to present information on a source of error that is of little or no interest, e.g., on analytical error when the relevant "error" for the treatment comparisons that are of interest arises from fruit to fruit variation.

4.6 Response Curves

Table 4.8 exhibits data that are strongly structured. The data are for an experiment in which a model car was released three times at each of four different distances up a 20° ramp. The experimenter recorded distances traveled from the bottom of the ramp across a concrete floor. This should be handled as a regression problem rather than as an analysis of variance problem.

Figure 4.9 shows a plot of these data.[23] What is the pattern of the response? An analysis that examines all pairwise comparisons does violence to the treatment structure, and confuses interpretation. In these data, the starting point effect is so strong that it will nevertheless

[23] `data(modelcars)     # From our DAAG package`
`   plot(distance.traveled ~ starting.point, pch=15, data=modelcars,`
`            xlab = "Distance up ramp", ylab="Distance traveled")`

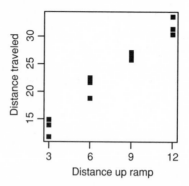

Figure 4.9: Distance traveled from bottom of ramp, relative to distance of starting point up a 20° ramp.

appear as highly significant. Response curve analyses should be used whenever appropriate in preference to comparison of individual pairs of means.

For example, data may show a clear pattern of changing insect weight with changing temperature. It may be possible to look at plots and decide that a quadratic equation type of response curve was needed. Where the data are not clear it may be necessary to go through the steps, in turn:

1. Does the assumption of a straight line form of response explain the data better than assuming a random scatter about a horizontal line?
2. Does a quadratic response curve offer any improvement?
3. Is there a suggestion of a response that looks like a cubic curve?

Notice that at this stage we are not concerned to say that a quadratic or cubic curve is a good description of the data. All we are examining is whether such a curve captures an important part of the pattern of change. If it does, but the curve is still not quite right, it may be worthwhile to proceed to look for a curve that does fit the data adequately.

A representation of the response curve in terms of coefficients of orthogonal polynomials provides information that makes it relatively easy to address questions 1–3. Consider for example a model that has terms in x and x^2. Orthogonal polynomials re-express this combination of terms in such a way that the coefficient of the "linear" term is independent of the coefficient of the "quadratic" term. Thus orthogonal polynomials have the property that the coefficient(s) of lower order terms (linear, . . .) do(es) not change when higher order terms are added. One model fit, with the highest order term present that we wish to consider, provides the information needed to make an assessment about the order of polynomial that is required. For details, see e.g., Steel et al. (1993).

4.7 Data with a Nested Variation Structure

Ten apples are taken from a box. A randomization procedure assigns five to one tester, and the other five to another tester. Each tester makes two firmness tests on each of their five fruit. Firmness is measured by the pressure needed to push the flat end of a piece of rod through the surface of the fruit. Table 4.9 gives the results, in N/m^2.

For comparing the testers, we have five experimental units for each tester, not ten. One way to do a t-test is to take means for each fruit. We then have five values (means) for one

Table 4.9: *Each tester made two firmness tests on each of five fruit.*

					Fruit				
	Tester 1					Tester 2			
1	2	3	4	5	6	7	8	9	10
6.8, 7.3	7.2, 7.3	7.4, 7.3	6.8, 7.6	7.2, 6.5	7.7, 7.7	7.4, 7	7.2, 7.6	6.7, 6.7	7.2, 6.8

Table 4.10: *This gives the mean for each cell from Table 4.9.*

					Fruit				
	Tester 1					Tester 2			
1	2	3	4	5	6	7	8	9	10
7.05	7.25	7.35	7.2	6.85	7.7	7.2	7.4	6.7	7.0

treatment, that we can compare with the five values for the other treatment. Thus we work with Table 4.10.

What happens if we ignore the data structure, and compare ten values for one tester with ten values for the other tester? This pretends that we have ten experimental units for each tester. The analysis will suggest that the treatment means are more accurate than is really the case. We obtain a pretend standard error that is not the correct standard error of the mean. We are likely to underestimate the standard error of the treatment difference.

4.7.1 Degrees of freedom considerations

For comparison of two means when the sample sizes n_1 and n_2 are small, it is important to have as many degrees of freedom as possible for the denominator of the t-test. It is worth tolerating possible bias in some of the calculated SEDs in order to gain extra degrees of freedom.

The same considerations arise in the one-way analysis of variance, and we pursue the issue in that context. It is illuminating to plot out, side by side, say 10 SEDs based on randomly generated normal variates, first for a comparison based on 2 d.f., then 10 SEDs for a comparison based on 4 d.f., etc.

A formal statistical test is thus unlikely, unless the sample is large, to detect differences in variance that may have a large effect on the result of the test. We therefore must rely on judgement. Both past experience with similar data and subject area knowledge may be important. In comparing two treatments that are qualitatively similar, differences in the population variance may be unlikely, unless the difference in means is at least of the same order of magnitude as the individual means. If the means are not much different then it is reasonable, though this is by no means inevitable, to expect that the variances will not be much different.

If the treatments are qualitatively different, then differences in variance may be expected. Experiments in weed control provide an example where it would be surprising to find a

Table 4.11: *These are the same data as in Table 3.2.*

	Pair #								
	1	2	3	4	5	6	7	8	9
Heated (mm)	244	255	253	254	251	269	248	252	292
Ambient	225	247	249	253	245	259	242	255	286
Difference	19	8	4	1	6	10	6	−3	6

common variance. There will be few weeds in all plots where there is effective weed control, and thus little variation. In control plots, or for plots given ineffective treatments, there may be huge variation.

If there do seem to be differences in variance, it may be possible to model the variance as a function of the mean. It may be possible to apply a variance-stabilizing transformation. Or the variance may be a smooth function of the mean. Otherwise, if there are just one or two degrees of freedom per mean, use a pooled estimate of variance unless the assumption of equal variance seems clearly unacceptable.

4.7.2 General multi-way analysis of variance designs

Generalization to multi-way analysis of variance raises a variety of new issues. If each combination of factor levels has the same number of observations, and if there is no structure in the *error* (or *noise*), the extension is straightforward. The extension is less straightforward when one or both of these conditions are not met. For unbalanced data from designs with a simple error structure, it is necessary to use the lm() (linear model) function. The lme() function in the *nlme* package is able to handle problems where there is structure in the error term, including data from unbalanced designs. See Chapter 9 for further details.

Data from various special types of experimental and/or sampling design may require special treatment. Fractional factorial designs are one example. Currently R has no special tools for handling such designs.

4.8* Resampling Methods for Tests and Confidence Intervals

There are many different resampling methods. All rely on the selection of repeated samples from a "population" that is constructed using the sample data. In general, there are too many possible samples to take them all, and we therefore rely on repeated random samples. In this section, we demonstrate permutation and bootstrap methods. We start with permutation tests, illustrating their use for the equivalent of one-sample and two-sample *t*-tests.

4.8.1 The one-sample permutation test

Consider the paired elastic band data of Table 3.2 again, reproduced here as Table 4.11.

If the treatment has made no difference, then an outcome of 244 for the heated band and 225 for the ambient band might equally well have been 225 for the heated band and

244 for the ambient band. A difference of 19 becomes a difference of -19. There are $2^9 = 512$ permutations, and a mean difference associated with each permutation. We then locate the mean difference for the data that we observed within this permutation distribution. The *p*-value is the proportion of values that are as large in absolute value as, or larger than, the mean for the data.

The absolute values of the nine differences are

Difference	19	8	4	1	6	10	6	3	6

In the permutation distribution, these each have an equal probability of taking a positive or a negative sign. There are 2^9 possibilities, and hence $2^9 = 512$ different values for $\bar{d}$. The possibilities that give a mean difference that is as large as or larger than in the actual sample, where the value for pair 8 has a negative sign, are

Difference	19	8	4	1	6	10	6	3	6
	19	8	4	-1	6	10	6	3	6
	19	8	4	1	6	10	6	-3	6

There are another three possibilities that give a mean difference that is of the same absolute value, but negative. Hence $p = 6/512 = 0.0117$.

In general, when the number of pairs is large, it will not be feasible to use such an enumeration approach to get information on relevant parts of the upper and lower tails of the distribution. We therefore take repeated random samples from the permutation distribution. The function `onet.permutation()` in our *DAAG* package may be used for this.

4.8.2 *The two-sample permutation test*

Suppose we have n_1 values in one group and n_2 in a second, with $n = n_1 + n_2$. The permutation distribution results from taking all possible samples of n_2 values from the total of n values. For each such sample, we calculate

mean of the n_2 values that are selected – mean of remaining n_1 values.

The permutation distribution is the distribution of all such differences of means. We locate the differences of means for the actual samples within this permutation distribution.

The calculation of the full permutation distribution is not usually feasible. We therefore take perhaps 1000 samples from this distribution. The function `twot.permutation()` that is in our *DAAG* package may be used for this repeated sampling.

Thus consider the data from Subsection 4.3.1.

Ambient: 254 252 239 240 250 256 267 249 259 269 (Mean = 253.5)
Heated: 233 252 237 246 255 244 248 242 217 257 254 (Mean = 244.1)

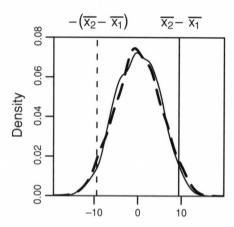

Figure 4.10: Density curves for two samples of 2000 each from the permutation distribution of the difference in means, for the two-sample elastic band data.

Figure 4.10 shows two estimates of the permutation distribution that were obtained by taking, in each instance, 2000 random samples from this distribution. The point where the difference in means falls with respect to this distribution ($253.5 - 244.1 = 9.4$) has been marked, as has minus this difference.[24]

The density estimate corresponding to the solid line gave a p-value of 0.051. The density estimate corresponding to the dashed line gave a p-value of 0.060. Use of a larger sample size will of course lead to more accurate p-values.

4.8.3 Bootstrap estimates of confidence intervals

The usual approach to constructing confidence intervals is based on a statistical theory that relies, in part, on the assumption of normally distributed observations. Sometimes this theory is too complicated to work out, and/or the normal assumption is not applicable. In such cases, the bootstrap may be helpful. We demonstrate the use of the methodology to calculate confidence intervals for the median and for the correlation.

A number of methods are available for computing bootstrap confidence intervals. We will demonstrate the so-called "percentile" method – applying it both to the median and to

---

[24]
```
## Draw just one of these curves
data(two65)                # From our DAAG package
x1 <- two65$ambient; x2 <- two65$heated; x <- c(x1, x2)
n1 <- length(x1); n2 <- length(x2); n <- n1+n2
dbar <- mean(x2)-mean(x1)
z <- numeric(2000)
for(i in 1:2000){
    mn <- sample(n, n2, replace=FALSE)
    dbardash <- mean(x[mn]) - mean(x[-mn])
    z[i]<- dbardash
    }
pval <- (sum(z > abs(dbar)) + sum (z< -abs(dbar)))/2000
plot(density(z), yaxs="i")
abline(v = dbar)
abline(v = -dbar, lty=2)
print(signif(pval,3))
rm(x1, x2, n1, n2, x, n, dbar, z, mn, dbardash, pval)
```

the correlation coefficient. This is one of the simplest bootstrap methods, which however works reasonably well for these applications.

The median

We will construct a 95% confidence interval for the median of the cuckoo egg lengths in wrens' nests. Just as for the computing of bootstrap standard errors, we calculate sample medians for a large number of resamples. The endpoints of the confidence interval are the 2.5 and 97.5 percentiles (in order to contain the middle 95% of the approximate sampling distribution of the median). Here is the R output:[25]

```
BOOTSTRAP CONFIDENCE INTERVAL CALCULATIONS
Based on 999 bootstrap replicates

CALL :
boot.ci(boot.out = wren.boot, type = "perc")

Intervals :
Level     Percentile
95%    (20.9, 22.1 )
Calculations and Intervals on Original Scale
```

The correlation coefficient

Bootstrap methods do not require bivariate normality. Independence between observations, i.e., between (x, y) pairs, is as important as ever. Note however that a correlation of e.g., 0.75 for a non-normal distribution may have quite different implications from a correlation of 0.75 when normality assumptions apply.

We will compute a 95% confidence interval for the correlation between chest and belly for the possum data frame:[26]

```
BOOTSTRAP CONFIDENCE INTERVAL CALCULATIONS
Based on 999 bootstrap replicates
CALL :
boot.ci(boot.out = possum.boot, type = "perc")

Intervals :
Level     Percentile
95%    ( 0.481,  0.708 )
Calculations and Intervals on Original Scale
```

[25] ## Use boot.wren, calculated on p.74, as a starting point
```
    boot.ci(wren.boot, type="perc")
```
[26]
```
data(possum)           # From our DAAG package
    possum.fun <- function(data, indices) {
        chest <- data$chest[indices]
        belly <- data$belly[indices]
        cor(belly, chest)
        }
    possum.boot <- boot(possum, possum.fun, R=999)
    boot.ci(possum.boot, type="perc")
```

The bootstrap – parting comments

Bootstrap methods are not a panacea. We must respect the structure of the data; any form of dependence in the data must be taken into account. There are contexts where the bootstrap is invalid and will mislead. As a rough guideline, we recommend against the bootstrap for statistics, including maximum, minimum and range, that are functions of sample extremes. The bootstrap is usually appropriate for statistics from regression analysis – means, variances, coefficient estimates, and correlation coefficients. It also works reasonably well for medians and quartiles, and other such statistics. See Davison and Hinkley (1997) or Efron and Tibshirani (1993). See also the references in the help page for the `boot()` function in the *boot* package.

4.9 Further Comments on Formal Inference

4.9.1 Confidence intervals versus hypothesis tests

Many researchers find hypothesis tests (significance tests) hard to understand. The methodology is too often abused. Papers that present a large number of significance tests are, typically, not making good use of the data. Also, it becomes difficult to know what to make of the results. Among a large number of tests, some will be significant as a result of chance. Misunderstandings are common in the literature, even among mature researchers. A p-value does not allow the researcher to say anything about the probability that either hypothesis, the null or its alternative, is true. Then why use them? Perhaps the best that can be said is that hypothesis tests often provide a convenient and quick answer to questions about whether effects seem to stand out above background noise. However if all that emerges from an investigation are a few p-values, we have to wonder what has been achieved.

Because of these problems, there are strong moves away from hypothesis testing and towards confidence intervals. Tukey (1991) argues strongly, and cogently, that confidence intervals are more informative and more honest than p-values. He argues

Statisticians classically asked the wrong question – and were willing to answer with a lie, one that was often a downright lie. They asked "Are the effects of A and B different?" and they were willing to answer "no".

All we know about the world teaches us that the effects of A and B are always different – in some decimal place – for every A and B. Thus asking "Are the effects different?" is foolish. What we should be answering first is "Can we tell the direction in which the effects of A differ from the effects of B?" In other words, can we be confident about the direction from A to B? Is it "up", "down", or "uncertain"? [Tukey, 1991]

Tukey argues that we should never conclude that we "accept the null hypothesis." Rather our uncertainty is about the direction in which A may differ from B. Confidence intervals do much better at capturing the nature of this uncertainty.

Here are guidelines on the use of tests of significance. Few scientific papers make more than half-a-dozen points that are of consequence. Any significance tests should be closely tied to these main points, preferably with just one or two tests for each point that is made. Keep any significance tests and p-values in the background. Once it is apparent that an effect is statistically significant, the focus of interest should shift to its pattern and magnitude, and to its scientific significance. For example, it is very poor practice to perform t-tests for each comparison between treatments when the real interest is (or should be) in the overall pattern of response. Where the response depends on a continuous variable, it is often pertinent to ask such questions as whether the response keeps on rising (falling), or whether it rises (falls) to a maximum (minimum) and then falls (rises).

Significance tests should give the researcher, and the reader of the research paper, confidence that the effects that are discussed are real! The focus should then move to the substantive scientific issues. Statistical modeling can be very helpful for this. The interest is often, finally, in eliciting the patterns of response that the data present.

4.9.2 If there is strong prior information, use it!

Any use of the hypothesis testing methodology that ignores strong prior information is inappropriate, and may be highly misleading. Diagnostic testing and criminal investigations provide cogent examples.

Suppose a diagnostic test for HIV infection has a specificity of 0.1%. This implies that, for every 1000 people tested, there will on average be one false positive. (This rate may be at the low end of the range of what is now achievable, and might apply to laboratories whose procedures do not follow the highest current standard.) Now suppose an adult Australian male who does not belong to any of the recognized risk categories has the test, with a positive result.

Using the hypothesis testing framework, we take the null hypothesis H_0 to be the hypothesis that the individual does not have HIV. Given this null hypothesis, the probability of a positive result is 0.001. Therefore the null hypothesis is rejected.

However, wait! There is strong prior information. The incidence of HIV in adult Australian males (15–49 years) may be 1 in 10 000. Assume that 10 001 people are tested. One person will, on average, have HIV and test positive. (We assume close to 100% specificity.) Among the remaining 10 000 who do not have HIV, 10 will test positive. Thus, if we incorporate the prior information, the odds that the person has HIV are 1:10, i.e., less than 10%.

Not infected	Infected
$10\,000 \times 0.001 = 10$ (false) positives	1 true positive

In a serious criminal case, the police might scrutinize a large number of potential perpetrators. A figure of 10 000 or more is entirely within the range of possibility. Suppose there is a form of incriminating evidence that is found in one person in 1000. Scrutiny of 10 000 potential perpetrators will on average net 10 suspects. Suppose one of these is later charged. The probability of such incriminating evidence, assuming that the defendant is innocent, is indeed 0.001. The police screening will net around 10 innocent people along with, perhaps,

Not the perpetrator	The perpetrator
$10\,000 \times 0.001 = 10$ (false) positives	1 true positive

the 1 perpetrator. This evidence leads to odds of 1:10 or worse, i.e., less than 10%, that the defendant is guilty. On its own, it should be discounted.

The interpretation of results from medical tests for AIDS is discussed in detail, with up to date probability estimates, in Gigerenzer (2002).

The calculations just given made an informal use of Bayesian methodology. Such an approach is essential when there is strong prior knowledge, and can be insightful as a commentary on classical methodology even when the prior assessment of probabilities is little more than a guess. For an exposition of a thorough-going use of the Bayesian approach, see, e.g., Gelman et al. (1995).

4.10 Recap

Dos and don'ts

- Do examine appropriate plots.
- Ensure that the analysis and graphs reflect any important structure in the data.
- Always present means, standard errors, and numbers for each group. Results from formal significance tests have secondary usefulness.
- The use of a large number of significance tests readily leads to data summaries that lack coherence and insight. Consider whether there is an alternative and more coherent analysis that would provide better insight.
- Reserve multiple range tests for unstructured data.
- Think about the science behind the data. What analysis will best reflect that science?
- The aim should be an insightful and coherent account of the data, placing it in the context of what is already known. The statistical analysis should assist this larger purpose.

4.11 Further Reading

On general issues of style and approach, see Wilkinson et al. (1999), Maindonald (1992), Krantz (1999) and Gigerenzer (1998 and 2002). See also the statistical good practice guidelines at the web site http://www.rdg.ac.uk/ssc/dfid/booklets.html. Miller (1986) has extensive comment on consequences of failure of assumptions, and on how to handle such failures. On the design of experiments, and on analysis of the resulting data, see Cox (1958) and Cox and Reid (2000). We include further brief discussion of the design and analysis of experiments in Chapter 9.

Formal hypothesis testing, which at one time had become almost a ritual among researchers in psychology, is now generating extensive controversy, reflected in the contributions to Harlow et al. (1997). See also Gigerenzer (1998), Wilkinson et al. (1999) and Nicholls (2000). Krantz (1999), which is a review of the Harlow et al. book, is a good guide to the controversy.

Gigerenzer et al. (1989) give interesting historical background to different styles and approaches to inference that have grown up in one or other area of statistical application. Gelman et al. (1995) is a careful account of a Bayesian inference; later chapters make relatively severe technical demands. There is a helpful brief summary of Bayesian methodology, including Bayesian modeling, in Chapter 3 of Baldi and Brunak (2001). A weakness of Baldi and Brunak's account is that there is no discussion of model criticism. Gigerenzer (2002) demonstrates the use of Bayesian arguments in several important practical contexts, including AIDS testing and screening for breast cancer.

References for further reading

Baldi, P. and Brunak, S. 2001. *Bioinformatics. The Machine Learning Approach*. MIT Press.

Cox, D.R. 1958. *Planning of Experiments*. Wiley, New York.

Cox, D.R. and Reid, N. 2000. *Theory of the Design of Experiments*. Chapman and Hall.

Gelman, A.B., Carlin, J.S., Stern, H.S. and Rubin, D.B. 1995. *Bayesian Data Analysis*. Chapman and Hall/CRC.

Gigerenzer, G. 1998. We need statistical thinking, not statistical rituals. *Behavioural and Brain Sciences* 21: 199–200.

Gigerenzer, G. 2002. *Reckoning with Risk: Learning to Live with Uncertainty*. Penguin Books.

Gigerenzer, G., Swijtink, Z., Porter, T., Daston, L., Beatty, J. and Krüger, L. 1989. *The Empire of Chance*. Cambridge University Press.

Harlow, L.L., Mulaik, S.A., and Steiger, J.H. (eds.) 1997. *What If There Were No Significance Tests?* Lawrence Erlbaum Associates.

Krantz, D.H. 1999. The null hypothesis testing controversy in psychology. *Journal of the American Statistical Association* 44: 1372–1381.

Maindonald J.H. 1992. Statistical design, analysis and presentation issues. *New Zealand Journal of Agricultural Research* 35: 121–141.

Miller R.G. 1986. *Beyond ANOVA, Basics of Applied Statistics*. Wiley.

Nicholls, N. 2000. The insignificance of significance testing. *Bulletin of the American Meteorological Society* 81: 981–986.

Tukey, J.W. 1991. The philosophy of multiple comparisons. *Statistical Science* 6: 100–116.

Wilkinson, L. and Task Force on Statistical Inference 1999. Statistical methods in psychology journals: guidelines and explanation. *American Psychologist* 54: 594–604.

4.12 Exercises

1. Draw graphs that show, for degrees of freedom between 1 and 100, the change in the 5% critical value of the *t*-statistic. Compare a graph on which neither axis is transformed with a graph on which the respective axis scales are proportional to log(*t*-statistic) and log(degrees of freedom). Which graph gives the more useful visual indication of the change in the 5% critical value of the *t*-statistic with increasing degrees of freedom?

2. Generate a random sample of 10 numbers from a normal distribution with mean 0 and standard deviation 2. Use `t.test()` to test the null hypothesis that the mean is 0. Now generate a random sample of 10 numbers from a normal distribution with mean 1.5 and standard deviation 2. Again use `t.test()` to test the null hypothesis that the mean is 0. Finally write a function that generates a random sample of *n* numbers from a normal distribution with mean μ and standard deviation 1, and returns the *p*-value for the test that the mean is 0.

3. Use the function that was created in exercise 2 to generate 50 independent *p*-values, all with a sample size $n = 10$ and with mean $\mu = 0$. Use `qqplot()`, with the parameter setting

x = qunif(ppoints(50)), to compare the distribution of the *p*-values with that of a uniform random variable, on the interval [0, 1]. Comment on the plot.

4. Here we generate random normal numbers with a sequential dependence structure.

```
library(ts)
y1 <- rnorm(51)
y <- y1[-1] + y1[-51]
acf(y1)    # acf is `autocorrelation function'
           # (see Ch. 9)
acf(y)
```

Repeat this several times. There should be no consistent pattern in the acf plot for different random samples y1. There will be a fairly consistent pattern in the acf plot for y, a result of the correlation that is introduced by adding to each value the next value in the sequence.

5. Create a function that does the calculations in the first two lines of the previous exercise. Put the calculation in a loop that repeats 25 times. Calculate the mean and variance for each vector y that is returned. Store the 25 means in the vector av, and store the 25 variances in the vector v. Calculate the variance of av.

6. Use mosaicplot() to display the table rareplants (Subsection 4.4.1) that was created using code in footnote 13. Annotate the mosaic plot to draw attention to the results that emerged from the analysis in Subsection 4.4.1.

7. *Type help(UCBAdmissions) to get information on the three-way table UCBAdmissions that is in the *ctest* package. The table gives admission rates, broken down by sex, for the six largest Berkeley departments in 1973. The third dimension of the table is department. What are the first two dimensions of the table? The following gives a table that adds the 2 × 2 tables of admission data over all departments:

```
data(UCBAdmissions)
UCBtotal <- apply(UCBAdmissions, c(1,2), sum)
```

(a) Create a mosaic plot, both for the data in UCBtotal and for the data for each faculty separately. (If necessary refer to the code given in the help page for UCBAdmissions.) Compare the table UCBtotal with the result from applying the function mantelhaen.test() to the table UCBAdmissions. Compare the two sets of results, and comment on the difference.

(b) The Mantel–Haenzel test is valid only if the male to female odds ratio for admission is similar across departments. The following code calculates the relevant odds ratios:

```
apply(UCBAdmissions, 3, function(x)
      (x[1,1]*x[2,2])/(x[1,2]*x[2,1]))
```

Is the odds ratio consistent across departments? Which department(s) stand(s) out as different? What is the nature of the difference?

8. *Tables 4.12 and 4.13 contain fictitious data that illustrate issues that arise in combining data across tables. To enter the data for Table 4.12, type in

```
admissions <- array(c(30,30,10,10,15,5,30,10),
                     dim=c(2,2,2))
```

Table 4.12: *In these data, addition over heterogeneous tables has created a spurious overall bias.*

	Engineering			Sociology			Total	
	Male	Female		Male	Female		Male	Female
Admit	30	10	Admit	15	30	Admit	45	40
Deny	30	10	Deny	5	10	Deny	35	20

Table 4.13: *In these data, biases that go in different directions in the two faculties have canceled in the table of totals.*

	Engineering			Sociology			Total	
	Male	Female		Male	Female		Male	Female
Admit	30	20	Admit	10	20	Admit	40	40
Deny	30	10	Deny	5	25	Deny	35	35

and similarly for Table 4.13. The third dimension in each table is faculty, as required for using faculty as a stratification variable for the Mantel–Haenzel test. From the help page for `man-telhaen.test()`, extract and enter the code for the function `woolf()`. Apply the function `woolf()`, followed by the function `mantelhaen.test()`, to the data of each of Tables 4.12 and 4.13. Explain, in words, the meaning of each of the outputs. Then apply the Mantel–Haenzel test to each of these tables.

9. Re-do the analysis of the plant root weight data presented in Figure 4.6 after taking logarithms of the weights. In the analysis, construct an analogue to Figure 4.6 as well as a new analysis of variance table.

10. The function `overlap.density()` in the *DAAG* package can be used to visualize the unpaired version of the *t*-test. Type in

```
data(two65)
attach(two65)
overlap.density(ambient, heated)      # Included with our
                                      # DAAG package
```

in order to observe estimates of the stretch distributions of the ambient (control) and heated (treatment) elastic bands.

11. *For constructing bootstrap confidence intervals for the correlation coefficient, it is advisable to work with the Fisher *z*-transformation of the correlation coefficient. The following lines of R code show how to obtain a bootstrap confidence interval for the *z*-transformed correlation between `chest` and `belly` in the `possum` data frame. The last step of the procedure is to apply the inverse of the *z*-transformation to the confidence interval to return it to the original scale. Run the following code and compare the resulting interval with the one computed without transformation. Is the *z*-transform necessary here?

```
z.transform <- function(r) .5*log((1+r)/(1-r))
z.inverse <- function(z) (exp(2*z)-1)/(exp(2*z)+1)
possum.fun <- function(data, indices) {
    chest <- data$chest[indices]
    belly <- data$belly[indices]
    z.transform(cor(belly, chest))}
possum.boot <- boot(possum, possum.fun, R=999)
z.inverse(boot.ci(possum.boot, type="perc")$percent[4:5])
  # See help(bootci.object). The 4th and 5th elements of
  # the percent list element hold the interval endpoints.
```

12. The 24 paired observations in the data set `mignonette` were from five pots. The observations
    are in order of pot, with the numbers 5, 5, 5, 5, 4 in the respective pots. Plot the data in a way that
    shows the pot to which each point belongs. Also do a plot that shows, by pot, the differences be-
    tween the two members of each pair. Do the height differences appear to be different for different
    pots?

5

Regression with a Single Predictor

Data for which the models of this chapter may be appropriate can be displayed as a scatterplot. Our focus is on the straight line model, though the use of transformations makes it possible to accommodate specific forms of non-linear relationship within this framework. The convention we follow is that the x-variable, plotted on the horizontal axis, has the role of explanatory variable. The y-variable, plotted on the vertical axis, has the role of response or outcome variable.

The issues that arise in these simple regression models are fundamental to any study of regression methods. There are various special applications of linear regression that raise their own specific issues. We consider one such application, to size and shape data.

Scrutiny of the scatterplot should precede the regression calculations that we will describe. Such a plot may indicate that the intended regression is plausible, or it may reveal unexpected features. If there are many observations, try fitting a smooth curve. If the smooth curve differs substantially from an intended line, then straight line regression may be inappropriate, as in Figure 2.6. The smooth curve may, if a line is inadequate, be all that is needed. For example, it will allow for the prediction of values of y at any given x that lies within the range of the data. Such smooth curves, however, lack the simple form of mathematical description that parametric regression offers.

5.1 Fitting a Line to Data

In the straight line regression model that we will describe,

- there is just one explanatory or predictor variable x,
- the values y are independent,
- given x, the response y has a normal distribution with mean given by a linear function of x.

Notice that the role of the variables is not symmetric. Our interest is in predicting y given x. The straight line regression model has the form

$$y = \alpha + \beta x + \varepsilon.$$

Subscripts are often used. Given observations (x_1, y_1), (x_2, y_2), ..., (x_n, y_n), we may write

$$y_i = \alpha + \beta x_i + \varepsilon_i.$$

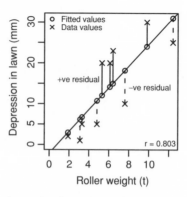

Figure 5.1: Lawn depression for various weights of roller, with fitted line.

In standard analyses, we assume that the ε_i are independently and identically distributed as normal variables with mean 0 and variance σ^2. In the terminology of Chapter 3, the $\alpha + \beta x$ term is the deterministic component of the model, and ε represents the random noise. Interest usually centers on the deterministic term. The R function $\mathtt{lm()}$ provides a way to estimate the slope β and the intercept α (the line is chosen so that the sum of squares of residuals is as small as possible). Given estimates (a for α and b for β), we can pass the straight line

$$\widehat{y} = a + bx$$

through the points of the scatterplot. Fitted or predicted values are calculated using the above formula, i.e.

$$\widehat{y}_1 = a + bx_1, \ \widehat{y}_2 = a + bx_2, \ldots.$$

By construction, the fitted values lie on the estimated line. The line passes through the cloud of observed values. Useful information about the noise can be gleaned from an examination of the residuals, which are the differences between the observed and fitted values,

$$e_1 = y_1 - \widehat{y}_1, \ e_2 = y_2 - \widehat{y}_2, \ldots.$$

In particular, a and b are estimated so that the sum of the squared residuals is as small as possible, i.e., the resulting fitted values are as close (in this "least squares" sense) as possible to the observed values.

5.1.1 Lawn roller example

Figure 5.1 shows, for the lawn roller data, details of the straight line regression. The depression measurements serve as the response values $y_1, y_2, \ldots$, and the weights are the values of the predictor variable $x_1, x_2, \ldots$. The regression model is

$$\text{depression} = \alpha + \beta \times \text{weight} + \text{noise}.$$

We use the R model formula $\mathtt{depression \sim weight}$ to supply the model information to the R function $\mathtt{lm()}$, thus:[1]

----

[1] ```
Previous to these statements, we specified:
   options(digits=4)   # This setting will be assumed in most later output
```
```

```
data(roller)          # DAAG package
roller.lm <- lm(depression ~ weight, data = roller)
summary(roller.lm)
```

The name `roller.lm` that we gave to the output of the calculation was chosen for mnemonic reasons – the object was the result of `lm` calculations on the `roller` data set. We have used the `summary.lm()` function to extract summary information. There are other such extractor functions that give more specialized information. Or we can examine `roller.lm` directly. Here is the output:

```
> summary(roller.lm)

Call:
lm(formula = depression ~ weight, data = roller)

Residuals:
   Min     1Q Median     3Q    Max
 -8.18  -5.58  -1.35   5.92   8.02

Coefficients:
             Estimate Std. Error t value Pr(>|t|)
(Intercept)    -2.09       4.75   -0.44   0.6723
weight          2.67       0.70    3.81   0.0052

Residual standard error: 6.74 on 8 degrees of freedom
Multiple R-Squared: 0.644,       Adjusted R-squared:  0.6
F-statistic: 14.5 on 1 and 8 DF,  p-value: 0.00518
```

Note that we have the global settings `options(digits=4)` and `options(show.signif.stars=FALSE)`. The setting `show.signif.stars=FALSE` suppresses interpretive features that, for our use of the output, are unhelpful.

The R output gives a numerical summary of the residuals as well as a table of the estimated regression coefficients with their standard errors. Here, we observe that the intercept of the fitted line is $a = -2.09$ (SE $= 4.75$), and the estimated slope is $b = 2.67$ (SE $= .70$). Also included in the table are t-statistics and p-values for individual tests of the hypotheses "true slope $= 0$" and "true intercept $= 0$". For this example, the p-value for the slope is small, which should not be surprising, and the p-value for the intercept is large. This is consistent with our intuition that depression should be proportional to weight. Thus, we could have fitted a model that did not contain an intercept term.[2] The standard deviation of the noise term is also estimated. In the R output, it is identified as the residual standard error (6.735 for this example). We defer comment on R^2 and the F-statistic until Subsection 5.1.4.

5.1.2 Calculating fitted values and residuals

Fitted values can be obtained from the regression output. For example, when `weight` has the value 1.9, the predicted value of `depression` is, rounding to two decimal places,

$$\widehat{y} = -2.09 + 2.67 \times 1.9 = 2.98.$$

[2] `lm(depression ~ -1 + weight, data = roller)`

The residual corresponding to this observation is the difference between the observed and fitted value:

$$e_1 = 2 - 2.98 = -0.98.$$

The fitted values can be obtained using the predict() function. The resid() function gives the residuals:[3]

```
    weight depression fitted.value residual
1      1.9          2         2.98    -0.98
2      3.1          1         6.18    -5.18
3      3.3          5         6.71    -1.71
4      4.8          5        10.71    -5.71
5      5.3         20        12.05     7.95
6      6.1         20        14.18     5.82
7      6.4         23        14.98     8.02
8      7.6         10        18.18    -8.18
9      9.8         30        24.05     5.95
10    12.4         25        30.98    -5.98
```

5.1.3 Residual plots

Model assumptions should be checked, as far as is possible. The common checks are:

- a plot of residuals versus fitted values – for example there may be a pattern in the residuals that suggests that we should be fitting a curve rather than a line;
- a normal probability plot of residuals – if residuals are from a normal distribution points should lie, to within statistical error, close to a line.

We will use the straight line analysis that we have been considering to illustrate the use of these plots. Our conclusion will be that the plots are consistent with the assumptions that underlie the analysis. Note however that, in such a small data set, departures from assumptions will be hard to detect.

Figure 5.2A shows the first of these plots, for the calculations we have just done.[4] There is a suggestion of clustering in the residuals, but no indication that there should be a curve rather than a line. The normal probability plot in Figure 5.2B allows, as indicated in the second bullet point above, an assessment whether the distribution of residuals follows a normal distribution.[5]

Interpretation of these plots may require a certain amount of practice. This is especially the case for normal probability plots, as noted in Subsection 3.4.2. Again here, we need

---

[3] data.frame(roller, fitted.value=predict(roller.lm), residual=resid(roller.lm))
[4] plot(roller.lm, which = 1)
    # if "which" is not specified, we get all four diagnostic plots
[5] plot(roller.lm, which = 2, pch=16)
    # requests the second of 4 diagnostic plots

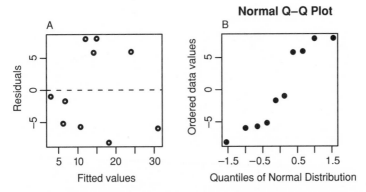

Figure 5.2: Diagnostic plots for the regression of Figure 5.1. Panel A is a plot of residuals against fitted values. Panel B is a normal probability plot of residuals. If residuals follow a normal distribution the points should fall close to a line.

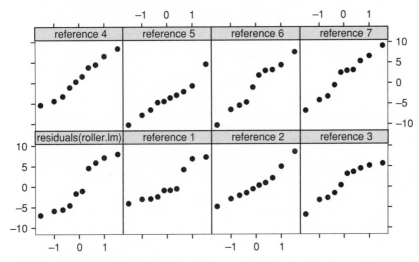

Figure 5.3: Normal probability plot for the regression of Figure 5.1, together with normal probability plots for computer-generated normal data.

to calibrate the eye by examining a number of independent plots from computer-generated normal data, as in Figure 5.3.[6]

Iron slag example: is there a pattern in the residuals?

We now consider an example where there does appear to be a clear pattern in the residuals. The data compare two methods for measuring the iron content in slag – a magnetic method

```
6 test <- residuals(roller.lm); n <- length(test)
  av <- mean(test); sdev <- sd(test)
  y <- c(test, rep(rnorm(n, av, sdev), 7))
  fac <- c(rep("residuals(roller.lm)",n), paste("reference", rep(1:7, rep(n,7))))
  fac <- factor(fac, levels=unique(fac))
  library(lattice)
  qqmath(~ y|fac, aspect=1, layout=c(4,2))
  ## Alternative, using the  function qreference(), from our DAAG package
  qreference(residuals(roller.lm), nrep=8, nrows=2)
```

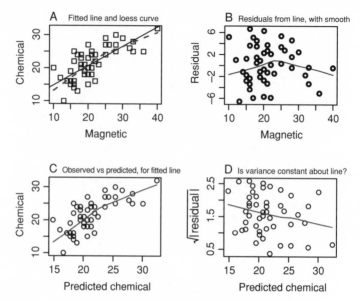

Figure 5.4: Chemical Test of Iron Content in Slag, versus Magnetic Test. The fitted curves used the lowess smooth. In D, the downward slope suggests lower variance for larger fitted values. See however Figure 5.5. (Exercise 6 on page 133 has R code for panels B and D.)

and a chemical method (data are from Roberts, 1974, p. 126). The chemical method requires greater effort and is presumably expensive, while the magnetic method is quicker and easier.

Figure 5.4A suggests that the straight line model is wrong. The smooth curve (shown with a dashed line) gives a better indication of the pattern in the data. Panel B shows the residuals from the straight line fit. The nonlinearity is now obvious. Panel C plots observed values against predicted values. Panel D is designed to allow a check on whether the error variance is constant. There are theoretical reasons for plotting the square root of absolute values of the residuals on the vertical axis in panel D.

Taken at face value, Figure 5.4D might be interpreted as a strong indication that the variance decreases with increasing predicted value. Note, however, that the residuals are from the straight line model, which we know to be wrong. Thus the plot in panel D may be misleading. We reserve judgement until we have examined the equivalent plot for residuals from the smooth curve.

Figure 5.5 shows the plot of residuals from the loess fit versus the predictor, together with the plot of square root of absolute values of residuals against predicted chemical test result. There is now no suggestion of heterogeneity. The heterogeneity or change of variance that was suggested by Figure 5.4D was an artifact of the use of a line, where the true response was curvilinear.

Where there is true heterogeneity of variance, we can obtain estimates for the variance at each x-value, and weight data points proportionately to the reciprocal of the variance. Getting an estimate to within some constant of proportionality is enough. It may be possible to guess at a suitable functional form for the change in variance with x or (equivalently, since $y = a + bx$) with y. For example, the variance may be proportional to y.

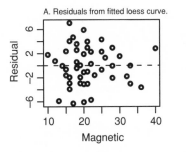
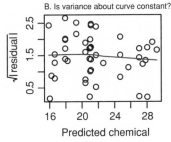

Figure 5.5: Residuals (A), and square root of absolute values of residuals (B), for the loess smooth for the data of Figure 5.4. (Exercise 7 on page 133 has the R code.)

5.1.4 The analysis of variance table

The analysis of variance table breaks the sum of squares about the mean, for the y-variable, into two parts: a part that is accounted for by the deterministic component of the model, i.e., by a linear function of `weight`, and a part attributed to the noise component or residual. For the lawn roller example, the analysis of variance table is[7]

```
Analysis of Variance Table

Response: depression
          Df Sum Sq Mean Sq F value   Pr(>F)
weight     1 657.97  657.97  14.503 0.005175
Residuals  8 362.93   45.37
```

The total sum of squares (about the mean) for the 10 observations is 1020.9 ($= 658.0 + 362.9$; we round to one decimal place). Including weight reduced this by 658.0, giving a residual sum of squares equal to 362.9. For comparing the reduction with the residual, it is best to look at the column headed `Mean Sq`, i.e., *mean square*. The mean square of the reduction was 658.0, giving 45.4 as the mean square for the residual. Note also the degrees of freedom: as 2 parameters (the slope and intercept) were estimated the residuals have $10 - 2 = 8$ degrees of freedom. For two points, both residuals would equal zero. Each additional point beyond 2 adds one further degree of freedom.

This table has the information needed for calculating R^2 (also known as the "coefficient of determination") and adjusted R^2. The R^2 statistic is the square of the correlation coefficient, and is the sum of squares due to weight divided by the total sum of squares:

$$R^2 = \frac{658.0}{1020.9} = 0.64,$$

while

$$\text{adjusted } R^2 = 1 - \frac{362.9/8}{1020.9/9} = 0.60.$$

Adjusted R^2 takes into account the number of degrees of freedom, and is preferable to R^2, though neither statistic is an effective summary of what has been achieved. They do not

[7] `anova(roller.lm)`

Table 5.1: *Volumes (≤m³) and weights (g), for eight softback books.*

volume	weight
412	250
953	700
929	650
1492	975
419	350
1010	950
595	425
1034	725

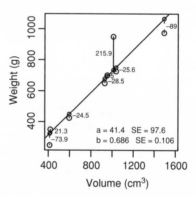

Figure 5.6: Weight versus volume for paperback books, with fitted line.

give any direct indication of how well the regression equation will predict when applied to a new data set.

Even for comparing models, adjusted R^2 and R^2 are in general unsatisfactory. Section 5.5 describes more appropriate measures.

5.2 Outliers, Influence and Robust Regression

The data in Table 5.1 are for a collection of eight softback books. Figure 5.6 displays these data. Also shown is information about the residuals.

Here is the output from the regression calculations:[8]

```
> summary(softback.lm)

Residuals:
    Min      1Q   Median      3Q      Max
-89.674 -39.888 -25.005   9.066 215.910
```

---

[8] data(softbacks)
  softback.lm <- lm(weight ~ volume, data=softbacks)

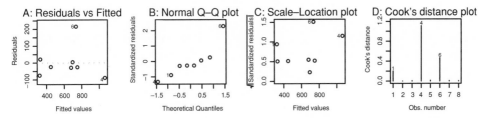

Figure 5.7: Diagnostic plots for Figure 5.6.

```
Coefficients:
              Estimate Std. Error t value Pr(>|t|)
(Intercept)    41.3725    97.5588   0.424 0.686293
volume          0.6859     0.1059   6.475 0.000644

Residual standard error: 102.2 on 6 degrees of freedom
Multiple R-Squared: 0.8748,      Adjusted R-squared: 0.8539
F-statistic: 41.92 on 1 and 6 degrees of freedom,
                       p-value: 0.0006445
```

Figure 5.7 shows regression diagnostics.[9]

We should already be familiar with plots A and B from Figure 5.7. For regression with one explanatory variable, plot A is equivalent to a plot of residuals against the explanatory variable. Plot C, which is of no interest for the present data, is designed for examining the constancy of the variance. Plot D identifies residuals that are influential in determining the form of the regression line. *Influential* points are those with Cook's distances that are close to, or greater than, one.

The largest two residuals in Figure 5.7D are identified with their row labels, which for the `softbacks` data frame are the observation numbers. The largest residual is that for observation 6. Cook's distance is a measure of *influence*; it measures the extent to which the line would change if the point were omitted. The most influential observation is observation 4, which corresponds to the largest value of `volume`. In part, it has this large influence because it is at the extreme end of the range of x-values. It is a high *leverage* point, exerting a greater pull on the regression line than observations closer to the center of the range. Since its y-value is lower than would be predicted by the line, it pulls the line downward.

Point 6 seems an outlier. There may then be two sets of results to report – the main analysis, and the results for any outlying observations. Here, with only eight points, we do not really want to omit any of them, especially as points 4 and 6 are both, for different reasons, candidates for omission.

Diagnostic plots, such as Figure 5.7, are not definitive. Rather, they draw attention to points that require further investigation. Following identification of an apparent outlier, it is good practice to do the analysis both with and without the outlier. If retention of an apparent outlier makes little difference to the practical use and interpretation of the results, then it should be retained in the main analysis.

---

[9] `plot(softback.lm)`

Robust regression

Robust regression offers a half-way house between including outliers and omitting them entirely. Rather than omitting outliers, it downweights them, reducing their influence on the fitted regression line. This has the additional advantage of making outliers stand out more strongly against the line. The *MASS* package has the function `rlm()` that may be used for robust regression. Note also the `lqs()` *resistant* regression function in the *lqs* package. For the distinction between robust regression and resistant regression, see the R help page for `lqs()`. We pursue the investigation of robust and resistant regression in the exercises at the end of this chapter.

5.3 Standard Errors and Confidence Intervals

The regression output gives the standard error of the regression slope. Setting up a confidence interval for the regression slope follows a similar format to that for a population mean.

In addition, we can ask for standard errors of predicted values, then make them the basis for calculating confidence intervals. A wide confidence interval for the regression slope implies that intervals for predicted values will be accordingly wide.

Recall that since two parameters (the slope and intercept) have been estimated, the error mean square is calculated with $n - 2$ degrees of freedom. As a consequence, the standard errors for the slope and for predicted values are calculated with $n - 2$ degrees of freedom. Both involve the use of the square root of the error mean square.

5.3.1 Confidence intervals and tests for the slope

A 95% confidence interval for the regression slope is

$$b \pm t_{.975}\text{SE}_b$$

where $t_{.975}$ is the 97.5% point of the t distribution with $n - 2$ degrees of freedom, and SE_b is the standard error of b.

We demonstrate the calculation for the `roller` data. From the second row of the `Coefficients` table in the summary output of Subsection 5.1.1, the slope estimate is 2.67, with standard error equal to 0.70. The t-critical value for a 95% confidence interval on $10 - 2 = 8$ degrees of freedom is $t_{.975} = 2.30$. Therefore, the 95% confidence interval is[10]

$$2.67 \pm 2.3 \times .7 = (1.1, 4.3).$$

The validity of such confidence intervals depends crucially upon the regression assumptions.

For a formal statistical test, the statistic

$$t = \frac{b}{\text{SE}_b}$$

---

[10] `SEb <- summary(roller.lm)$coefficients[2, 2]`
`coef(roller.lm)[2] + qt(c(0.025,.975), 8)*SEb`

Table 5.2: *Observed and fitted values of* depression *at the given* weight *values, together with two different types of SE. The column headed SE gives the precision of the predicted value. The column headed SE.OBS gives the prediction of a new observation.*

	Predictor weight	Observed depression	Fitted	SE	SE.OBS	
1	1.9	2	3.0	3.6	7.6	$\sqrt{3.6^2 + 6.74^2}$
2	3.1	1	6.2	3.0	7.4	$\sqrt{3.0^2 + 6.74^2}$
3	3.3	5	6.7	2.9	7.3	
4	4.8	5	10.7	2.3	7.1	
5	5.3	20	12.0	2.2	7.1	
6	6.1	20	14.2	2.1	7.1	
7	6.4	23	15.0	2.1	7.1	
8	7.6	10	18.2	2.4	7.1	
9	9.8	30	24.0	3.4	7.5	
10	12.4	25	31.0	4.9	8.3	

is used to check whether the true regression slope differs from 0. Again, this is compared with a t-critical value based on $n - 2$ degrees of freedom. An alternative is to check whether the 95% confidence interval for b contains 0.

5.3.2 SEs and confidence intervals for predicted values

There are two types of predictions: the prediction of the points on the line, and the prediction of a new data value. The SE estimates of predictions for new data values take account both of uncertainty in the line and of the variation of individual points about the line. Thus the SE for prediction of a new data value is larger than that for prediction of points on the line.

Table 5.2 shows expected values of the depression, with SEs, for various roller weights.[11] The column that is headed SE indicates how precisely points on the line are determined. The column headed SE.OBS indicates the precision with which new observations can be predicted. For determining SE.OBS, there are two sources of uncertainty: the standard error for the fitted value (estimated at 3.6 in row 1) and the noise standard error (estimated at 6.74) associated with a new observation.

Figure 5.8 shows 95% pointwise confidence bounds for the fitted line.[12]

---

[11]
```
## Code to obtain fitted values and standard errors (SE, then SE.OBS)
fit.with.se <- predict(roller.lm, se.fit=TRUE)
fit.with.se$se.fit                                       # SE
sqrt(fit.with.se$se.fit^2+fit.with.se$residual.scale^2)  # SE.OBS
```
[12]
```
plot(depression ~ weight, data=roller, xlab = "Weight of Roller (tonnes)",
     ylab = "Depression in Lawn (mm)", pch = 16)
roller.lm <- lm(depression ~ weight, data = roller)
abline(roller.lm$coef, lty = 1)
xy <- data.frame(weight = pretty(roller$weight, 20))
yhat <- predict(roller.lm, newdata = xy, interval="confidence")
ci <- data.frame(lower=yhat[,"lwr"],upper=yhat[,"upr"])
lines(xy$weight, ci$lower, lty = 2, lwd=2, col="grey")
lines(xy$weight, ci$upper, lty = 2, lwd=2, col="grey")
```

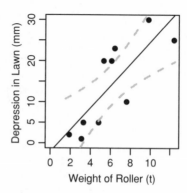

Figure 5.8: Lawn depression, for various weights of roller, with fitted line and showing 95% pointwise confidence bounds for points on the fitted line.

It bears emphasizing that the validity of these calculations depends crucially on the appropriateness of the fitted model for the given data.

5.3.3* Implications for design

The point of this subsection is to emphasize that the choice of location of the x-values, which is a design issue, is closely connected with sample size considerations. As often, increasing the sample size is not the only, or necessarily the best, way to improve precision.

The estimated variance of the slope estimate is

$$\mathrm{SE}_b{}^2 = \frac{s^2}{ns_x^2},$$

where we define

$$s_x^2 = \frac{\sum_i (x_i - \bar{x})^2}{n}.$$

Here s^2 is the error mean square, i.e., s is the estimated SD for the population from which the residuals are taken.

Now consider two alternative ways that we might reduce SE_b by a factor of (e.g.) 2:

- If we fix the configuration of x-values, but multiply by 4 the number of values at each discrete x-value, then s_x is unchanged. As n has been increased by a factor of 4, the effect is to reduce SE_b by a factor of 2.
- Alternatively, we can reduce SE_b by pushing the x-values further apart, thus increasing s_x. Increasing the average separation between x-values by a factor of 2 will reduce SE_b by a factor of 2.

Providing the relationship remains close to linear, and we have enough distinct well-separated x-values that we can check for linearity, pushing the x-values further apart is the better way to go. The caveats are important – it is necessary to check that the relationship is linear over the extended range of x-values.

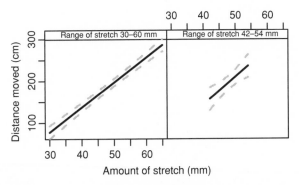

Figure 5.9: Two rubber band experiments, with different ranges of x-values. The dashed curves are pointwise 95% confidence bounds for points on the fitted line. Note that, for the panel on the right, the axis labels appear above the panel, as is done for *lattice* plots.

Reducing SE_b does at the same time reduce the standard error of points on the fitted line. Figure 5.9 shows the effect of increasing the range of x-values (the code for both panels is a ready adaptation of the code for Figure 5.8). Both experiments used the same rubber band. The first experiment used a much wider range of values of x (= amount by which the rubber band was stretched). For the left panel of Figure 5.9, $s_x = 10.8$, while for the right panel $s_x = 4.3$.

5.4 Regression versus Qualitative anova Comparisons

Issues of power

An analysis that fails to take advantage of structure in the data may fail to find what is there. Figure 5.10 shows six sets of data that have been simulated to follow a linear trend.

The first p-value tests for linear trend, while the second p-value tests for qualitative differences between treatment effects, ignoring the fact that the levels are quantitative (note that the test for linear trend is equivalent to the test for a linear contrast from the aov() function that is available when the explanatory term is an ordered factor). A test for linear trend is more powerful than an analysis of variance that treats the five levels as qualitatively different levels. In repeated simulations of Figure 5.10, the p-values in the test for linear trend will on average be smaller than in the analysis of variance that makes qualitative comparisons between the five levels.

To get a clear indication of the effect we need a more extensive simulation. Figure 5.11 plots results from 200 simulations. Both axes use a scale of $\log(p/(1-p))$. On the vertical axis are the p-values for a test for linear trend, while the horizontal axis plots p-values for an aov test for qualitative differences. The majority of points (for this simulation, 91%) lie below the line $y = x$.

The function simulate.linear() in our DAAG package allows readers to experiment with such simulations. Write the p-values for a test for linear trend as p_l, and the p-values for the analysis of variance test for qualitative differences as p_a. Specifying type="density" gives overlaid plots of the densities for the two sets of p-values, both on a scale of $\log(p/(1-p))$, together with a plot of the density of

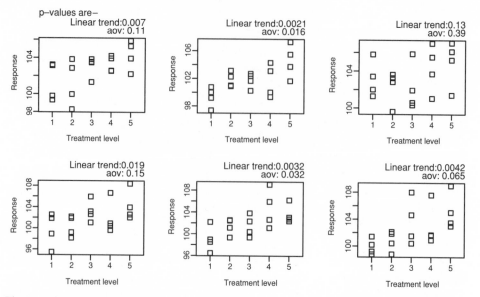

Figure 5.10: Test for linear trend, versus analysis of variance comparison that treats the levels as qualitatively different. The six panels are six different simulations from the straight line model with slope 0.8, SD = 2, and 4 replications per level.

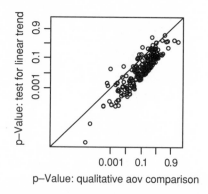

Figure 5.11: This plot compares *p*-values in a test for linear trend with *p*-values in an analysis of variance test for qualitative differences, in each of 200 sets of simulated results. We show also the line $y = x$.

$\log(p_l/(1 - p_l)) - \log(p_a/(1 - p_a))$. As the data are paired, this last plot is the preferred way to make the comparison.

The pattern of change

There are other reasons for fitting a line or curve, where this is possible, rather than fitting an analysis of variance type model that has a separate parameter for each separate level of the explanatory variable. Fitting a line (or a curve) allows interpolation between successive levels of the explanatory variable. It may be reasonable to hazard prediction a small distance beyond the range of the data. The nature of the response curve may give scientific insight.

5.5 Assessing Predictive Accuracy

The *resubstitution* estimate of predictive accuracy, derived by direct application of model predictions to the data from which the regression relationship was derived, gives in general an optimistic assessment. There is a mutual dependence between the model prediction and the data used to derive that prediction. The problem becomes more serious as the number of parameters that are estimated increases, i.e., as we move from the use of one explanatory variable to examples (in later chapters) where there may be many explanatory variables. The simple models discussed in the present chapter are a good context in which to begin discussion of the issue.

5.5.1 Training/test sets, and cross-validation

An ideal is to assess the performance of the model on a new data set. The data that are used to develop the model form the *training* set, while the data on which predictions are tested form the *test* set. This is a safe and generally valid procedure, provided that the test set can be regarded as a random sample of the population to which predictions will be applied. If there are too few data to make it reasonable to divide data into training and test sets, then the method of cross-validation can be used.

Cross-validation extends the training/test set approach. As with that approach, it estimates predictive accuracy for data that are sampled from the population in the same way as the existing data. The data are divided into k sets (or *folds*), where k is typically in the range 3 to 10. Each of the k sets becomes in turn the test set, with the remaining data forming the training set. The average of predictive accuracy assessments from the k folds is used as a measure of the predictive performance of the model. This may be done for several different measures of predictive performance.

5.5.2 Cross-validation – an example

We present an example of the use of cross-validation with a small data set. In order to simplify the discussion, we will use 3-fold validation only. Cross-validation has special importance in contexts where believable model-based estimates of predictive accuracy are not available.

Table 5.3 shows data on floor area and sale price for 15 houses in a suburb of Canberra, in 1999. Rows of data have been numbered from 1 to 15. For demonstrating cross-validation, we use a random number sampling system to divide the data up into three equal groups.[13]

---

[13]
```
rand <- sample(1:15)%%3 + 1
    # a%%3 is the remainder of a, modulo 3
    # It is the remainder after subtracting from a the
    # largest multiple of 3 that is <= a
  (1:15)[rand == 1] # Observation numbers for the first group
  (1:15)[rand == 2] # Observation numbers for the second group
  (1:15)[rand == 3] # Observation numbers for the third group.
```

Table 5.3: *Floor area and sale price, for 15 houses
in Aranda, a suburb of Canberra, Australia.*

Row number	area	bedrooms	sale.price
1	694	4	192.0
2	905	4	215.0
	. . .	. . .	. . .
15	1191	6	375.0

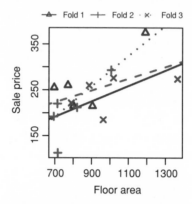

Figure 5.12: Graphical summary of 3-fold cross-validation for the house sale data (Table 5.3). The
line is fitted leaving out the corresponding "test" set of points. Predictions for these omitted points
are used to assess predictive accuracy.

The observation numbers for three groups we obtain are

```
2  3  12  13  15
1  5  7  8  14
4  6  9  10  11
```

Re-running the calculations will of course lead to a different division into three
groups.

   At the first pass (fold 1) the first set of rows will be set aside as the test data, with remaining
rows making up the training data. Each such division between training data and test data is
known as a *fold*. At the second pass (fold 2) the second set of rows will be set aside as the
test data, while at the third pass (fold 3) the third set of rows will be set aside as the test
data. A crucial point is that at each pass the data that are used for testing are separate from
the data used for prediction. Figure 5.12 is a visual summary.[14]

   The following summary of the cross-validation results includes, for each fold, estimates
of the mean square error.

```
fold 1
Observations in test set: 2 3 12 13 15
```

---

[14] `data(houseprices)`   `# DAAG package`
   `houseprices.lm <- lm(sale.price ~ area, data=houseprices)`
   `cv.lm (houseprices, houseprices.lm)`

```
Floor area        905.0 802.00 696.0 771.0 1191
Predicted price 225.9 208.63 190.9 203.4  274
Observed price  215.0 215.00 255.0 260.0  375
Residual        -10.9   6.37  64.1  56.6  101

Sum of squares = 17719    Mean square = 3544    n = 5

fold 2
Observations in test set: 1 5 7 8 14

Floor area        694.0   716 821.0 714.00 1006.0
Predicted price 222.4   225 238.6 224.97  262.2
Observed price  192.0   113 212.0 220.00  293.0
Residual        -30.4 -113 -26.6  -4.97   30.8

Sum of squares = 15269    Mean square = 3054    n = 5

fold 3
Observations in test set: 4 6 9 10 11

Floor area        1366 963.0 1018.0 887.00 790.00
Predicted price   412 278.4  296.6 253.28 221.22
Observed price    274 185.0  276.0 260.00 221.50
Residual         -138 -93.4  -20.6   6.72   0.28

Sum of squares = 28127    Mean square = 5625    n = 5
Overall ms
      4074
```

At each fold, the training set consists of the remaining rows of data.

To obtain the estimate of the error mean square, take the total of the sums of squares and divide by 15. This gives

$$s^2 = (17\,719 + 15\,269 + 28\,127)/15 = 4074.$$

Actually, what we have is an estimate of the error mean square when we use only 2/3 of the data. Thus we expect the cross-validated error to be larger than the error if all the data could be used. We can reduce the error by doing 10-fold rather than 3-fold cross-validation. Or we can do leave-one-out cross-validation, which for these data is 15-fold cross-validation.

Contrast $s^2 = 4074$ with the estimate $s^2 = 2323$ that we obtained from the model-based estimate in the regression output for the total data set.[15]

5.5.3* Bootstrapping

We first indicate how resampling methods can be used to estimate the standard error of slope of a regression line. Recalling that the standard error of the slope is the standard deviation of the sampling distribution of the slope, we need a way of approximating this sampling

---

[15] summary(houseprices.lm)$sigma^2

distribution. One way of obtaining such an approximation is to resample the observations or cases directly. For example, suppose five observations have been taken on a predictor x and response y:

$$(x_1, y_1), (x_2, y_2), (x_3, y_3), (x_4, y_4), (x_5, y_5).$$

Generate five random numbers with replacement from the set $\{1, 2, 3, 4, 5\}$: 3, 5, 5, 1, 2, say. The corresponding resample is then

$$(x_3, y_3), (x_5, y_5), (x_5, y_5), (x_1, y_1), (x_2, y_2).$$

Note we are demonstrating only the so-called case-resampling approach. Another approach involves fitting a model and resampling the residuals. Details for both methods are in Davison and Hinkley (1997, Chapter 6). A regression line can be fit to the resampled observations, yielding a slope estimate. Repeatedly taking such resamples, we obtain a distribution of slope estimates, the bootstrap distribution.

As an example, consider the regression relating `sale.price` to `area` in the `houseprices` data. For comparison purposes, we compute the estimate given by `lm()`.[16]

```
> summary(houseprices.lm)$coef
```

```
              Estimate Std. Error t value Pr(>|t|)
(Intercept)   70.750     60.3477    1.17    0.2621
area           0.188      0.0664    2.83    0.0142
```

The `lm()` estimate of the standard error of the slope is thus .0664.

In order to use the `boot()` function, we need a function that will evaluate the slope for the bootstrap resamples:

```
houseprices.fn <- function (houseprices, index){
    house.resample <- houseprices[index, ]
    house.lm <- lm(sale.price ~ area, data=house.resample)
    coef(house.lm)[2]      # slope estimate for resampled data
    }
```

Using 999 resamples, we obtain a standard error estimate that is almost 40% larger than the `lm()` estimate:[17]

```
Bootstrap Statistics :
    original  bias     std. error
t1*    0.188  0.0169      0.0916
```

By changing the statistic argument in the `boot()` function appropriately, we can compute standard errors and confidence intervals for fitted values. Here we use the `predict()` function to obtain predictions for the given area:

```
houseprices1.fn <- function (houseprices, index){
    house.resample <- houseprices[index, ]
```

---

[16] `houseprices.lm <- lm(sale.price ~ area, data=houseprices)`
   `summary(houseprices.lm)$coef`
[17] `set.seed(1028)`   `# only necessary for replicating our results exactly`
   `library(boot)`   `# ensure that the boot package is loaded`
   `houseprices.boot <- boot(houseprices, R=999, statistic=houseprices.fn)`

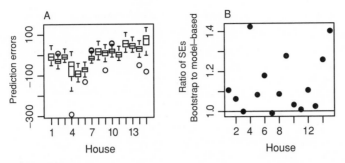

Figure 5.13: (A) Plot of bootstrap distributions of prediction errors for regression relating sale.price to area, each based on 200 bootstrap estimates of the prediction error. (B) Ratios of bootstrap prediction standard errors to model-based prediction standard errors.

```
house.lm <- lm(sale.price ~ area, data=house.resample)
predict(house.lm, newdata=data.frame(area=1200))
}
```

For example, a 95% confidence interval for the expected sale price of a house (in Aranda) having an area of 1200 square feet is (249 000, 363 000).[18]

The bootstrap procedure can be extended to gain additional insight into how well a regression model is making predictions. Regression estimates for each resample are used to compute predicted values at all of the original values of the predictor. The differences (i.e., the prediction errors) between the observed responses and these resampled predictors can be plotted against observation number. Repeating this procedure a number of times gives a distribution of the prediction errors at each observation. Figure 5.13A displays a prediction error plot for the houseprices data.[19] Note the large variability in the prediction error associated with observation 4. We can use the same bootstrap output to estimate the standard errors. These can be compared with the usual estimates obtained by lm. Figure 5.13B displays ratios of the bootstrap standard errors to the model-based standard errors.[20] In this case, the model-based standard errors are generally smaller than the bootstrap standard errors. A cautious data analyst might prefer the bootstrap standard errors.

---

[18] houseprices1.boot <- boot(houseprices, R=999, statistic=houseprices1.fn)
    boot.ci(houseprices1.boot, type="perc") # "basic" is an alternative to "perc"
[19] houseprices2.fn <- function (houseprices, index)
    {
       house.resample <- houseprices[index,]
       house.lm <- lm(sale.price ~ area, data=house.resample)
       houseprices$sale.price - predict(house.lm, houseprices)   # resampled prediction
                                                                   # errors
    }
    n <- length(houseprices$area)
    R <- 200
    houseprices2.boot <- boot(houseprices, R=R, statistic=houseprices2.fn)
    house.fac <- factor(rep(1:n, rep(R, n)))
    plot(house.fac, as.vector(houseprices2.boot$t), ylab="Prediction Errors", xlab=
        "House")
[20] bootse <- apply(houseprices2.boot$t, 2, sd)
    usualse <- predict.lm(houseprices.lm, se.fit=TRUE)$se.fit
    plot(bootse/usualse, ylab="Ratio of Bootstrap SE's to Model-Based SE's",
        xlab="House", pch=16)
    abline(1, 0)

We can also compute an estimate of the aggregate prediction error, as an alternative to the cross-validation estimate obtained in the previous subsection. There are a number of ways to do this, and some care should be taken. We refer the interested reader to Davison and Hinkley (1997, Section 6.4).

Commentary

The cross-validation and bootstrap estimates of mean square error are valid, provided we can assume a homogeneous variance. This is true even if data values are not independent. However, the estimate of predictive error applies only to data that have been sampled in the same way as the data that are used as the basis for the calculations. In the present instance, the estimate of predictive accuracy applies only to 1999 house prices in the same city suburb.

Such standard errors may have little relevance to the prediction of house prices in another suburb, even if thought to be comparable, or to prediction for more than a very short period of time into the future. This point has relevance to the use of regression methods in business "data mining" applications. A prediction that a change will make cost savings of $500 000 in the current year may have little relevance to subsequent years. The point will have special force if the change will take six or twelve months to implement. A realistic, though still not very adequate, assessment of accuracy may be derived by testing a model that is based on data from previous years on a test set that is formed from the current year's data. Predictions based on the current year's data may, if other features of the business environment do not change, have a roughly comparable accuracy for prediction a year into the future. If the data series is long enough, we might, starting at a point part-way through the series, compare predictions one year into the future with data for that year. The estimated predictive accuracy would be the average accuracy for all such predictions. A more sophisticated approach might involve incorporation of temporal components into the model, i.e., use of a time series model.

5.6* A Note on Power Transformations

Among the more common transformations for continuous data are:

- **Logarithmic**: This is often the right transformation to use for size measurements (linear, surface, volume or weight) of biological organisms. Some data may be too skewed even for a logarithmic transformation. Some types of insect counts provide examples.
- **Square root** or **cube root**: These are milder than the logarithmic transformation. If linear measurements on insects are normally distributed, then we might expect the cube root of weight to be approximately normally distributed. The square root is useful for data for counts of "rare events". The power transformation generalizes the transformations that we have just discussed. Examples of power transformations are y^2, $y^{0.5}$, y^3, etc. Figure 5.14 shows a number of response curves, and describes the particular power transformation that would make the relationship linear.

If the ratio of largest to smallest data value is greater than 10, and especially if it is more than 100, then the logarithmic transformation should be tried as a matter of course. Check this advice against the response curves shown in Figure 5.14.

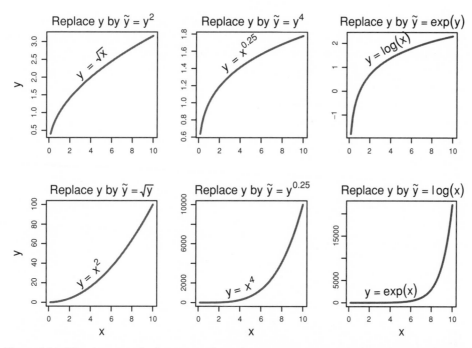

Figure 5.14: The above panels show some alternative response curves. The formula for $\tilde{y}$ gives the power family transformation of y that will make $\tilde{y}$ a linear function of x. Thus, if $y = \log(x)$, then the transformation $\tilde{y} = \exp(y)$ will make $\tilde{y}$ a linear function of x.

We have so far mentioned only transformation of y. We might alternatively transform x. In practice, we will use whichever of these alternatives leads to the more homogeneous variance. Or, we may need to transform both x and y.

General power transformations

For $\lambda \neq 0$, the power transformation replaces a value y by y^λ. The logarithmic transformation corresponds to $\lambda = 0$. In order to make this connection, a location and scale correction is needed. The transformation is then

$$y(\lambda) = \frac{y^\lambda - 1}{\lambda}, \quad \text{if } \lambda \neq 0,$$
$$y(\lambda) = \log(y), \quad \text{if } \lambda = 0.$$

- If the small values of a variable need to be spread, make λ smaller.
- If the large values of a variable need to be spread, make λ larger.

This is called the Box–Cox transformation, as proposed in Box and Cox (1964).

The R function `boxcox()`, whose syntax is similar to that of `lm()`, can be used to obtain data-driven estimates of λ. We pursue investigation of `boxcox()` in an exercise at the end of the chapter.

5.7 Size and Shape Data

In animal growth studies, there is an interest in the rate of growth of one organ relative to another. For example, heart growth may be related to increase in body weight. As the animal must be sacrificed to measure the weights of organs, it is impossible to obtain data on growth profiles for individual animals. For seals and dolphins and some other protected marine species, the main source of information is animals who have died, e.g., snared in trawl nets, as an unintended consequence of commercial fishing. For each animal, the data provide information at just one point in time, when they died. At best, if conditions have not changed too much over the lifetimes of the animals in the sample, the data may provide an indication of the average of the population growth profiles. If, e.g., sample ages range from 1 to 10 years, it is pertinent to ask how food availability may have changed over the past 10 years, and whether this may have had differential effects on the different ages of animal in the sample.

5.7.1 Allometric growth

The allometric growth equation is

$$y = ax^b$$

where x may, e.g., be body weight and y heart weight. It may alternatively be written

$$\log y = \log a + b \log x,$$

i.e.,

$$Y = A + bX,$$

where

$$Y = \log y, \quad A = \log a, \quad \text{and } X = \log x.$$

If $b = 1$, then the two organs (e.g. heart and body weight) grow at the same rate. Thus, we have an equation that can be fitted by linear regression methods, allowing us to predict values of Y given a value for X.

Stewardson et al. (1999) present such data for the Cape fur seal. Figure 5.15 shows the plot of heart weight against body weight for 30 seals.

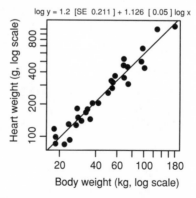

Figure 5.15: Heart weight versus body weight, for 30 Cape fur seals.

Here is the R output for the calculations that fit the regression line.[21]

```
> cfseal.lm <- lm(log(heart) ~ log(weight), data=cfseal)
> summary(cfseal.lm)

Call:
lm(formula = log(heart) ~ log(weight), data = cfseal)

Residuals:
      Min        1Q    Median        3Q       Max
-3.15e-01 -9.18e-02  2.47e-05  1.17e-01  3.21e-01

Coefficients:
             Estimate Std. Error t value Pr(>|t|)
(Intercept)    1.2043     0.2113     5.7  4.1e-06
log(weight)    1.1261     0.0547    20.6  < 2e-16

Residual standard error: 0.18 on 28 degrees of freedom
Multiple R-Squared: 0.938,       Adjusted R-squared: 0.936
F-statistic:  424 on 1 and 28 DF,  p-value: <2e-16
```

Note that the estimate of the exponent b ($= 1.126$) differs from 1.0 by 2.3 ($= 0.126/0.0547$) times its standard error. Thus for these data, the relative rate of increase seems slightly greater for heart weight than for body weight. We have no interest in the comparison between b and zero, for which the t-statistic and p-value in the regression output are appropriate (authors sometimes present p-values that focus on the comparison with zero, even though their interest is in the comparison with 1.0. See Table 10 and other similar tables in Gihr and Pilleri, 1969, p. 43). For an elementary discussion of allometric growth, see Schmidt-Nielsen (1994).

5.7.2 There are two regression lines!

At this point, we note that there are two regression lines – a regression line for y on x, and a regression line for x on y. It makes a difference which is the explanatory variable, and which the dependent variable. The two lines are quite different if the correlation is small. Figure 5.16 illustrates the point for two other data sets.

An alternative to a regression line

Here we note that there are yet other possibilities. A point of view that makes good sense for the seal organ growth data is that there is an underlying functional relationship. The analysis assumes that observed values of log(organ weight) and log(body weight) differ from the values for this underlying functional relationship by independent random amounts. The line that is obtained will lie between the regression line for y on x and the line for x on y.

---

[21] `data(cfseal)    # From our DAAG package`
  `cfseal.lm <- lm(log(heart) ~ log(weight), data=cfseal); summary(cfseal.lm)`

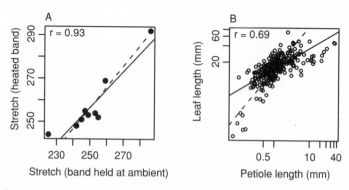

Figure 5.16: In each plot we show both the regression line for y on x (solid line), and the regression line for x on y (dotted line). In A the lines are quite similar, while in B where the correlation is smaller, the lines are quite different. Plot A is for the data of Table 3.2, while B is for a leaf data set.

5.8 The Model Matrix in Regression

This section presents the calculations of earlier sections from a point of view that moves us closer to the mathematical operations that are needed to handle the calculations. It will to an extent test the understanding of the content of earlier sections. We include it at this point because it will be important for understanding some of the extensions of straight line regression that we will consider in later chapters. Straight line regression is a simple context in which to introduce these ideas.

The formal mathematical structure of model matrices will allow us to fit different lines to different subsets of the data. It will allow us to fit polynomial curves. It allows the handling of multiple regression calculations. It will make it straightforward to fit qualitative effects that are different for different factor levels. The model matrix, which is mostly optional for understanding the examples that we consider in Chapter 6, will be of central importance in Chapter 7.

In straight line regression, the model or X matrix has two columns – a column of 1s and a column that holds values of the explanatory variable x. As fitted, the straight line model is

$$\widehat{y} = a + bx$$

which we can write as

$$\widehat{y} = 1 \times a + x \times b.$$

For an example, we return to the lawn roller data. The model matrix, with the y-vector alongside, is given in Table 5.4.[22]

For each row, we take some multiple of the value in the first column, another multiple of the value in the second column, and add them. Table 5.5 shows how calculations proceed given the estimates of a and b obtained earlier.

Note also the simpler (no intercept) model. For this

$$\widehat{y} = bx.$$

In this case the model matrix has only a single column, containing the values of x.

---

[22] `model.matrix(roller.lm)`

Table 5.4: *The model matrix, for the lawn roller data, with the vector of observed values in the column to the right.*

X		y
weight (t)		depression (mm)
1	1.9	2
1	3.1	1
1	3.3	5
1	4.8	5
1	5.3	20
1	6.1	20
1	6.4	23
1	7.6	10
1	9.8	30
1	12.4	25

Table 5.5: *The use of the model matrix for calculation of fitted values and residuals, in fitting a straight line to the lawn roller data.*

Model matrix		Multiply and add to yield fitted value $\widehat{y}$	Compare with observed y	Residual = $y - \widehat{y}$
$\times -2.09$	$\dfrac{\times\ 2.67}{\text{weight}}$			
1	1.9	$-2.1 + 2.67 \times\ 1.9 =\ 2.98$	2	$2 - 2.98$
1	3.1	$-2.1 + 2.67 \times\ 3.1 =\ 6.18$	1	$1 - 6.18$
1	3.3	$-2.1 + 2.67 \times\ 3.3 =\ 6.71$	5	$5 - 6.71$
1	4.8	$-2.1 + 2.67 \times\ 4.8 = 10.71$	5	$5 - 10.71$
1	5.3	$-2.1 + 2.67 \times\ 5.3 = 12.05$	20	$20 - 12.05$
1	6.1	$-2.1 + 2.67 \times\ 6.1 = 14.18$	20	$20 - 14.18$
1	6.4	$-2.1 + 2.67 \times\ 6.4 = 14.98$	23	$23 - 14.98$
1	7.6	$-2.1 + 2.67 \times\ 7.6 = 18.18$	10	$10 - 18.18$
1	9.8	$-2.1 + 2.67 \times\ 9.8 = 24.05$	30	$30 - 24.05$
1	12.4	$-2.1 + 2.67 \times 12.4 = 30.98$	25	$25 - 30.98$

5.9 Recap

In exploring the relationships in bivariate data, the correlation coefficient can be a crude and unduly simplistic summary measure. Keep in mind that it measures linear association. Wherever possible, use the richer and more insightful regression framework.

The model matrix, together with the coefficients, allows calculation of predicted values. The coefficients give the values by which the values in the respective columns must be multiplied. These are then summed over all columns. In later chapters, we will use the model matrix formulation to fit models that are not inherently linear.

In the study of regression relationships, there are many more possibilities than regression lines. If a line is adequate, use that. Simple alternatives to straight line regression using the original data are:

- Transform x and/or y.
- Use polynomial regression.
- Fit a smoothing curve.

Following the calculations:

- Plot residuals against fitted values.
- If it seems necessary, do a plot that checks homogeneity of variance.
- Plot residuals against the explanatory variable or, better still, do a component plus residual plot.

For size and shape data, the equation that assumes allometric variation is a good starting point. Relationships between the logarithms of the size variables are linear.

The line for the regression of y on x is different from the line for the regression of x on y. The difference between the two lines is most marked when the correlation is small.

5.10 Methodological References

We refer the reader to the suggestions for further reading at the end of Chapter 6.

5.11 Exercises

1. Here are two sets of data that were obtained using the same apparatus, including the same rubber band, as the data frame elasticband. For the data set elastic1, the values are

   ```
   stretch (mm): 46, 54, 48, 50, 44, 42, 52
   distance (cm): 183, 217, 189, 208, 178, 150, 249.
   ```

   For the data set elastic2, the values are

   ```
   stretch (mm): 25, 45, 35, 40, 55, 50, 30, 50, 60
   distance (cm): 71, 196, 127, 187, 249, 217, 114, 228, 291.
   ```

   Using a different symbol and/or a different color, plot the data from the two data frames elastic1 and elastic2 on the same graph. Do the two sets of results appear consistent?

2. For each of the data sets elastic1 and elastic2, determine the regression of stretch on distance. In each case determine
   (i) fitted values and standard errors of fitted values and
   (ii) the R^2 statistic. Compare the two sets of results. What is the key difference between the two sets of data?
   Use the robust regression function rlm() from the *MASS* package to fit lines to the data in elastic1 and elastic2. Compare the results with those from use of lm(). Compare regression coefficients, standard errors of coefficients, and plots of residuals against fitted values.

3. Using the data frame cars (in the *base* package), plot distance (i.e. stopping distance) versus speed. Fit a line to this relationship, and plot the line. Then try fitting and plotting a quadratic

curve. Does the quadratic curve give a useful improvement to the fit? [Readers who have studied the relevant physics might develop a model for the change in stopping distance with speed, and check the data against this model.]

4. In the data set `pressure` (*base* package), examine the dependence of pressure on temperature. [The relevant theory is that associated with the Claudius-Clapeyron equation, by which the logarithm of the vapor pressure is approximately inversely proportional to the absolute temperature. For further details of the Claudius-Clapeyron equation, search on the internet, or look in a suitable reference text.]

5. *Look up the help page for the function `boxcox()` from the *MASS* package, and use this function to determine a transformation for use in connection with Exercise 4. Examine diagnostics for the regression fit that results following this transformation. In particular, examine the plot of residuals against temperature. Comment on the plot. What are its implications for further investigation of these data?

6. The following code gives panels B and D of Figure 5.4. Annotate this code, explaining what each function does, and what the parameters are:

```
library(DAAG);  data(ironslag)
attach(ironslag)
par(mfrow=c(2,2))
ironslag.lm <- lm(chemical ~ magnetic)
chemhat <- fitted(ironslag.lm)
res <- resid(ironslag.lm)
## Figure 5.4B
plot(magnetic, res, ylab = "Residual", type = "n")
panel.smooth(magnetic, res, span = 0.95)
## Figure 5.4D
sqrtabs <- sqrt(abs(res))
plot(chemhat, sqrtabs, xlab = "Predicted chemical",
     ylab = expression(sqrt(abs(residual))), type = "n")
panel.smooth(chemhat, sqrtabs, span = 0.95)
detach(ironslag)
```

7. The following code gives the values that are plotted in the two panels of Figure 5.5.

```
library(modreg)
ironslag.loess <- loess(chemical ~ magnetic, data=ironslag)
chemhat <- fitted(ironslag.loess)
res2 <- resid(ironslag.loess)
sqrtabs2 <- sqrt(abs(res2))
```

Using this code as a basis, create plots similar to Figure 5.5A and 5.5B. Why have we preferred to use `loess()` here, rather than `lowess()`? [Hint: Is there a straightforward means for obtaining residuals from the curve that `lowess()` gives? What are the x-values, and associated y-values, that `lowess()` returns?]

6

Multiple Linear Regression

In straight line regression, a response variable y is regressed on a single explanatory variable x. Multiple linear regression generalizes this methodology to allow multiple explanatory or predictor variables. The focus may be on accurate prediction. Or it may, alternatively or additionally, be on the regression coefficients themselves. We warn the reader that interpreting the regression coefficients is not as straightforward as it might appear.

6.1 Basic Ideas: Book Weight and Brain Weight Examples

The book weight example has two x-variables in the regression equation. Table 6.1 is an extension of the data from Section 5.2. To the softback books, we now add 7 books with hardback covers.

Figure 6.1 plots the data.[1]

The explanatory variables are the volume of the book ignoring the covers, and the total area of the front and back covers. We might expect that

$$\text{weight of book} = b_0 + b_1 \times \text{volume} + b_2 \times \text{area of covers}.$$

The intercept, b_0, may not be needed. However, we will retain it for the moment. Later, we can decide whether to set it to zero.

Here is the regression output:[2]

```
> summary(allbacks.lm)


Call:
lm(formula = weight ~ volume + area, data = allbacks)

Residuals:
    Min     1Q Median     3Q    Max
 -104.1  -30.0  -15.5   16.8  212.3
```

---

[1] ## Simplified version of plot
```
  attach(allbacks)          # allbacks is from our DAAG package
  volume.split <- split(volume, cover)
  weight.split <- split(weight, cover)
  plot(weight.split$hb ~ volume.split$hb, pch=16, xlim=range(volume),
       ylim=range(weight), ylab="Weight (g)", xlab="Volume (cc)")
  points(weight.split$pb ~ volume.split$pb, pch=16, col=2)
  detach(allbacks)
```
[2] `allbacks.lm <- lm(weight ~ volume+area, data=allbacks)`
```
  summary(allbacks.lm)
  # for coefficient estimates and SEs, without other output enter
  # summary(allbacks.lm)$coef
```

Table 6.1: *Weights, volumes, and area of the cover, for 7 hardback (cloth) and 8 softback books.*

	volume (cm^3)	area (cm^2)	weight (g)	cover
1	885	382	800	hb
2	1016	468	950	hb
3	1125	387	1050	hb
4	239	371	350	hb
5	701	371	750	hb
6	641	367	600	hb
7	1228	396	1075	hb
8	412	0	250	pb
9	953	0	700	pb
10	929	0	650	pb
11	1492	0	975	pb
12	419	0	350	pb
13	1010	0	950	pb
14	595	0	425	pb
15	1034	0	725	pb

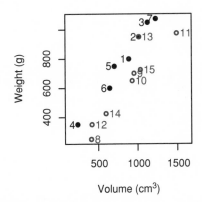

Figure 6.1: Weight versus volume, for 15 books.

```
Coefficients:
            Estimate Std. Error t value Pr(>|t|)
(Intercept)  22.4134    58.4025    0.38  0.70786
volume        0.7082     0.0611   11.60  7e-08
area          0.4684     0.1019    4.59  0.00062

Residual standard error: 77.7 on 12 degrees of freedom
Multiple R-Squared: 0.928,      Adjusted R-squared: 0.917
F-statistic: 77.9 on 2 and 12 DF,  p-value: 1.34e-007
```

The coefficient estimates are $b_0 = 22.4$, $b_1 = .708$ and $b_2 = .468$. Standard errors and p-values are provided for each estimate. Note that the p-value for the intercept suggests that it cannot be distinguished from 0, as we guessed earlier. The p-value for volume tests $b_1 = 0$, in the equation that has both volume and area as explanatory variables.

The estimate of the noise standard deviation (the residual standard error) is 77.7. There are now $15 - 3 = 12$ degrees of freedom for the residual; we start with 15 observations and estimate 3 parameters. In addition, there are two versions of R^2. We defer comment on these statistics for a later section. The F-statistic allows an overall test of significance of the regression. The null hypothesis for this test is that all coefficients (other than the intercept) are 0. Here, we obviously reject this hypothesis and conclude that at least one coefficient differs from 0.

The output is geared towards various tests of hypotheses. The t-statistics and associated p-values should however be used informally, rather than as a basis for formal tests of significance. Even in this simple example, the output has four p-values. This may not be too bad, but what if we have six or eight p-values? There are severe problems in interpreting results from such a multiplicity of formal tests, with varying amounts of dependence between the various tests.

The information on individual regression coefficients can readily be adapted to obtain a confidence interval for the coefficient. The 5% critical value for a t-statistic with 12 degrees of freedom is 2.18.[3] Thus, a 95% confidence interval is $0.708 \pm 2.18 \times 0.0611$, i.e., it ranges from 0.575 to 0.841.

The analysis of variance table is[4]

```
Analysis of Variance Table

Response: weight
          Df Sum Sq Mean Sq F value   Pr(>F)
volume     1 812132  812132   134.7    7e-08
area       1 127328  127328    21.1  0.00062
Residuals 12  72373    6031
```

Note that this is a sequential analysis of variance table. It gives the contribution of volume after fitting the overall mean, and then the contribution of area after fitting the overall mean and then volume. The p-value for area in the anova table must agree with that in the main regression output, since in both cases, we are testing the significance of the coefficient of area after including volume in the model. The p-values for volume need not necessarily agree. They will differ if there is a correlation between volume and other explanatory variables included in the model. They agree in this instance (to a fair approximation), because area is almost uncorrelated with volume. (The correlation is 0.0015.)

Finally, note the model matrix that has been used in the least square calculations:

```
> model.matrix(allbacks.lm)

  (Intercept) volume area
1           1    885  382
2           1   1016  468
3           1   1125  387
```

---

[3] qt(0.975, 12)
[4] anova(allbacks.lm)

4	1	239	371
5	1	701	371
6	1	641	367
7	1	1228	396
8	1	412	0
9	1	953	0
10	1	929	0
11	1	1492	0
12	1	419	0
13	1	1010	0
14	1	595	0
15	1	1034	0

Predicted values are given by multiplying the first column by b_0 (=22.4), the second by b_1 (=0.708), the third by b_2 (=0.468), and adding.

6.1.1 Omission of the intercept term

We now investigate the effect of leaving out the intercept. Here is the regression output:[5]

```
> summary(allbacks.lm0)

Call:
lm(formula = weight ~ -1 + volume + area, data = allbacks)

Residuals:
    Min     1Q Median     3Q     Max
 -112.5  -28.7  -10.5   24.6  213.8

Coefficients:
         Estimate Std. Error t value Pr(>|t|)
volume     0.7289     0.0277   26.34  1.1e-12
area       0.4809     0.0934    5.15  0.00019

Residual standard error: 75.1 on 13 degrees of freedom
Multiple R-Squared: 0.991,       Adjusted R-squared: 0.99
F-statistic:  748 on 2 and 13 DF,  p-value: 3.8e-014
```

Notice that the regression coefficients now have smaller standard errors. This is because, in the model that included the intercept, there was a substantial negative correlation between the estimate of the intercept and the coefficient estimates. The reduction in standard error is greater for the coefficient of volume, where the correlation was −0.88, than for area, where the correlation was −0.32.[6]

---

[5] allbacks.lm0 <- lm(weight ~ -1+volume+area, data=allbacks)
[6] ## By setting corr=TRUE output correlations between estimates
   summary(allbacks.lm, corr=TRUE)

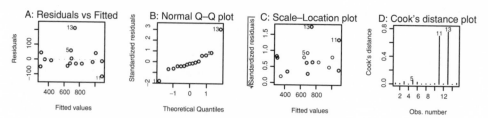

Figure 6.2: Diagnostic plots for Figure 6.1.

6.1.2 Diagnostic plots

Figure 6.2 displays useful information for checking on the adequacy of the model fit to the allbacks data.[7]

Should we omit observation 13? The first task is to check the data. The data are correct. The book is a computing book, with a smaller height to width ratio than any of the other books. It is clearly constructed from a heavier paper, though differences in the paper are not obvious to the inexpert eye. It may be legitimate to omit it from the main analysis, but noting that there was one book that seemed to have a much higher weight to volume ratio than any of the books that were finally included. If we omit observation 13, here is what we find:[8]

```
> summary(allbacks.lm13)

Call:
lm(formula = weight ~ -1 + volume + area, data = allbacks[-13, ])

Residuals:
    Min      1Q Median     3Q    Max
 -61.72 -25.29   3.43  31.24  58.86

Coefficients:
         Estimate Std. Error t value Pr(>|t|)
volume     0.6949     0.0163    42.6   1.8e-14
area       0.5539     0.0527    10.5   2.1e-07

Residual standard error: 41 on 12 degrees of freedom
Multiple R-Squared: 0.997,       Adjusted R-squared: 0.997
F-statistic: 2.25e+003 on 2 and 12 DF,  p-value: 3.33e-016
```

The residual standard error is substantially smaller in the absence of observation 13. Observation 11 stands out in the Cook's distance plot, but not in the plot of residuals. There is no reason to be unduly concerned, though we might want to examine the relevant row from the lm.influence()$coef or dfbetas() output that we now describe.

---

[7] par(mfrow = c(2,2))
   plot(allbacks.lm0)
   par(mfrow = c(1,1))
[8] allbacks.lm13 <- lm(weight ~ -1+volume+area, data=allbacks[-13,])

There will be further discussion of the interpretation of these and other diagnostic plots in Section 6.2.

6.1.3 Further investigation of influential points

In addition to examining the diagnostic plots, it may be interesting and useful to examine, for each data point in turn, how removal of that point affects the regression coefficients. The functions lm.influence() and dfbetas() give this information. We demonstrate the use of the lm.influence() function for the regression model, without the intercept term, that we fitted to the allbacks data set. The changes in the coefficients are[9]

```
volume    area
  0.73    0.48
```

```
          [,1]   [,2]
 [1,]   0.00  -0.01
 [2,]   0.00  -0.01
 [3,]   0.00   0.01
 [4,]   0.00   0.00
 [5,]   0.00   0.03
 [6,]   0.00  -0.02
 [7,]   0.00   0.00
 [8,]   0.00   0.01
 [9,]   0.00   0.00
[10,]   0.00   0.01
[11,]  -0.03   0.07
[12,]   0.00  -0.01
[13,]   0.03  -0.07
[14,]   0.00   0.00
[15,]   0.00   0.01
```

Rounding to two decimal places makes it easier to see where the changes are most substantial. The big changes are in observations 11 and 13.

The function dfbetas() gives a standardized version of the changes. Changes are divided by a standard error estimate for the change. Such standardization may assist visual scrutiny of the output. If the distributional assumptions are satisfied, standardized changes that are larger than 2 can be expected, for a specified coefficient, in about 1 observation in 20. For most diagnostic purposes, the output from dfbetas() should thus be used, rather than the output that we have given.

The Cook's distances, and examination of the drop-1 regression coefficients, provide useful guidance when there is just one influential residual that is distorting the regression estimates. Dynamic regression graphics, as in Cook and Weisberg (1999), may be a useful recourse when two or more residuals are jointly influential in distorting the regression estimates.

---

[9] `round(coef(allbacks.lm0), 2)   # Baseline for changes`
`round(lm.influence(allbacks.lm0)$coef, 2)`

Table 6.2: *Range of values, for the three variables in the data set* litters.

	lsize	bodywt	brainwt
Minimum	3.00	5.45	0.37
Maximum	12.00	9.78	0.44

6.1.4 Example: brain weight

The data set litters holds data on the variables lsize (litter size), bodywt (body weight) and brainwt (brain weight), for 20 mice. There were two mice from each litter size. Table 6.2 gives summary information on the range of each of the variables (these data relate to Wainright et al., 1989, and Matthews and Farewell, 1996, p. 121).

At this point, we introduce the scatterplot matrix (Figure 6.3).[10] The diagonal panels give the *x*-variable names for all plots in the column above or below, and the *y*-variable names for all plots in the row to the left or right. Note that the vertical axis labels alternate between the axis on the extreme left and the axis on the extreme right, and similarly for the horizontal axis labels. This avoids a proliferation of axis labels on the extreme left and lower axes.

The first point to check is whether the relationship(s) between explanatory variables is (are) approximately linear. If so, then examination of the panels in the row and column for the dependent variable gives a good indication whether the dependent variable may need transformation.

Figure 6.3 suggests that the relationship between lsize and bodywt is close to linear with little scatter. This may lead to difficulties in separating out the effect of litter size from that of body weight. The relationship between brainwt and bodywt may not be quite linear. Brain weight and litter size seem linearly related, apart from two points (litter sizes 9 and 11).

Figure 6.4 presents results both from the individual regressions and from the multiple regression.[11] Notice how the coefficient for litter size has a different sign (−ve versus +ve) in the individual regression from the multiple regression. Both coefficients are significant ($p < 0.05$). The coefficients have different interpretations:

- In the individual regressions, we are examining the variation of brainwt with litter size, regardless of bodywt. No adjustment is made because, in this sample, bodywt

---

¹⁰
```
data(litters)                                        # DAAG package
pairs(litters, labels=c("lsize\n\n(litter size)", "bodywt\n\n(Body Weight)",
       "brainwt\n\n(Brain Weight)"))
   # pairs(litters) gives a scatterplot matrix with less adequate labeling
   ## Alternative, using the lattice function splom()
   library(lattice);  splom(~litters)
```
¹¹
```
## Code for the several regressions
mice1.lm <- lm(brainwt ~ lsize, data = litters)    # Regress on lsize
mice2.lm <- lm(brainwt ~ bodywt, data = litters)   # Regress on bodywt
mice12.lm <- lm(brainwt ~ lsize + bodywt, data = litters)  # Both the above
## Now obtain coefficients and SEs, etc., for the first of these regressions
summary(mice1.lm)$coef
## Similarly for the other regressions.
```

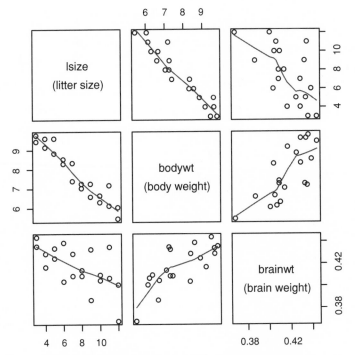

Figure 6.3: Scatterplot matrix for the litters data set.

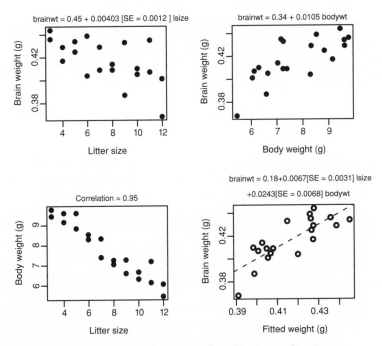

Figure 6.4: Brain weight as a function of `lsize` and `bodywt`.

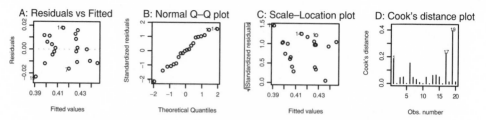

Figure 6.5: Diagnostic plots for the regression of `brainwt` on `bodywt` and `lsize`.

increases as `lsize` decreases. Thus, at small values of `lsize` we are looking at the `brainwt` for individuals with large values of `bodywt`, while at large values of `lsize` we are looking at `brainwt` for individuals who on the whole have low values of `bodywt`.

- In the multiple regression, the coefficient for `lsize` size estimates the change in `brainwt` with `lsize` when `bodywt` is held constant. For any particular value of `bodywt`, `brainwt` increases with `lsize`. This was a significant finding for the purposes of the study.

The results are consistent with the biological concept of brain sparing, whereby the nutritional deprivation that results from large litter sizes has a proportionately smaller effect on brain weight than on body weight.

Finally, Figure 6.5 displays the diagnostic plots from the regression calculations. There is perhaps a suggestion of curvature in the plot of residuals against fitted values in panel A.

6.2 Multiple Regression Assumptions and Diagnostics

We have now illustrated issues that arise in multiple regression calculations. At this point, as a preliminary to setting out a general strategy for fitting multiple regression models, we describe the assumptions that underpin multiple regression modeling. Given the explanatory variables $x_1, x_2, \ldots, x_p$, the assumptions are that:

- The expectation $E[y]$ is some linear combination of $x_1, x_2, \ldots, x_p$:

$$E[y] = \alpha + \beta_1 x_1 + \beta_2 x_2 + \cdots + \beta_p x_p.$$

- The distribution of y is normal with mean $E[y]$ and constant variance, independently between observations.

In general, the assumption that $E[y]$ is a linear combination of the xs is likely to be false. In particular, note that if the joint distribution of $(y, x_1, x_2, \ldots, x_p)$ is multivariate normal, then for given $x_1, x_2, \ldots, x_p$, the above assumptions will be satisfied. The best we can hope is that it is a good approximation. In addition, we can often do some simple checks that will indicate how the assumption may be failing. This assumption may be the best approximation we can find when the range of variation of each explanatory variable is sufficiently small, relative to the noise component of the variation in y, that it is impossible to detect any non-linearity in the effects of explanatory variables. Even where there are indications that it may not be quite adequate, a simple form of multiple regression model may be a reasonable starting point for analysis, on which we can then try to improve.

6.2.1 Influential outliers and Cook's distance

The Cook's distance statistic is a measure, for each observation in turn, of the extent of change in model estimates when that observation is omitted. Any observation for which the Cook's distance is close to one or more, or that is substantially larger than other Cook's distances, requires investigation. Data points that seriously distort the fitted response may lead to residuals that are hard to interpret or even misleading. Thus it is wise to remove any highly influential data points before proceeding far with the analysis. An alternative is to use a form of robust regression. There is a brief discussion of robust regression in Subsection 6.2.4.

Outliers may or may not be influential points. Influential outliers are of the greatest concern. They should never be disregarded. Careful scrutiny of the original data may reveal an error in data entry that can be corrected. If the influential outliers appear to be genuine, they should be set aside for separate scrutiny. Occasionally, it may turn out that their exclusion was a result of use of the wrong model, so that with the right model they can be re-incorporated. If they remain excluded from the final fitted model, they must be noted in the eventual report or paper. They should be included, separately identified, in graphs.

* Leverage and the hat matrix

Identifying high leverage points in multiple regression is less obvious than in the single predictor case, since the definition of extreme value in higher dimensions is ambiguous. However, the model matrix furnishes a means of identifying such points. More precisely, the leverage of the ith observation is the ith diagonal element of the so-called hat matrix that can be derived from the model matrix. (The name of the hat matrix H is due to the fact that the vector of fitted values ("hat" values) is related to the observed response vector by $\widehat{y} = Hy$.) Large values represent high leverage (to obtain the leverage values for `mice12.lm`, type in `hat(model.matrix(mice12.lm))`).

The Cook's distance statistic combines the effects of leverage and of the magnitude of the residual. Careful data analysts may prefer a plot of residuals against leverage values, thus disentangling the two components of the effect. This plot will in due course be available as part of the output from `plot.lm()`.[12]

6.2.2 Component plus residual plots

Suppose there are just two explanatory variables, x_1 and x_2, and we want to examine the contribution of each in turn to the model. The fitted model is

$$\widehat{y} = b_0 + b_1x_1 + b_2x_2. \tag{6.1}$$

Then

$$y = b_0 + b_1x_1 + b_2x_2 + E.$$

---

[12] `## Here is a way to obtain such a plot`
`plot(hat(model.matrix(mice12.lm)), residuals(mice12.lm))`

We denote the residual by E, to indicate that it may contain a systematic component that the model has not explained. Suppose that the "correct" model is

$$y = b_0 + t_1(x_1) + b_2 x_2 + e \qquad (6.2)$$

where $t_1(x_1)$ is a non-linear function of x_1. We can check this from the component plus residual (C+R) plot for x_1. (Component plus residual plots are also called partial residual plots.) This is a plot of $b_1 x_1 + E$ against x_1. If (6.1) is the correct model, then $b_1 x_1 + E$ will, apart from the noise, be linear in x_1. In fact, we can then write

$$t_1(x_1) = b_1 x_1$$

and

$$e = E.$$

If (6.1) is not the correct model, then there will be a residual pattern in E. Our task is to determine $t(x_1)$ such that

$$b_1 x_1 + E = t_1(x_1) + e$$

where e consists only of noise. We estimate the functional form of $t_1(x_1)$ from the plot of $b_1 x_1 + E$ against x_1. A first step may be to use a smoothing method to determine the trend. If the curve appears to follow some identifiable algebraic form, then we can fit an equation that follows that form, and compare it with the smooth curve that we fitted to $b_1 x_1 + E$.

We can do the analogous check for x_2, by plotting $b_2 x_2 + E$ against x_2. If this is not linear, then we can use the plot to determine another functional form $t_2(x_2) + e$ such that e contains only noise. What if, in fact,

$$y = t_1(x_1) + t_2(x_2) + e?$$

In this case, we need to examine both C+R plots, and choose whichever shows the smallest variability about whatever pattern is apparent. Suppose that we decide to start with $t_1(x_1)$. We get a working approximation to $t_1(x_1)$. Then, with this transformation applied to x_1, we obtain a working approximation to $t_2(x_2)$. This is the "backfitting" algorithm that is described, e.g., in Hastie and Tibshirani (1990).

Figure 6.6 shows the two C+R plots for the brain weight data. Here $x_1 = $ lsize, and $x_2 = $ bodywt. These plots both suggest a slight departure from linearity. However, we cannot at this point tell whether one or both of the explanatory variables should perhaps be transformed, or whether perhaps it is brainwt that should be transformed.

Component + residual plots have not been explicitly implemented in the standard R distribution, but it is easy to set them up.[13]

---

[13]
```
## Component + residual plots for the litters analysis
mice12.lm <- lm(brainwt ~ lsize+bodywt, data=litters)
par(mfrow = c(1,2))
res <- residuals(mice12.lm)
b1 <- mice12.lm$coef[2]; b2 <- mice12.lm$coef[3]
plot(litters$lsize, b1*litters$lsize+res, xlab="Litter size",
    ylab="Component + Residual")
panel.smooth(litters$lsize, b1*litters$lsize+res)    # Overlay curve
plot(litters$bodywt, b2*litters$bodywt+res, xlab="Body weight",
    ylab="Component + Residual")
panel.smooth(litters$bodywt, b2*litters$bodywt+res)  # Overlay curve
par(mfrow = c(1,1))
```

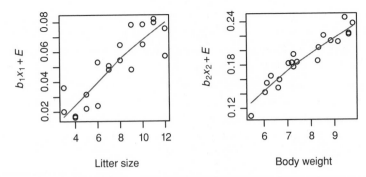

Figure 6.6: Component + residual (C+R) plots, for the regression of `brainwt` on `bodywt` & `lsize`. (Data are from Table 6.2.)

6.2.3 Further types of diagnostic plot*

The functions in the *car* package greatly extend the range of available diagnostic plots. See the examples and references included on the help pages for this package. As an indication of what is available, try

```
library(car)
leverage.plots(allbacks.lm, term.name="volume",
               identify.points=FALSE)
```

6.2.4 Robust and resistant methods

Robust and resistant methods can assist model diagnosis. Outliers show up most clearly when the fitting criterion is designed to downweight them. We refer the reader back to our earlier brief discussion, in Section 5.2. We pursue further investigation of these methods, now in connection with the examples in this chapter, in an exercise.

6.3 A Strategy for Fitting Multiple Regression Models

6.3.1 Preliminaries

Careful graphical scrutiny of the explanatory variables is an essential first step. This may lead to any or all of:

- Transformation of some or all variables.
- Replacement of existing variables by newly constructed variables that are a better summary of the information. For example, we might want to replace variables x_1 and x_2 by the new variables $x_1 + x_2$ and $x_1 - x_2$.
- Omission of some variables.

Here are steps that are reasonable to follow. They involve examination of the distributions of values of explanatory variables, and of the pairwise scatterplots.

- Examine the distribution of each of the explanatory variables, and of the dependent variable. Look for any instances where distributions are highly skew, or where there are outlying values. Check whether any outlying values may be mistakes.

- Examine the scatterplot matrix involving all the explanatory variables. (Including the dependent variable is, at this point, optional.) Look first for evidence of non-linearity in the plots of explanatory variables against each other. Look for values that appear as outliers in any of the pairwise scatterplots.

- Note the ranges of each of the explanatory variables. Do they vary sufficiently to affect values of the dependent variable? If values of a variable are essentially constant, there can be no apparent effect from varying levels of that variable.

- Where the distribution is skew, consider transformations that may lead to a more symmetric distribution. If some pairwise plots show evidence of non-linearity, examine the possible use of transformation(s) to give more nearly linear relationships. Surprisingly often, standard forms of transformation both give more symmetric distributions and lead to scatterplots where the relationships appear more nearly linear. It is the latter requirement, i.e., linear pairwise linear relationships, that is the more important.

- Look for pairs of explanatory variables that seem highly correlated. Do scientific considerations help in judging whether both variables should be retained? For example, the two members of the pair may measure what is essentially the same quantity. On the other hand, there will be other instances where the difference, although small, is important. Section 8.2 has an example.

If relationships between explanatory variables are approximately linear, perhaps after transformation, it is then possible to interpret plots of predictor variables against the response variable with confidence. Contrary to what might be expected, this is more helpful than looking at the plots of the response variable against explanatory variables. Look for non-linearity, for values that are outlying in individual bivariate plots, for clustering, and for various unusual patterns of scatter. If there is non-linearity in these plots, examine whether a transformation of the dependent variable would help.

It is not always possible to obtain pairwise linear scatterplots. This can create problems both for the interpretation of the diagnostic plots for the fitted regression relationship and for the interpretation of the coefficients in the fitted regression equation.

6.3.2 Model fitting

Here is a suggested procedure for fitting and checking the regression relationship:

- Fit the multiple regression equation.
- Examine the Cook's distance statistics. If it seems helpful, examine standardized versions of the drop-1 coefficients directly, using `dfbetas()`. It may be necessary to delete influential data points and refit the model. Influential points may be more troublesome than points that, although residuals are large, have little influence on the fitted model.
- Plot residuals against fitted values. Check for patterns in the residuals, and for the fanning out (or in) of residuals as the fitted values change. (Do not plot residuals against observed values. This is potentially deceptive; there is an inevitable positive correlation.)

Table 6.3: *Distance* (`dist`)*, height climbed* (`climb`)*, and record times* (`time`)*, for some of the 35 Scottish hill races.*

	dist (mi)	climb (ft)	time (h)
1	2.40	650	0.27
2	6.0	2500	0.81
3	6.0	900	0.56
...	...	...	...
18	3.0	350	1.31
...	...	...	...
35	20.0	5000	2.66

- For each explanatory variable, do a component plus residual plot, in order to check whether any of the explanatory variables require transformation.

We leave the discussion of variable selection issues until Subsection 6.6.1.

6.3.3 An example – the Scottish hill race data

The data set `hills`, from which Table 6.3 has selected observations, and which is in the Venables and Ripley *MASS* package (we have included a version of it in our *DAAG* package with the time variable converted from minutes to hours), gives distances (`dist`), height climbed (`climb`), and `time` taken, for some of the 35 hill races held in Scotland (Atkinson 1986, 1988). The times are the record times in 1984.

We begin with the scatterplot matrix (Figure 6.7).[14] Apart from a possible outlier, the relationship between `dist` and `climb` seems approximately linear. If we can assume such a linear relationship, this makes it straightforward to interpret the panels for `time` and `climb`, and for `time` and `dist`.

The following are reasons for at least investigating the taking of logarithms.

- The range of values of `time` is large (3.4:0.27, i.e., > 10:1), and similarly for `dist` and `climb`. The times are bunched up towards zero, with a long tail.
- Figure 6.7 hints at nonlinearity for the longer races.
- It seems unlikely that `time` would be a linear function of `dist` and `climb`. We would expect `time` to increase more than linearly at long times, and similarly for `climb`. This suggests looking at a power relationship

$$\text{time} = A(\text{dist})^{b_1}(\text{climb})^{b_2}.$$

Taking logarithms gives

$$\log(\text{time}) = a + b_1 \log(\text{dist}) + b_2 \log(\text{climb}),$$

where $a = \log(A)$. We hope that the use of a logarithmic scale will stabilize the variance and alleviate the problem with outliers at long times.

---

[14] `library(DAAG); data(hills)`
`pairs(hills, labels=c("dist\n\n(miles)", "climb\n\n(feet)", "time\n\n(hours)"))`
`## Alternative, using the lattice function splom()`
`library(lattice); splom(~hills)`

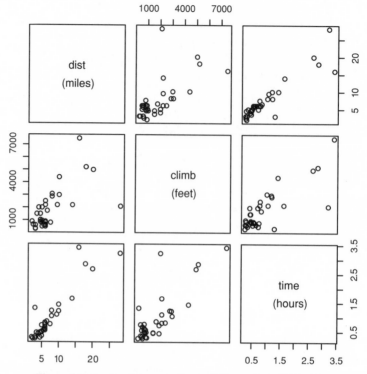

Figure 6.7: Scatterplot matrix for the hill race data (Table 6.3).

Figure 6.8 shows the scatterplot matrix when we take logarithms.[15] This does seem marginally more consistent with the assumption of a linear relationship between predictors. Apart from one outlier the relationship between `climb` and `dist` seems linear. Apart from the one outlier the relationships involving `time` then also seem linear, and the variability does not now increase at the longer times. Hence, we pursue the analysis using the log transformed data. We will leave the troublesome observation 18 in place for the moment and note how it affects the diagnostic plots. Figure 6.9 shows the plots.[16]

Observation 18 (Knock Hill), which we had already identified, emerges as the main problem outlier. It gives a time of 1.3 hours for a distance of three miles. The two other three-mile races had times of 0.27 hours and 0.31 hours. The time of 1.3 hours is undoubtedly a mistake. Atkinson (1988) corrected the figure to 0.3 hours. Omission of observation 18

---

[15] `pairs(log(hills), labels=c("dist\n\n(log(miles))", "climb\n\n(log(feet))",`
`      "time\n\n(log(hours))"))`
`## Alternative, using the lattice function splom()`
`library(lattice);`
`splom(~ log(hills), varnames=c("logdist", "logclimb", "logtime"))`
[16] `hills0.loglm <- lm(log(time) ~ log(dist) + log(climb), data = hills)`
`par(mfrow=c(2,2))`
`plot(hills0.loglm)`
`par(mfrow=c(1,1))`
`# The 0 in hills0.loglm! reminds us that this is a preliminary analysis.`

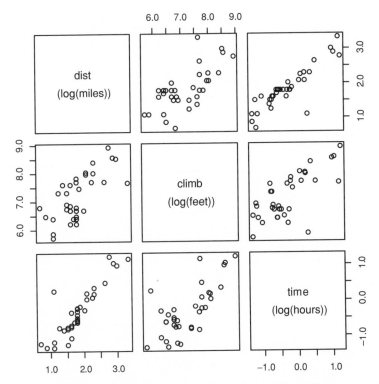

Figure 6.8: Scatterplot matrix, hill race data set, log scales.

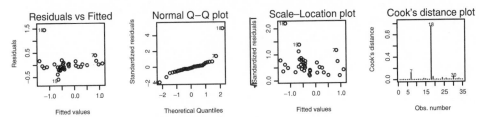

Figure 6.9: Diagnostic plots, from the regression of `log(time)` on `log(climb)` and `log(dist)`.

leads to the diagnostic information in Figure 6.10.[17] Observe that problems with outliers have now largely disappeared. Black Hill is now something of an outlier. However, it is not an influential point. Thus it does contribute substantially to increasing the error mean square, but does not distort the regression equation.

We now check to see whether it is advantageous to include the term `log(dist)` × `log(climb)`. We fit a further model that has this interaction (product) term. The following

---

[17]
```
hills.loglm <- lm(log(time) ~ log(dist) + log(climb), data = hills[-18, ])
summary(hills.loglm) # Get a summary of the analysis results.
  # Note that we have not displayed this output.
par(mfrow = c(2,2))
plot(hills.loglm)
par(mfrow = c(1,1))
```

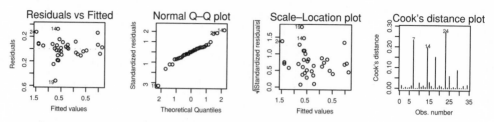

Figure 6.10: Diagnostic plots, from the regression of `log(time)` on `log(climb)` and `log(dist)`, omitting Knock Hill.

is the R output that compares the two models:[18]

```
Analysis of Variance Table

Model 1: log(time) ~ log(dist) + log(climb)
Model 2: log(time) ~ log(dist) + log(climb)
         + log(dist):log(climb)
  Res.Df    RSS Df Sum of Sq      F Pr(>F)
1     31 0.798
2     30 0.741  1     0.057   2.31   0.14
```

The additional term is not statistically significant. None of the panels in the diagnostic plot, which we do not show, are greatly different from Figure 6.10. As often, however, our real interest is not in whether the additional coefficient is statistically significant at the 5% or any other level, but in whether adding the extra term is likely to improve the predictive power of the equation. The Akaike Information Criterion (AIC), discussed in more detail in the next section, is designed to decrease as the estimated predictive power increases. We may compare the AIC value with and without the interaction term as follows.

```
> step(hills2.loglm)

Start:  AIC= -122.1
 log(time) ~ log(dist) + log(climb) + log(dist):log(climb)

                          Df Sum of Sq   RSS    AIC
<none>                                    0.7 -122.1
- log(dist):log(climb)     1       0.1   0.8 -121.6

Call:
lm(formula = log(time) ~ log(dist) + log(climb)
    + log(dist): log(climb), data = hills[-18, ])

Coefficients:
 (Intercept)    log(dist)    log(climb)   log(dist):log(climb)
     -2.4755       0.1685        0.0672                 0.0993
```

---

[18] `hills2.loglm <- lm(log(time) ~ log(dist)+log(climb)+log(dist):log(climb),`
                `data=hills[-18, ])`
  `anova(hills.loglm, hills2.loglm)`

The output examines changes from the model that includes the interaction term. No change, i.e., leaving the interaction in, gives the better expected predictive power, though the difference is small. On its own, the AIC value is pretty meaningless. What matters is the comparison between the values of the statistic for the different models.

The model without the interaction term

Now, examine the model that lacks the interaction term.

```
> summary(hills.loglm, corr=TRUE)$coef
```

	Estimate	Std. Error	t value	Pr(>|t|)
(Intercept)	-3.88	0.2826	-13.73	9.99e-15
log(dist)	0.91	0.0650	13.99	6.22e-15
log(climb)	0.26	0.0484	5.37	7.33e-06

The estimated equation is

$$\log(\texttt{time}) = -3.88[\text{SE}=0.28] + 0.909[\text{SE }0.065] \times \log(\texttt{dist}) + 0.260[\text{SE }0.048]$$
$$\times \log(\texttt{climb})$$

By noting that $\exp(-3.88) = 0.0206$, this can be rewritten as

$$\texttt{time} = 0.0206 \times \texttt{dist}^{0.909} \times \texttt{climb}^{0.260}.$$

We note the implications of this equation. For a given height of climb, the time taken is smaller for the second three miles, faster than for the first three miles. This follows because $0.909 < 1$. Is this plausible?

Note first that the coefficient for log(dist) has a substantial standard error, i.e., 0.065. A 95% confidence interval is $(.78, 1.04)$.[19] A more important point is that the values of dist are subject to substantial error. Most distances are given to the nearest half mile. This leads to what has been called a regression dilution effect; the coefficient in the regression equation is reduced relative to its value in the underlying theoretical relationship. Subsection 5.7.2 demonstrated the effect for the case where there is a single explanatory variable.

The model with the interaction term

The summary information is[20]

```
> summary(hills2.loglm)$coef
```

	Estimate	Std. Error	t value	Pr(>|t|)
(Intercept)	-2.47552	0.9653	-2.5645	0.01558
log(dist)	0.16854	0.4913	0.3431	0.73394
log(climb)	0.06724	0.1354	0.4966	0.62307
log(dist):log(climb)	0.09928	0.0653	1.5205	0.13886

---

[19] .909 + c(-.065, .065) * qt(.975, 31)
[20] summary(hills2.loglm)$coef

The estimated equation for this model is

$$\log(\texttt{time}) = -2.48[\text{SE } 0.97] + 0.168[\text{SE } 0.491] \times \log(\texttt{dist}) + 0.067[\text{SE } 0.135]$$
$$\times \log(\text{climb}) + 0.0993[\text{SE } 0.0653] \times \log(\texttt{dist}) \times \log(\text{climb}).$$

Noting that $\exp(-2.48) = 0.084$, this can be rewritten

$$\texttt{time} = 0.084 \times \texttt{dist}^{0.168 + 0.0993 \times \log(\texttt{climb})} \times \texttt{climb}^{0.067}.$$

Thus, we have a more complicated equation. Notice that the coefficients are now much smaller than are the SEs. Even though the model can be expected to give good predictions, individual coefficients are uninterpretable.

What happens if we do not transform?

If we avoid transformation (see Exercise 6.10.7 for further development), we find several outlying observations, some of which are influential. It is unsatisfactory to omit many points. Note, though, that we might be able to put one or more of these points back at a later stage. When individual outlying points distort the fitted regression plane, this may make other points appear to be outliers. Once the effect of a genuine outlier has been removed, other points that were initially identified as outliers may cease to be a problem.

It is, largely, the cluster of three points for long races that causes the difficulty for the untransformed data. The model that uses the untransformed variables will inevitably identify two out of three of them as outliers, as can be readily verified.

6.4 Measures for the Comparison of Regression Models

6.4.1 R^2 and adjusted R^2

Both R^2 and adjusted R^2 are included in the output that comes from `summary.lm()`; refer back to Section 6.1. One or other of these statistics is often used to give a rough sense of the adequacy of a model. Whereas R^2 is the proportion of the sum of squares about the mean that is explained, adjusted R^2 is the proportion of the mean sum of squares (mss, or variance) that is explained. In those contexts where such a statistic perhaps has a place, adjusted R^2 is preferable to R^2.

Neither statistic should be used for making comparisons between different studies, where the range of values of the explanatory variables may be different. Both are likely to be largest in those studies where the range of values of the explanatory variables is greatest. For making comparisons between different models on the same data, the "information measure" statistics that we now consider are preferable.

6.4.2 AIC and related statistics

The "information measure" statistics that we discuss here are designed to choose, from among a small number of alternatives, the model with the best predictive power. Statistics that are in use include the Akaike Information Criterion (AIC), Mallows' C_p, and various elaborations of the AIC. The AIC criterion and related statistics can be used for more general models than multiple regression. We give the version that is used in R.

The model that gives the smallest value is preferred, whether for the AIC or for the C_p statistic. The variance estimate $\widehat{\sigma}^2$ is often determined from the full model.

Another statistic is the Bayesian Information Criterion (BIC), which is claimed as an improvement on the AIC. The AIC is inclined to overfit, i.e., to choose too complex a model.

Calculation of R's version of the AIC

Let n be the number of observations, let p be the number of parameters that have been fitted, and let RSS denote the residual sum of squares. Then, if the variance is known, R takes the AIC statistic as

$$\text{AIC} = \frac{\text{RSS}}{\sigma^2} + 2p.$$

In the more usual situation where the variance is not known, R takes the AIC statistic as

$$\text{AIC} = n \log\left(\frac{\text{RSS}}{n}\right) + 2p.$$

The BIC statistic, which replaces $2p$ by $\log(n) \times p$, penalizes models with many parameters more strongly.

The C_p statistic differs from the AIC statistic only by subtraction of n; it is

$$C_p = n \log\left(\frac{\text{RSS}}{n}\right) + 2p - n.$$

6.4.3 How accurately does the equation predict?

The best test of how well an equation predicts is the accuracy of its predictions. We can measure this theoretically, by examining, e.g., 95% confidence intervals for predicted values. Alternatively we can use cross-validation, or another resampling procedure, to assess predictive accuracy.

As always, we need to be cautious. Theoretical assessments of the confidence intervals rely on the validity of that theory. If the theoretical assumptions are wrong, perhaps because of clustering or other forms of dependence, then these prediction intervals will also be wrong. Both methods (theoretical and cross-validation) for assessing predictive accuracy assume that the sample is randomly drawn from the population that is of interest, and that any new sample for which predictions are formed is from that same population. If the sample for which predictions are made is really from a different population, or is drawn with a different sampling bias, the assessments of predictive accuracy will be wrong.

The following table gives 95% coverage (confidence) intervals for predicted times in the regression of `log(time)` on `log(climb)` and `log(dist)`. Displayed are the intervals

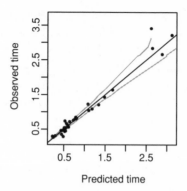

Figure 6.11: Observed versus predicted values, for the hill race data. The line $y = x$ is shown. The horizontal limits of the bounding curves are 95% pointwise confidence limits for predicted times.

for the first few and last observations from the hill race data:[21]

```
      fit   lwr   upr
1    0.25  0.22  0.28
2    0.80  0.74  0.88
3    0.62  0.58  0.66
 . . .
35   2.88  2.54  3.27
```

Figure 6.11 shows what these mean graphically. The 95% pointwise intervals apply to the fitted values, i.e., to the values on the horizontal axis.

This assessment of predictive accuracy has two limitations:

1. It is crucially dependent on the model assumptions – independence of the data points, homogeneity of variance, and normality. Departures from normality are likely to be the least of our problems.
2. It applies only to the population from which these data have been sampled. Thus it would be hazardous to use it to predict winning times for hill races in England or Mexico or Tasmania.

Point 2 can be addressed only by testing the model against data from these other locations. Or, we might take data from different locations and use them to develop a model that allows for variation between locations as well as variation within locations.

We might consider cross-validation, or another resampling method. Cross-validation uses the data we have to simulate the effect of having a new sample, sampled in the same way as the present data were sampled. The bootstrap is essentially a generalization of cross-validation. Cross-validation and other resampling methods can cope, to a limited extent and depending on how they are used, with lack of independence. We still end up with a predictive accuracy estimate that is valid only for data that were sampled in the same

---

[21] `hills.loglm <- lm(log(time) ~ log(dist)+log(climb), data=hills[-18, ])`
`hills.ci <- exp(predict(hills.loglm, interval="confidence"))`
`   # Use of exp() undoes the logarithmic transformation.`
`print(hills.ci)`

way as the original data. Note that neither this nor any other method can address sampling bias.

Heterogeneity of variance, leading to predictions for some values of the explanatory variables that are more accurate than predictions for other values, remains a problem. Thus for the hill race data:

1. If we use the untransformed data, cross-validation will exaggerate the accuracy for long races and be slightly too pessimistic for short races.
2. If we use the log transformed data, cross-validation will exaggerate the accuracy for short races and underrate the accuracy for long races.

Heterogeneity causes the same problems as for model-based assessments of accuracy.

6.4.4 An external assessment of predictive accuracy

As noted above, we have been discussing assessments of predictive accuracy that are internal to these data. Ideally we would like an assessment that is external to these data. For that we need new data, sampled in a manner that matches the way that data will be obtained when we apply the model.

Now in fact data are available, for the year 2000, that update these earlier data. We have included them in our *DAAG* package, in the data set `hills2000`. In some cases there is more up to date information for the same races. Other results are for venues that were not in the original data set. Thus we do have new data. We leave investigations that involve these data to the exercises.

6.5 Interpreting Regression Coefficients – the Labor Training Data

With observational data involving a number of explanatory variables, interpretation of coefficient estimates can be hazardous. In any discussion of how far it is reasonable to press the use of regression methods, we should note the important study of Lalonde (1986), revisited more recently by Dehejia and Wahba (1999).

The US National Supported Work (NSW) Demonstration program operated in the mid-1970s to provide work experience for a period of 6–18 months to individuals who had faced economic and social hardship. Over the period 1975–1977, an experiment randomly assigned individuals who met the eligibility criteria either to a treatment group that participated in the NSW training program, or to a control group that did not participate. The results for males, because they highlight methodological problems more sharply, have been studied more extensively than the corresponding results for females. Participation in the training gave an increase in male 1978 earnings, relative to those in the control group, by an average of $886 [SE $472].

Lalonde had, in addition, data from two further studies that examined other groups of individuals who did not participate in any training program. These were

1. The Panel Study of Income Dynamics study (PSID: 2490 males),
2. Westat's Matched Current Population Survey – Social Security Administration file (CPS: 16 289 males).

These data are pertinent to an investigation of the way that earnings changed, between 1974–1975 and 1978, in the absence of training. Here, our interest is in a methodological issue, whether regression adjustments allow useful estimates of the treatment effect when the comparison is between a treatment group and a non-experimental control group where covariates (such as age, 1975 income, ethnicity, etc.) may not be closely matched.

Lalonde's comparison between male experimental and regression results

We follow Lalonde in replacing the NSW experimental control group with either the PSID or the CPS non-experimental group, then using regression methods to estimate the effect on earnings. The data frame `nsw74psid1` combines data for the experimental treatment group (185 observations), using as control data results from the PSID (Panel Study of Income Dynamics) study (2490 observations). (The data that formed the basis of the Dehejia and Wahba (1999) paper are available from the web site http://www.columbia.edu/ rd247/nswdata.html. The data frame `nsw74psid1` is formed from two of the data sets that appear on that web site.) Variables are

```
trt (0 = control 1=treatment)
age (years)
educ (years of education)
black (0=white 1=black)
hisp (0=non-hispanic 1=hispanic)
marr (0 = not married 2=married)
nodeg (0=completed high-school 1=dropout)
re74 (real earnings in 1974, available for a subset
  of the data)
re75 (real earnings in 1975)
re78 (real earnings in 1978)
```

Lalonde examined several different measures of the treatment effect. We limit attention to the subset of the data for which 1974 earnings are available. We give our own analyses of these data, always adjusting for covariates. For comparability with Lalonde's results, we do not investigate transformation of the dependent variable. For the model that estimates 1978 earnings, using all pre-intervention variables as covariates, we have[22]

```
> summary(nsw74psid1.lm)$coef
```

	Estimate	Std. Error	t value	Pr(>|t|)
(Intercept)	-129.743	1.69e+03	-0.0768	9.39e-01
trt	751.946	9.15e+02	0.8216	4.11e-01
age	-83.566	2.08e+01	-4.0149	6.11e-05
educ	592.610	1.03e+02	5.7366	1.07e-08
re74	0.278	2.79e-02	9.9598	0.00e+00
re75	0.568	2.76e-02	20.6130	0.00e+00
black	-570.928	4.95e+02	-1.1530	2.49e-01

---

[22] `data(nsw74psid1)      # DAAG package`
   `nsw74psid1.lm <- lm(re78~ trt+ (age + educ + re74 + re75) +`
                   `(black + hisp + marr + nodeg), data = nsw74psid1)`

```
hisp          2163.281    1.09e+03  1.9805 4.78e-02
marr          1240.520    5.86e+02  2.1160 3.44e-02
nodeg          590.467    6.47e+02  0.9129 3.61e-01
```

Notice that the treatment effect estimate is smaller than its standard error. The diagnostic plots are a concern. There are some large outliers that we might consider removing. This has been a cavalier analysis.

A more careful analysis

The 1974 and 1975 earnings span a huge range of values. We examine various quantiles of the continuous variables:[23]

```
      age educ   re74    re75    re78
25%  25.0   10   8817    7605    9243
50%  32.0   12  17437   17008   19432
75%  43.5   14  25470   25584   28816
95%  53.0   17  41145   41177   45810
100% 55.0   17 137149  156653  121174
```

The quartiles of re74 and re75 (and re78) are roughly equal distances either side of the median. However, the distributions have very long tails. Note especially the huge range of values of re75, with the largest value over \$150 000. Such very large values will have huge leverage in the analysis, and are likely to cause serious distortion of results.

The quantiles for the treated group alone are much smaller. Here are the details:[24]

```
      age educ  re74   re75     re78
25%    20    9     0      0    485.2
50%    25   11     0      0   4232.3
75%    29   12  1291   1817   9643.0
95%    42   13 12342   6855  18774.7
100%   48   16 35040  25142  60307.9
```

In Figure 6.12, we restrict the data to observations for which re74 $\leq$ 15 000 and re75 $\leq$ 15 000. We plot the densities, separately for treated and control observations, of the continuous variables educ, age, re74, re75 and re78.

We note the large differences between the two groups. Differences in the distributions are much smaller, though still substantial, for educ and age. In order to achieve some modest comparability, we limit attention to age $\leq$ 40, re74 $\leq$ 5000 and re75 $\leq$ 5000. We also note that the distribution for values of re78 is reasonably symmetric, but with a long tail. However, we prefer to look at the scatterplot matrix for the reduced data before proceeding. This leads to Figure 6.13.

---

[23] options(digits=4)
    sapply(nsw74psid1[, c(2,3,8,9,10)], quantile, prob=c(.25,.5,.75,.95,1))
[24] attach(nsw74psid1)
    sapply(nsw74psid1[trt==1, c(2,3,8,9,10)], quantile, prob=c(.25,.5,.75,.95,1))

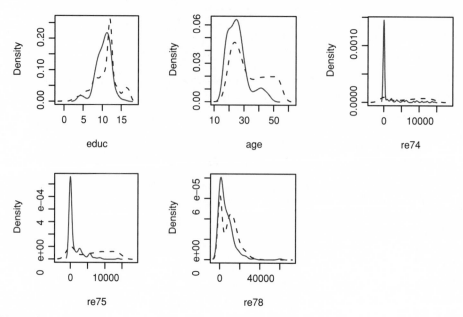

Figure 6.12: Density plots, separately for the treated and control data, for the variables educ, age, re74, re75 and re78. The solid curves are for controls, while the dashed curves are for treated individuals.

We now fit the regression equation, including only main effects. We add the further restriction re78 $\leq$ 30 000, in order to remove values that are hugely different from the main body of the data. We find:[25]

```
  0    1
119  133
```

	Estimate	Std. Error	t value	Pr(>|t|)
(Intercept)	-2093.441	4253.834	-0.4921	0.6231
trt	2942.967	1169.943	2.5155	0.0125
age	71.200	77.139	0.9230	0.3569
educ	245.989	250.688	0.9813	0.3274
re74	0.314	0.400	0.7846	0.4335
re75	0.872	0.383	2.2780	0.0236
black	905.595	1135.158	0.7978	0.4258
hisp	3723.038	2211.820	1.6832	0.0936
marr	24.722	1104.606	0.0224	0.9822
nodeg	-883.452	1294.208	-0.6826	0.4955

---

[25] here <- age <= 40 & re74 <= 5000 & re75 <= 5000 & re78 < 30000
    nsw74psidA <- nsw74psid1[here,]
    detach(nsw74psid1)
    table(nsw74psidA$trt)
    A1.lm <- lm(re78 ~ trt + (age + educ + re74 + re75) + (black +
              hisp + marr + nodeg), data = nsw74psidA)
    summary(A1.lm)$coef

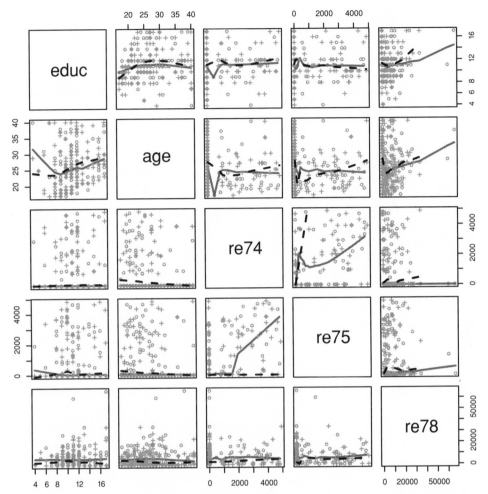

Figure 6.13: Scatterplot matrix, for data where age $\leq$ 40, re74 $\leq$ 5000 and re75 $\leq$ 5000, for the variables educ, age, re74, re75 and re78. We use o for control, and + for treated individuals. The solid fitted smooth curves are for controls, while the dashed curves are for treated individuals.

This gives a plausible estimate for the treatment effect. Figure 6.14 shows the contribution of individual terms. There is nothing that calls for obvious attention.[26]

We try a further model, including interactions between the binary variables and the continuous variables:[27]

```
Analysis of Variance Table
. . . .
  Res.Df      RSS  Df Sum of Sq     F Pr(>F)
1    242 9.72e+09
2    226 9.08e+09  16  6.47e+08  1.01    0.45
```

[26] `par(mfrow = c(3,3))`
`termplot(A1.lm, partial=TRUE)`
`par(mfrow = c(1,1))`
[27] `A2.lm <- lm(re78 ~ trt + (age + educ + re74 + re75) * (black +`
`        hisp + marr + nodeg), data = nsw74psidA)`
`anova(A1.lm, A2.lm)`

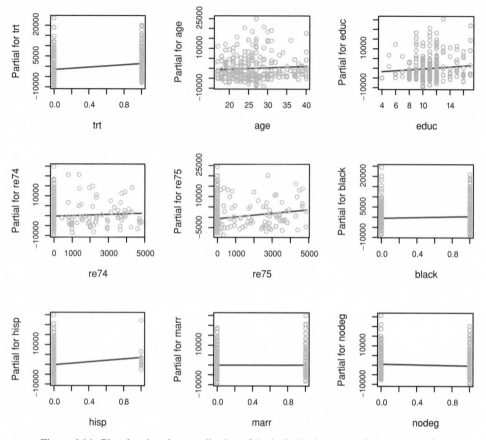

Figure 6.14: Plot showing the contribution of the individual terms to the regression fit.

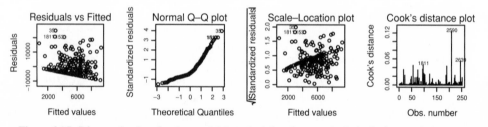

Figure 6.15: Diagnostic plots for the model (A1.1m) that does not include the interaction terms.

The interactions do not improve the fit.

Figure 6.15 plots the diagnostic information for the model (A1.1m) that does not include the interaction terms.[28] Note that the residuals are far from normal.

Chapter 11 will investigate these data further, using discriminant analysis to identify similarity or dissimilarity between the two groups.

[28]
```
par(mfrow = c(2,2))
plot(A1.lm)
par(mfrow = c(1,1))
```

Lalonde's results – further comments

Dehejia and Wahba (1999) revisited Lalonde's study, using his data. They used the propensity score methodology, as expounded, e.g., in Rosenbaum and Rubin (1983). These papers warn that coefficients in regression equations can be highly misleading. Regression relationships must model within group relationships acceptably well, and in addition they must model differences between groups. These two demands can be in conflict. Even where the groups are reasonably well matched on relevant variables, the methodology is unlikely to handle these conflicting demands well. Further complications arise because some effects are too far from linear for linear regression adjustments to be satisfactory. Where the ranges of some variables are widely different in the different groups, the task is even more difficult.

6.6 Problems with Many Explanatory Variables

Variable selection is an issue when the aim is to obtain the best prediction possible. Be sure to distinguish the variable selection problem from that of determining which variables have greatest explanatory power. If the interest is in which variables have useful explanatory power, then the choice of variables will depend on which quantities are to be held constant when the effects of other quantities are estimated. There should in any case be an initial exploratory investigation of explanatory variables, as described in Section 6.3, leading perhaps to transformation of one or more of the variables.

One suggested rule is that there should be at least 10 times as many observations as variables, before any use of variable selection takes place. For any qualitative factor, subtract one from the number of levels, and count this as the number of variables contributed by that factor. This may be a reasonable working rule when working with relatively noisy data.

For an extended discussion of state of the art approaches to variable selection, we refer the reader to Harrell (2001) and to Hastie et al. (2001, Chapter 3). We will limit discussion here to widely used approaches that do not take us much beyond the methodology discussed earlier in this chapter. We begin by noting strategies that are designed, broadly, to keep to a minimum the number of different models that will be compared. The following strategies may be used individually, or in combination.

1. A first step may be an informed guess as to what variables/factors are likely to be important. An extension of this approach classifies explanatory variables into a small number of groups according to an assessment of scientific "importance". Fit the most important variables first, then add the next set of variables as a group, checking whether the fit improves from the inclusion of all variables in the new group.
2. Interaction effects are sometimes modeled by including pairwise multiples of explanatory variables, e.g. $x_1 \times x_2$ as well as x_1 and x_2. Use is made of an omnibus check for all interaction terms, rather than checking for interaction effects one at a time.
3. Use principal components analysis to look for a small number of components, i.e., combinations of the explanatory variables, that together account for most of the variation in the explanatory variables. If we are fortunate, the principal components may

suggest simple forms of summary of the original variables. The components, or the new explanatory variables that they suggest, then become the explanatory variables. See Chapter 11 for further details.

4. Discriminant analysis can sometimes be used to identify a summary variable. There is an example in Chapter 11.

6.6.1 Variable selection issues

We caution against giving much credence to output from conventional automatic variable selection techniques – various forms of stepwise regression, and best subsets regression. The resulting regression equation may have poorer genuine predictive power than the regression that includes all explanatory variables. The standard errors and t-statistics typically ignore the effects of the selection process; estimates of standard errors, p-values and F-statistics will be optimistic. Estimates of regression coefficients are biased upwards in absolute value – positive coefficients will be larger than they should be, and negative coefficients will be smaller than they should be. See Harrell (2001) for further discussion.

Repeated simulation of a regression problem where the data consist entirely of noise will demonstrate the extent of the problem. In each regression there are 41 vectors of 100 numbers that have been generated independently and at random from a normal distribution. In these data:[29]

1. The first vector is the response variable y
2. The remaining 40 vectors are the variables $x_1, x_2, \ldots, x_{40}$.

If we find any regression relationships in these data, this will indicate faults with our methodology. (In computer simulation, we should not however totally discount the possibility that a quirk of the random number generator will affect results. We do not consider this an issue for the present simulation!)

We perform a best subsets regression in which we look for the three x-variables that best explain y (This subsection has an example that requires access to the *leaps* package for R. Before running the code, be sure that it is installed.)[30]

```
Call:
lm(formula = y ~ -1 + xx[, subvar])


Residuals:
    Min    1Q Median     3Q     Max
-2.363 -0.655  0.131  0.618  2.215
```

---

[29] y <- rnorm(100)
    xx <- matrix(rnorm(4000), ncol = 40)
    dimnames(xx)<- list(NULL, paste("X",1:40, sep=""))
[30] library(leaps)
    xx.subsets <- regsubsets(xx, y, method = "exhaustive", nvmax = 3, nbest = 1)
    subvar <- summary(xx.subsets)$which[3,-1]
    best3.lm <- lm(y ~ -1+xx[, subvar])
    print(summary(best3.lm, corr = FALSE))
    rm(xx, y, xx.subsets, subvar, best3.lm)
      # For once, we note objects that should be removed.
      ## Note also our function bestset.noise(). The following
      ## call is effectively equivalent to the above.
    bestset.noise(m=100, n=40)

```
Coefficients:
                  Estimate Std. Error t  value Pr(>|t|)
xx[, subvar]X12    0.2204     0.0896    2.46   0.0156
xx[, subvar]X23   -0.1431     0.0750   -1.91   0.0593
xx[, subvar]X28    0.2529     0.0927    2.73   0.0076

Residual standard error: 0.892 on 97 degrees of freedom
Multiple R-Squared: 0.132,        Adjusted R-squared: 0.105
F-statistic: 4.93 on 3 and 97 DF,  p-value: 0.00314
```

Note that two of the three variables selected have p-values less than .05.

When we repeated this experiment ten times, the outcomes were as follows. Categories are exclusive:

	Instances
All three variables selected were significant at $p < 0.01$	1
All three variables had $p < 0.05$	3
Two out of three variables had $p < 0.05$	3
One variable with $p < 0.05$	3
Total	10

In the modeling process there are two steps:

1. Select variables.
2. Do a regression and determine SEs and p-values, etc.

The p-values have been calculated assuming that step 2 was the only step in building the model. Therefore, they are wrong, as demonstrated by our ability to find "significance" in data sets that consist only of noise. Cross-validation is one way to determine realistic standard errors and p-values. At each cross-validation step, we repeat both of steps 1 and 2 above, i.e., both the variable selection step and the comparison of predictions from the regression equation with data different from that used in forming the regression equation.

Cross-validation will not however deal with the bias in the regression coefficients (note, in passing, that the cross-validation SEs for these coefficients are similarly biased upwards). Regression on principal components, which we discuss briefly in the next subsection and in more detail in Chapter 11, may sometimes be a useful recourse that can avoid this problem. Hastie et al. (2001) discuss principal components regression alongside a number of other methods that are available. Note especially the "shrinkage" methods, which directly shrink the coefficients.

6.6.2 Principal components summaries

Principal components analysis, to be discussed in Chapter 11, can, for some data sets, be a useful tool. It seeks a small number of variables (or *components*) that summarize most of the information in the data. Note that we do this separately from response variable information, so that the selection process does not have implications for the assessment of statistical significance. Principal components analysis is then a "dimension reduction" mechanism. The new variables (components) may be hard to interpret, or may require modification to make them interpretable. An important application of the methodology is to data from carefully designed survey "instruments".

6.7 Multicollinearity

Some later variables may be exactly, or very nearly, linear combinations of earlier variables. Technically, this is known as multicollinearity. For each multicollinear relationship, there is one redundant variable.

The approaches that we have advocated – careful thinking about the background science, careful initial scrutiny of the data, and removal of variables whose effect is already accounted for by other variables – will generally avoid the more extreme effects of multicollinearity that we will illustrate. Milder consequences are pervasive, especially for observational data. Our admittedly contrived example may help alert the reader to these effects. While it is contrived, we have from time to time seen comparable effects in computer output that researchers have brought us for scrutiny. Indeed, we have ourselves, on occasion, generated data sets that have exhibited similar effects.

6.7.1 A contrived example

We have formed the data set `carprice` from observations for cars of US origin in the data set `Cars93` that is in the *MASS* package. The data set includes the factor `Type`, three price measurements and the variables `MPG.city` and `MPG.highway` from `Cars93`. We have added two further variables. The variable `Range.Price` is the difference between `Max.Price` and `Min.Price`. The variable `RoughRange` has been obtained by adding to `Range.Price` random values from a normal distribution with mean 0 and standard deviation 0.01. The variable `gpm100` is the number of gallons used in traveling 100 miles. For this, we averaged `MPG.city` and `MPG.highway`, took the inverse, and multiplied by 100.

Figure 6.16 shows the scatterplot matrix for some of the continuous variables in this data set.[31] Notice that the relationships between `gpm100` and the price variables are approximately linear. We leave it to the reader to verify that the relationship with `MPG.city` or `MPG.highway` is not at all linear.

We will now fit two models that have `gpm100` as the response variable. The first uses `Type, Min.Price, Price, Max.Price` and `Range.Price` as explanatory variables. The second replaces `Range.Price` with `RoughRange`. We will, for the time being, keep our attention fixed well away from the columns that hold the *t*-value and the *p*-value, lest the nonsense of the calculations in which we are engaged should bear too strongly on our

---

[31] `data(carprice)        # DAAG package`
`pairs(carprice[,-c(1,8,9)])`
`## Alternative, using the lattice function splom()`
`splom(~carprice[,-c(1,8,9)])`

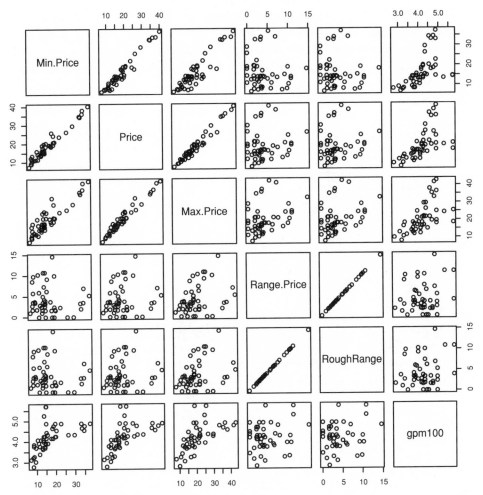

Figure 6.16: Scatterplot matrix for the continuous variables in the constructed `carprice` data set.

attention! Here are the calculations for the first model:[32]

```
> round(summary(carprice1.lm)$coef, 3)
```

	Estimate	Std. Error	t value	Pr(>|t|)
(Intercept)	3.287	0.153	21.467	0.000
TypeLarge	0.324	0.173	1.867	0.069
TypeMidsize	0.185	0.166	1.114	0.272
TypeSmall	-0.389	0.168	-2.317	0.026
TypeSporty	0.205	0.175	1.176	0.247
TypeVan	1.349	0.193	6.997	0.000
Min.Price	0.700	0.989	0.707	0.484
Price	-1.377	1.983	-0.695	0.491
Max.Price	0.711	0.994	0.715	0.479

[32] `carprice1.lm <- lm(gpm100 ~ Type+Min.Price+Price+Max.Price+Range.Price, data=carprice)`

Notice that the variable `Range.Price` has been silently omitted. We constructed it as a linear combination (in fact a difference) of `Min.Price` and `Max.Price`, and it adds no information additional to what those variables contain. The `alias()` function can be a useful recourse in such circumstances; it will identify the linear relationship(s). In this instance we find[33]

```
> alias(carprice1.lm)

Model :
gpm100 ~ Type + Min.Price + Price + Max.Price + Range.Price

Complete :
            (Intercept) TypeLarge TypeMidsize TypeSmall
Range.Price  0           0          0           0
            TypeSporty TypeVan
             0          0
            Min.Price Price Max.Price
Range.Price -1         0    1
```

This tells us what in this instance we already knew, that `Range.Price` is the result of multiplying `Min.Price` by -1, `Max.Price` by 1, and adding.

Here now is the result that emerges when we replace `Price.Range` with `RoughRange` which differs ever so slightly. (Recall that to get `Rough.Range`, we added random values from a normal distribution with mean 0 and standard deviation 0.01.) Here is the result:[34]

```
> round(summary(carprice2.lm)$coef, 2)

              Estimate Std. Error t value Pr(>|t|)
(Intercept)    3.29        0.16    21.19     0.00
TypeLarge      0.33        0.18     1.84     0.07
TypeMidsize    0.18        0.17     1.09     0.28
TypeSmall     -0.39        0.17    -2.29     0.03
TypeSporty     0.21        0.18     1.16     0.25
TypeVan        1.35        0.20     6.90     0.00
Min.Price      0.26        4.28     0.06     0.95
Price         -1.35        2.02    -0.67     0.51
Max.Price      1.13        4.04     0.28     0.78
RoughRange    -0.43        4.04    -0.11     0.92
```

Notice first that `RoughRange` is now included as an explanatory variable. The coefficients of two of the price variables have increased substantially in magnitude. The standard errors of all three price variables have increased very substantially. This is a strong version of a so-called *multicollinearity* effect, where there are near linear relationships between the explanatory variables. The differences between the two sets of results serve only to highlight the point that having once generated nonsensical coefficient estimates, very small changes

---

[33] `alias(carprice1.lm)`
[34] `carprice2.lm <- lm(gpm100 ~ Type+Min.Price+Price+Max.Price+RoughRange,`
     `data=carprice)`

in the input data lead to new nonsense that looks superficially different. Nonsense is not all of one kind!

Note however that the fitted values and standard errors of predicted values for these different models will all be quite similar. It is the coefficients that are used to generate these predicted values that are problematic.

An analysis that makes modest sense

We can sensibly fit only one price variable, in addition to Type. Here is the analysis:[35]

```
> round(summary(carprice.lm)$coef, 3)
```

	Estimate	Std. Error	t value	Pr(>|t|)
(Intercept)	3.305	0.148	22.357	0.000
TypeLarge	0.329	0.168	1.955	0.057
TypeMidsize	0.196	0.162	1.207	0.234
TypeSmall	-0.365	0.162	-2.254	0.030
TypeSporty	0.243	0.163	1.490	0.144
TypeVan	1.392	0.180	7.712	0.000
Price	0.033	0.007	4.482	0.000

The standard error for the coefficient of Price (.007) should be compared with the corresponding standard error (1.98) when we fitted all three price variables; see above under the summary information for carprice1.lm. The "residual standard error" estimates for the two models are very similar:[36] 0.306 and 0.301. Almost all the increase in the standard error is due to the introduction of two further variables – Min.Price and Max.Price – with which Price is strongly correlated.

6.7.2 The variance inflation factor (VIF)

The variance inflation factor (VIF) quantifies the effect of correlation with other variables in inflating the standard error of a regression coefficient. We can best explain it by showing its values for the coefficient of the variable Price in the models above, using the function vif() in our *DAAG* package for the calculations. The variance inflation factor reflects changes in the amount by which the residual variance is multiplied to give the variance of the coefficient of Price. It does not reflect changes in the residual variance. Thus, it depends only on the model matrix.

We start with the model that includes Price as the only explanatory variable. This model is the baseline, relative to which the variance inflation factors for other models are calculated. As we might have hoped, the variance inflation factor for this model is 1. (the variance inflation factor is always 1 when there is a single predictor variable!)

```
> library (DAAG)        # Be sure to use vif() from the DAAG package
> vif(lm(gpm100 ~ Price, data=carprice))  # Baseline
  Price 1
```

---

[35] carprice.lm <- lm(gpm100 ~ Type + Price, data = carprice)

[36] summary(carprice1.lm)$sigma   # residual se when fitting all 3 price variables
   summary(carprice.lm)$sigma   # residual standard error when only price is used

```
> vif(carprice1.lm)      # has Min.Price, Price & Max.Price
   TypeLarge TypeMidsize    TypeSmall  TypeSporty      TypeVan
    2.72e+00    2.33e+00     1.80e+00    2.17e+00     1.78e+00
   Min.Price        Price   Max.Price
    2.99e+04    1.21e+05     3.34e+04
> vif(carprice2.lm)      # has Min.Price, Price,
  TypeLarge TypeMidsize    TypeSmall  TypeSporty
   2.78e+00    2.35e+00     1.82e+00    2.18e+00
      Price   Max.Price   RoughRange
   1.22e+05    5.38e+05     9.62e+04
  Max.Price & RoughRange
    TypeVan    Min.Price
   1.78e+00    5.45e+05
> vif(carprice.lm)        # Price alone
  TypeLarge TypeMidsize TypeSmall TypeSporty  TypeVan   Price
       2.65        2.30      1.74       1.96     1.62    1.73
```

6.7.3 Remedying multicollinearity

As we noted at the beginning of the section, the careful initial choice of variables, based on scientific knowledge and on careful scrutiny of relevant exploratory plots of explanatory variables, will often avert the problem. In some situations, it may be possible to find additional data that will reduce correlations among the explanatory variables.

Ridge regression is one of several approaches that may be used to alleviate the effects of multicollinearity, in inflating coefficients and their standard errors. We refer the reader to the help page for the function lm.ridge() in the *MASS* package and to the discussions in Myers (1990) and Harrell (2001). For a less elementary and more comprehensive account, see Hastie et al. (2001). Use of the function lm.ridge() requires the user to choose the tuning parameter lambda. Typically, the analyst tries several settings and chooses the value that, according to one of several available criteria, is optimal. Principal components regression is another possible remedy.

6.8 Multiple Regression Models – Additional Points

6.8.1 Confusion between explanatory and response variables

As an example, we return to the hill races data. The equation for predicting time given dist and climb (without interaction) is

$$\log(\text{predicted time}) = -3.88 + 0.909 \log(\text{dist}) + 0.260 \log(\text{climb}).$$

It is necessary to note that, to obtain the equation for predicting dist given time and climb, we need to regress dist on climb and time. We cannot find it by re-arranging the above equation. The equation for predicting dist given time and climb is[37]

$$\log(\text{predicted dist}) = 3.44 + 0.95 \log(\text{time}) - 0.178 \log(\text{climb}),$$

which is equivalent to

$$\log(\text{time}) = -3.62 + 1.053 \log(\text{predicted dist}) + 0.187 \log(\text{climb}).$$

Compare this with

$$\log(\text{predicted time}) = -3.88 + 0.909 \log(\text{dist}) + 0.260 \log(\text{climb}).$$

Only if the relationship is exact, so that predicted time is the same as observed time, will the equations be exactly the same.

For examples where this issue arises in earth sciences, see Williams (1983).

Unintended correlations

This highlights another aspect of the desirability of maintaining a clear distinction between explanatory and response variables. Suppose that x_i $(i = 1, 2, \ldots, n)$ are results from a series of controls, while y_i $(i = 1, 2, \ldots, n)$ are the results from the corresponding treated group. It is tempting to plot $y - x$ versus x. Unfortunately, there is inevitably a negative correlation between $y - x$ and x. See the example given in Sharp et al. (1996).

6.8.2 Missing explanatory variables

Here the issue is use of the wrong model for the expected value. With the right "balance" in the data, the expected values are unbiased or nearly unbiased. Where there is serious imbalance, the bias may be huge. This has been discussed earlier.

6.8.3 The use of transformations*

Often there are scientific reasons for transformations. Thus, suppose we have weights w of individual apples, but the effects under study are more likely to be related to surface area. We should consider using $x = w^{\frac{2}{3}}$ as the explanatory variable. If the interest is in studying relative, rather than absolute, changes, it may be best to work with the logarithms of measurements.

Statisticians use transformations for one or more of the following reasons.

1. To form a straight line or other simple relationship.
2. To ensure that the "scatter" of the data is similar for all categories, i.e., to ensure that the boxplots all have a similar shape. In more technical language, the aim is to achieve homogeneity of variance.
3. To make the distribution of data more symmetric and closer to normal.

---

[37] ## For convenience, here again are the R commands:
```
hills.loglm <- lm(log(dist) ~ log(time)+log(climb), data=hills[-18, ])
summary(hills.loglm)
```

If we can find a transformation that deals with all these issues at once, we are fortunate. It may greatly simplify the analysis.

A log transformation may both remove an interaction and give more nearly normal data. It may on the other hand introduce an interaction where there was none before. Or a transformation may reduce skewness while increasing heterogeneity. The availability of direct methods for fitting special classes of model with non-normal errors, for example the generalized linear models that we will discuss in Chapter 8, has reduced the need for transformations.

6.8.4* Non-linear methods – an alternative to transformation?

This is a highly important area for which, apart from the present brief discussion, we have not found room in the present monograph. We will investigate the use of the R `nls()` function to shed light on the loglinear model that we used for the hill race data in Subsection 6.3.3.

The analysis of Subsection 6.3.3 assumed additive errors on the transformed logarithmic scale. This implies, on the untransformed scale, multiplicative errors. We noted that the assumption of homogeneity of errors was in doubt. One might alternatively assume that the noise term is additive on the untransformed scale, leading to the non-linear model

$$y = x_1^{\alpha} x_2^{\beta} + \varepsilon$$

where $y = $ time, $x_1 = $ dist, and $x_2 = $ climb.

Because we could be taking a square or cube of the climb term, we prefer to work with the variable climb.mi that is obtained by dividing climb by 5280:

```
hills$climb.mi <- hills$climb/5280
```

We will use the `nls()` non-linear least squares function to estimate α and β. The procedure used to solve the resulting non-linear equations is iterative, requiring starting values for α and β. We use the estimates from the earlier loglinear regression as starting values.

```
library(nls)
hills.nls0 <- nls(time ~ (dist^alpha)*(climb.mi^beta), start =
                c(alpha = .909, beta = .260), data = hills[-18,])
summary(hills.nls0)
plot(residuals(hills.nls0) ~ predict(hills.nls0)) # residual plot
```

Output from the `summary()` function includes the following:

```
Parameters:
      Estimate Std. Error t value Pr(>|t|)
alpha  0.35642    0.01939  18.379  < 2e-16
beta   0.65846    0.06744   9.764 4.05e-11
```

Note that these parameter estimates differ substantially from those obtained under the assumption of multiplicative errors. This is not an unusual occurrence; the non-linear least squares problem has an error structure that is different from the linearized least-squares problem solved earlier. A glance at the residual plot shows the 11th observation as an extreme outlier. Other residuals exhibit a non-linear pattern.

Another possibility, that allows `time` to increase nonlinearly with `climb.mi`, is

$$y = \alpha + \beta x_1 + \gamma x_2^\delta + \varepsilon.$$

We then fit the model, using a fairly arbitrary starting guess:

```
hills.nls <- nls(time ~ alpha + beta*dist +
    gamma*(climb.mi^delta), start=c(alpha = 1, beta = 1,
    gamma = 1, delta = 1), data=hills[-18, ])
summary(hills.nls)
plot(residuals(hills.nls) ~ predict(hills.nls))
    # residual plot
```

The result is

```
Parameters:
      Estimate Std. Error t value Pr(>|t|)
alpha -0.05310    0.02947   -1.80    0.082
beta   0.11076    0.00398   27.82  < 2e-16
gamma  0.77277    0.07523   10.27  2.4e-11
delta  2.18414    0.22706    9.62  1.1e-10

Residual standard error: 0.0953 on 30 degrees of freedom
```

There are no outliers on the residual plot. In addition, there is no indication of an increase in the variance as the fitted values increase. Thus, a variance-stabilizing transformation or the use of weighted least squares is unnecessary. The model is highly interpretable, indicating that a racer should allow roughly 5.5 minutes per mile plus about 45 minutes for every squared mile of `climb`. Using this rough guide, the prediction is that the missing race (18), a distance of 3 miles with a climb of .066 miles, has a time of 16.7 minutes.

The intercept `alpha`, which implies a negative time for very short races, should probably be removed from the model.

6.9 Further Reading

Weisberg (1985) offers a relatively conventional approach. Cook and Weisberg (1999) rely heavily on graphical explorations to uncover regression relationships. Venables and Ripley (2002) is a basic reference for using R for regression calculations. Hastie et al. (2001) offers wide-ranging challenges for the reader who would like to explore well beyond the forms of analysis that we have described.

On variable selection, which warrants more attention than we have given it, see Harrell (2001), Hastie et al. (2001) and Venables (1998). There is certain to be, in the next several years, substantial enhancement of what R packages offer in this area.

In our examples, there have been two categories of variables – explanatory variables and response variables. More generally there may be three or more categories. We may, for example, have explanatory variables, intermediate variables, and dependent variables, where intermediate are dependent with respect to one or more of the explanatory variables, and explanatory variables with respect to one or more of the dependent variables.

Cox and Wermuth (1996) and Edwards (2000) describe approaches that use regression methods to elucidate the relationships. Cox and Wermuth is useful for the large number of examples that it gives and for its illuminating comments on practical issues, while Edwards now has a more up to date account of the methodology.

Bates and Watts (1988) discuss non-linear models in detail. A more elementary presentation is given in one of the chapters of Myers (1990).

References for further reading

Bates, D.M. and Watts, D.G. 1988. *Non-linear Regression Analysis and Its Applications.* Wiley.

Cook, R.D. and Weisberg, S. 1999. *Applied Regression Including Computing and Graphics.* Wiley.

Cox, D.R. and Wermuth, N. 1996. *Multivariate Dependencies: Models, Analysis and Interpretation.* Chapman and Hall.

Edwards, D. 2000. *Introduction to Graphical Modelling*, 2nd ed. Springer-Verlag.

Harrell, F.E. 2001. *Regression Modelling Strategies, with Applications to Linear Models, Logistic Regression and Survival Analysis.* Springer-Verlag.

Hastie, T., Tibshirani, R. and Friedman, J. 2001. *The Elements of Statistical Learning. Data Mining, Inference and Prediction.* Springer-Verlag.

Myers, R.H. 1990. *Classical and Modern Regression with Applications*, 2nd edn. Brooks Cole.

Venables, W.N. 1998. Exegeses on linear models. Proceedings of the 1998 international S-PLUS user conference. Available from www.stats.ox.ac.uk/pub/MASS3/Compl.html

Venables, W.N. and Ripley, B.D. 2002. *Modern Applied Statistics with S-PLUS*, 4th edn. Springer-Verlag.

Weisberg, S. 1985. *Applied Linear Regression*, 2nd edn. Wiley.

6.10 Exercises

1. The data set `cities` lists the populations (in thousands) of Canada's largest cities over 1992 to 1996. There is a division between Ontario and the West (the so-called "have" regions) and other regions of the country (the "have-not" regions) that show less rapid growth. To identify the "have" cities we can specify

```
data(cities)              # DAAG package
cities$have <- factor((cities$REGION=="ON")|
                      (cities$REGION=="WEST"))
```

Plot the 1996 population against the 1992 population, using different colors to distinguish the two categories of city, both using the raw data and taking logarithms of data values, thus:

```
plot(POP1996 ~ POP1992, data=cities,
     col=as.integer(cities$have))
```

```
plot(log(POP1996) ~ log(POP1992), data=cities,
                         col=as.integer(cities$have))
```

Which of these plots is preferable? Explain.
Now carry out the regressions

```
cities.lm1 <- lm(POP1996 ~ have+POP1992, data=cities)
cities.lm2 <- lm(log(POP1996) ~ have+log(POP1992),
                  data=cities)
```

and examine diagnostic plots. Which of these seems preferable? Interpret the results.

2. In the data set cement (*MASS* package), examine the dependence of y (amount of heat produced) on x1, x2, x3 and x4 (which are proportions of four constituents). Begin by examining the scatterplot matrix. As the explanatory variables are proportions, do they require transformation, perhaps by taking $\log(x/(100 - x))$? What alternative strategies might be useful for finding an equation for predicting heat?

3. The data frame hills2000 in our *DAAG* package has data, based on information from the Scottish Running Resource web site, that updates the 1984 information in the data set hills. Fit a regression model, for men and women separately, based on the data in hills2000. Check whether it fits satisfactorily over the whole range of race times. Compare the equation that you obtain with that based on the hills data frame.

4. For each covariate, compare the (NSW) treatment group in nsw74psid1 with the (PSID) control group. Use overlaid density plots to compare the continuous variables, and two-way tables to compare the binary (0/1) variables. Where are the greatest differences?

5. Repeat the analysis in Section 6.5, using the data set nsw74psid1, but now working with log(re78) and log(re75). What difference does the use of the logarithms of the income variables have for the interpretation of the results?
    [Hint: For each observation, determine predicted values. Then exp(predicted values) gives predicted incomes in 1978. Take exp(estimated treatment effect) to get an estimate of the factor by which a predicted income for the control group must be multiplied to get a predicted income for the experimental group, if covariate values are the same.]

6. Section 6.1 used lm() to analyze the allbacks data that are presented in Figure 6.1. Repeat the analysis using (1) the function rlm() in the *MASS* package, and (2) the function lqs() in the *lqs* package. Compare the two sets of results with the results in Section 6.1.

7. This exercise illustrates what happens if a transformation is not used for the hills data analysis.

    (a) Fit two models, one a model that is linear in dist and climb, and one that has, additionally, dist × climb, i.e.

    ```
    hills.lm <- lm(time ~ dist+climb, data=hills)
    hills2.lm <- lm(time ~ dist+climb+dist:climb,
                      data=hills)
    anova(hills.lm, hills2.lm)
    ```

    (b) Using the *F*-test result, make a tentative choice of model, and proceed to examine diagnostic plots. Are there any problematic observations? What happens if these points are removed? Re-fit both of the above models, and check the diagnostics again.

8. Check the variance inflation factors for `bodywt` and `lsize` for the model `brainwt ~ bodywt + lsize`, fitted to the `litters` data set. Comment.

9. Compare the ranges of `dist` and `climb` in the data frames `hills` and `hills2000`. In which case would you expect it to be more difficult to find a model that fits well? For each of these data frames, fit both the model based on the formula

   `log(time) ~ log(dist) + log(climb)`

   and the model based on the formula

   `time ~ alpha*dist + beta*I(climb^2)`

   Is there one model that gives the best fit in both cases?

7

Exploiting the Linear Model Framework

The model matrix is fundamental to all calculations for a linear model. The model matrix carries the information needed to calculate the fitted values that correspond to any particular choice of coefficients. There is one coefficient for each column of X.

For calculating fitted values, multiply the first column by the first coefficient (by default the intercept), the second column by the second coefficient, and so on across all the columns. Then add up the total. The regression calculations find the set of coefficients that best predicts the observed values. In the classical theory, the best choice is the choice of coefficients that minimizes the sum of squares of residuals and this is the definition of "best" that we will use.

In Chapter 6, the columns of the model matrix were the explanatory variables, although we did allow ourselves the option of transforming them in the process of creating a column for the model matrix. Here, we will explore new ways to use the columns of the model matrix. We can use vectors of zeros and ones to handle factor levels. For modeling a quadratic form of response, we take values of x as one of the columns, and values of x^2 as another.

In R, as in many other statistical software packages, a model formula specifies the model. Terms in model formulae may include factors and interactions of factors with other terms. Given this information, R then sets up the model matrix, without further intervention from the user. Where there are factors, there are various alternatives to R's default for setting up the relevant columns of the model matrix. In technical language, there are alternative factor parameterizations, or alternative choices of factor *contrasts*. Users need to understand the implications of these alternative choices for the interpretation of the resulting regression coefficients.

7.1 Levels of a Factor – Using Indicator Variables

7.1.1 Example – sugar weight

Table 7.1 displays data from an experiment that compared an unmodified wild type plant with three different genetically modified forms (data are in the data set `sugar` in our *DAAG* package). The measurements are weights (mg) of sugar that were obtained by breaking down the cellulose. There is a single explanatory factor (treatment), with one level for each of the different control agents. For convenience, we will call the factor levels Control, A (Modified 1), B (Modified 2) and C (Modified 3). Figure 7.1 shows the data.

Table 7.1: *Comparison of weights (`weight`) of sugar in a control (wild type) plant and in three different genetically modified plant types.*

Control (Wild Type)	A (Modified 1)	B (Modified 2)	C (Modified 3)
82.0	58.3	68.1	50.7
97.8	67.9	70.8	47.1
69.9	59.3	63.6	48.9
Mean = 83.2	61.8	67.5	48.9

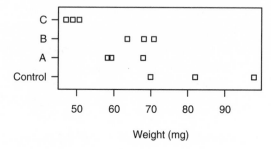

Weight (mg)

Figure 7.1: Weight of sugar (`weight`) extracted from four different types of plant: Control (wild type), and modified types A, B and C.

We could reduce the apparent difference in variability between treatments by working with the `log(weight)`. For present illustrative purposes, we will however work with the variable `weight`, leaving as an exercise for the reader the analysis that works with `log(weight)`.

We have a choice whether to treat this as a regression problem, or as an analysis of variance problem. Depending on the choice we make, the statements used for analysis may be different, and the default output will be different. As our focus is on regression, we will formulate it as a regression problem.

For any problem that involves factor(s), there are several different ways to set up the model matrix. The default, for R and for many other computer programs, is to set up one of the treatment levels as a baseline or reference, with the effects of other treatment levels then measured from the baseline. Here it makes sense to set `Control` (Wild) as the baseline. With `Control` as baseline, the model matrix for the data in Table 7.1 is given in Table 7.2.[1]

In Table 7.3, the multiples determined by least squares calculations are shown above each column. Also shown is $\hat{y}$, which is the predicted value.[2] Residuals can be obtained by

---

[1] `## Code to obtain the model matrix`
    `data(sugar)              # DAAG package`
    `sugar.aov <- aov(weight ~ trt, data=sugar)`
    `model.matrix(sugar.aov)`
    `# Responses are sugar$weight`
[2] `fitted.values(sugar.aov)`

Table 7.2: *Model matrix for the analysis of variance calculation for the data in Table 7.1. The values of the outcome variable are in the right column.*

Control (baseline)	A	B	C	weight
1	0	0	0	82.0
1	0	0	0	97.8
1	0	0	0	69.9
1	1	0	0	58.3
1	1	0	0	67.9
1	1	0	0	59.3
1	0	1	0	68.1
1	0	1	0	70.8
1	0	1	0	63.6
1	0	0	1	50.7
1	0	0	1	47.1
1	0	0	1	48.9

Table 7.3: *At the head of each column is the multiple, as determined by least squares, that is taken in forming the fitted values.*

Control: 83.2	A: −21.4	B: −15.7	C: --34.3	Fitted value
1	0	0	0	83.2
1	0	0	0	83.2
1	0	0	0	83.2
1	1	0	0	61.8
1	1	0	0	61.8
1	1	0	0	61.8
1	0	1	0	67.5
1	0	1	0	67.5
1	0	1	0	67.5
1	0	0	1	48.9
1	0	0	1	48.9
1	0	0	1	48.9

subtracting the predicted values ($\widehat{y}$) in Table 7.3 from the observed values (y) in Table 7.2. Here is the output from R:

```
> summary.lm(sugar.aov)    # NB: summary.lm(),
                           # not summary() or summary.aov()


Call:
aov(formula = weight ~ trt, data = sugar)
```

```
Residuals:
    Min      1Q   Median      3Q      Max
-13.333  -2.783  -0.617   2.175   14.567
Coefficients:
              Estimate Std. Error t value Pr(>|t|)
(Intercept)      83.23       4.47   18.61  7.2e-08
trtA            -21.40       6.33   -3.38  0.00960
trtB            -15.73       6.33   -2.49  0.03768
trtC            -34.33       6.33   -5.43  0.00062

Residual standard error: 7.75 on 8 degrees of freedom
Multiple R-Squared: 0.791,        Adjusted R-squared: 0.713
F-statistic: 10.1 on 3 and
8 degrees of freedom,                p-value: 0.00425
```

The row labeled (Intercept) gives the estimate (= 83.23) for the baseline, in our case Control. The interpretations of the remaining coefficients are:

A: weight changed by −21.40.
B: weight changed by −15.73.
C: weight changed by −34.33.

All of these changes (differences from control) are significant at the conventional 5% level. In this example, the estimate for each treatment is the treatment mean. Regression calculations have given us a relatively complicated way to calculate the treatment means! The example is intended to demonstrate a standard approach to working with factors. It does not pretend to be the best way to analyze these data.

7.1.2 Different choices for the model matrix when there are factors

In the language used in the R help pages, different choices of *contrasts* are available, with each different choice leading to a different model matrix. These different choices lead to different regression parameters, but the same fitted values, and the same analysis of variance table.

The default (*treatment*) choice of *contrasts* uses the initial factor level as baseline, as we have noted. Different choices of the baseline or reference level lead to different versions of the model matrix. The other common choice, i.e., *sum* contrasts, uses the average of treatment effects as the baseline. The choice of contrasts may call for careful consideration, in order to obtain the output that will be most helpful for the problem in hand. Or, more than one run of the analysis may be necessary, in order to gain information on all effects that are of interest.

Here is the output when the baseline is the average of the treatment effects, i.e., from using the *sum* contrasts:[3]

```
> summary.lm(sugar.aov)
```

[3] options(contrasts=c("contr.sum", "contr.poly"))
   # The mean over all treatment levels is now the baseline
  sugar.aov <- aov(formula = weight ~ trt, data = sugar)
  summary.lm(sugar.aov)

```
Call:
aov(formula = weight ~ trt, data = sugar)


Residuals:
    Min      1Q  Median      3Q     Max
-13.333  -2.783  -0.617   2.175  14.567


Coefficients:
            Estimate Std. Error t value Pr(>|t|)
(Intercept)    65.37       2.24   29.23  2.0e-09
trt1           17.87       3.87    4.61   0.0017
trt2           -3.53       3.87   -0.91   0.3883
trt3            2.13       3.87    0.55   0.5968


Residual standard error: 7.75 on 8 degrees of freedom
Multiple R-Squared: 0.791,        Adjusted R-squared: 0.713
F-statistic: 10.1 on 3 and
8 degrees of freedom,          p-value: 0.00425
```

Notice how it differs from the output from the default choice of model matrix. The baseline, labeled (Intercept), is now the treatment mean. This equals 65.37. Remaining coefficients are differences, for Control and for treatment levels A and B, from this mean. The sum of the differences for all three treatments is zero. Thus the difference for C is (rounding up)

$$-(17.87 - 3.53 + 2.13) = -16.5.$$

The estimates (means) are:

Control: $65.37 + 17.87 = 83.2$.
A: $65.37 - 3.53 = 61.8$.
B: $65.37 + 2.13 = 67.5$.
C: $65.37 - 16.5 = 48.9$.

Note finally the possibility of using *helmert* contrasts (these are the S-PLUS default). We refer the reader to Sections 12.4 and the Appendix for further details.

7.2 Polynomial Regression

Polynomial regression provides a straightforward way to model simple forms of departure from linearity. The simplest case is where the response curve has a simple cup up or cup down shape. For a cup down shape, the curve has some part of the profile of a path that follows the steepest slope up a rounded hilltop towards the summit and down over the other side. For a cup up shape the curve passes through a valley. Such cup down or cup up shapes can often be modeled quite well using quadratic, i.e., polynomial with degree 2, regression. For this the analyst uses x^2 as well as x as explanatory variables. If a straight line is not adequate, and the departure from linearity suggests a simple cup up or cup down form of response, then it is reasonable to try quadratic regression.

Table 7.4: *Number of grains*
*per head of barley (*grain*),*
*versus seeding rate (*rate*).*

	rate	grain
1	50	21.2
2	75	19.9
3	100	19.2
4	125	18.4
5	150	17.9

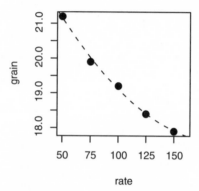

Figure 7.2: Plot of number of grains per head versus seeding rate, for the barley seeding rate data, with fitted quadratic curve. Data relate to McLeod (1982).

The calculations are formally identical to those for multiple regression. To avoid numerical problems, it is often preferable to use orthogonal polynomial regression. The orthogonal polynomial coefficients must be translated back into coefficients of powers of x, if these are required. Interested readers may wish to pursue for themselves the use of orthogonal polynomial regression, perhaps using as a starting point Exercise 7.7.13.

Table 7.4 gives number of grains per head (averaged over eight replicates), for different seeding rates of barley. Figure 7.2 shows a quadratic curve that has been fitted to these data.[4] The quadratic regression appears, from visual inspection, to be a good fit to the data. The fitted model may be written

$$\widehat{y} = a + b_1 x_1 + b_2 x_2$$

---

[4]
```
data(seedrates)        # DAAG package
seedrates.lm2 <- lm(grain ~ rate + I(rate^2), data = seedrates)
   # The wrapper function I() ensures that the result from
   # calculating rate^2 is treated as a variable in its own right.
plot(grain ~ rate, data = seedrates, pch = 16,
     xlim = c(50, 160),  cex=1.4)
new.df <- data.frame(rate = (1:14) * 12.5) # for plotting the fitted curve
hat2 <- predict(seedrates.lm2, newdata = new.df, interval="predict",
             coverage = 0.95)
lines(new.df$rate, hat2[,"fit"], lty = 2, lwd=2)
```

Table 7.5: *The model matrix, for fitting a quadratic curve to the seeding rate data of Table 7.4.*

(Intercept)	rate	rate^2
1	50	2 500
1	75	5 625
1	100	10 000
1	125	15 625
1	150	22 500

where $x_1 = x$, and $x_2 = x^2$. Thus, the model matrix (Table 7.5) has a column of 1s, a column of values of x, and a column that has values of x^2.

Here is the output from R:

```
> summary.lm(seedrates.lm2, corr=TRUE)

Call:
lm(formula = grain ~ rate + I(rate^2), data = seedrates)

Coefficients:
              Value Std. Error t value Pr(>|t|)
(Intercept)  24.060   0.456     52.799   0.000
       rate  -0.067   0.010     -6.728   0.021
  I(rate^2)   0.000   0.000      3.497   0.073

Residual standard error: 0.115 on 2 degrees of freedom
Multiple R-Squared: 0.996
F-statistic: 256 on 2 and 2 degrees of freedom,
the p-value is 0.0039

Correlation of Coefficients:
          (Intercept)    rate
     rate -0.978
I(rate^2)  0.941        -0.989
```

Observe the high correlations between the coefficients. Note in particular the large negative correlation between the coefficients for rate and I(rate^2). Forcing the coefficient for rate to be high would lead to a low coefficient for I(rate^2), and so on.

7.2.1 Issues in the choice of model

The coefficient of the x^2 term in the quadratic model fell short of statistical significance at the 5% level. Fitting the x^2 leaves only two degrees of freedom for error. For prediction our interest is likely to be in choosing the model that is on balance likely to give the more accurate predictions; for this, use of the model that includes the quadratic term may be preferred.

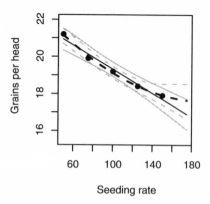

Figure 7.3: Number of grains per head versus seeding rate, with fitted line (solid) and fitted quadratic curve (dashed). Also shown are 95% pointwise confidence bounds.

Figure 7.3 shows both a fitted line and a fitted curve, in both cases with 95% confidence bounds.[5] It shows a quadratic curve (dashed line) as well as a line (solid line). In addition, the graph shows 95% pointwise confidence bounds, both about the line and about the curve.

The curve is a better fit to the data than the line. For a short distance beyond the final data point, it almost certainly gives a better estimate than does the line. Notice that the confidence bounds for the curve are much wider, beyond a rate of about 170, than the line. Not only does the line almost certainly give a biased estimate, it also gives unrealistically narrow bounds for that estimate. If the model is wrong, it will give wrong estimates of predictive accuracy. This is especially serious for extrapolation beyond the limits of the data.

Beyond the limits of the data, it would be unwise to put much trust in either the line or the curve. Our point is that the bounds for the quadratic curve do better reflect uncertainty in the curve that ought to be fitted. We could try other, single parameter, models. Selecting a model from a number of choices that allow for the curvature may not however be much different, in its effect on the effective degrees of freedom, from adding an x^2 term. The wider confidence bounds for the quadratic model reflect this uncertainty in choice of model, better than results from any individual model that has one parameter additional to the intercept.

---

[5]
```
## Code to fit the line, the curve, determine the pointwise coverage
## bounds and create the plots
plot(grain ~ rate, data=seedrates, xlim=c(50,180), ylim=c(15.5,22), axes=FALSE)
new.df <- data.frame(rate=(2:8)*25)
seedrates.lm1 <- lm(grain ~ rate, data=seedrates)
seedrates.lm2 <- lm(grain ~ rate+I(rate^2), data=seedrates)
hat1 <- predict(seedrates.lm1, newdata=new.df, interval="confidence")
hat2 <- predict(seedrates.lm2, newdata=new.df, interval="confidence")
axis(1, at=new.df$rate); axis(2); box()
z1 <- spline(new.df$rate, hat1[,"fit"])
z2 <- spline(new.df$rate, hat2[,"fit"])
rate <- new.df$rate; lines(z1$x, z1$y)
lines(spline(rate, hat1[,"lwr"]), lty=1, col=3)
lines(spline(rate, hat1[,"upr"]), lty=1, col=3)
lines(z2$x, z2$y, lty=4)
lines(spline(rate, hat2[,"lwr"]), lty=4, col=3)
lines(spline(rate, hat2[,"upr"]), lty=4, col=3)
```

Table 7.6: *Selected rows, showing values of*
CO2level, vapPress *and* tempDiff, *from the*
data set leaftemp.

CO2level	vapPress	tempDiff
low	1.88	1.36
low	2.20	0.60
low	1.75	0.23
...	...	...
medium	2.38	1.94
medium	2.72	0.83
medium	2.21	−0.11
...	...	...
high	2.56	1.50
high	2.55	0.85
high	2.17	−0.04
...	...	...

We can in fact fit the data well by modeling grain as a linear function of log(rate). This model seems intuitively more acceptable; the fitted value of grain continues to decrease as the rate increases beyond the highest rate used in the experiment. This is perhaps the model that we should have chosen initially on scientific grounds.

7.3 Fitting Multiple Lines

Multiple regression can be used to fit multiple lines. In the example that follows (Table 7.6), we have measurements of vapor pressure (vapPress) and of the difference between leaf and air temperature (tempDiff), for three different levels of carbon dioxide.

Possibilities we may want to consider are:

- Model 1 (constant response): $y = a$.
- Model 2 (a single line): $y = a + bx$.
- Model 3 (three parallel lines): $y = a_1 + a_2 z_2 + a_3 z_3 + bx$.
  (For the low CO_2 group ($z_2 = 0$ and $z_3 = 0$) the constant term is a_1; for the medium CO_2 group ($z_2 = 1$ and $z_3 = 0$) the constant term is $a_1 + a_2$, while for the high CO_2 group ($z_2 = 0$ and $z_3 = 1$) the constant term is $a_1 + a_3$.)
- Model 4 (three separate lines): $y = a_1 + a_2 z_2 + a_3 z_3 + b_1 x + b_2 z_2 x + b_3 z_3 x$.
  (Here, z_2 and z_3 are as in Model 3 (Panel B). For the low CO_2 group ($z_2 = 0$ and $z_3 = 0$) the slope is b_1; for the medium CO_2 group ($z_2 = 1$ and $z_3 = 0$) the slope is $b_1 + b_2$, while for the high CO_2 group ($z_2 = 0$ and $z_3 = 1$) the slope is $b_1 + b_3$.)

Selected rows from the model matrices for Model 3 and Model 4 are displayed in Tables 7.7 and 7.8, respectively.

The statements used to fit the four models are

```
data(leaftemp)       # DAAG package
leaf.lm1 <- lm(tempDiff ~ 1 , data = leaftemp)
```

Table 7.7: *Model matrix for fitting three parallel lines (Model 3) to the data of Table 7.6. The y-values are in the separate column to the right.*

(Intercept)	Medium	High	vapPress	tempDiff
1	0	0	1.88	1.36
1	0	0	2.2	0.6
1	0	0	1.75	0.23
1	0	0	1.85	0.48
...	...	...	...	...
1	1	0	2.38	1.94
1	1	0	2.72	0.83
1	1	0	2.21	−0.11
1	1	0	1.67	0.85
...	...	...	...	...
1	0	1	2.56	1.5
1	0	1	2.55	0.85
1	0	1	2.17	−0.04
1	0	1	1.64	1.25

Table 7.8: *Model matrix for fitting three separate lines (Model 4), with y-values in the separate column to the right.*

(Intercept)	Medium	High	vapPress	Medium: vapPress	High: vapPress	tempDiff
1	0	0	1.88	0	0	1.36
1	0	0	2.2	0	0	0.6
1	0	0	1.75	0	0	0.23
1	0	0	1.85	0	0	0.48
...	...	...	...	...	...	...
1	1	0	2.38	2.38	0	1.94
1	1	0	2.72	2.72	0	0.83
1	1	0	2.21	2.21	0	−0.11
1	1	0	1.67	1.67	0	0.85
...	...	...	...	...	...	...
1	0	1	2.56	0	2.56	1.5
1	0	1	2.55	0	2.55	0.85
1	0	1	2.17	0	2.17	−0.04
1	0	1	1.64	0	1.64	1.25

```
leaf.lm2 <- lm(tempDiff ~ vapPress, data = leaftemp)
leaf.lm3 <- lm(tempDiff ~ CO2level + vapPress,
               data = leaftemp)
leaf.lm4 <- lm(tempDiff ~ CO2level + vapPress
               + vapPress:CO2level, data = leaftemp)
```

Table 7.9: *Analysis of variance information. The starting point is a model that has only an intercept or "constant" term. The entries in rows 1–3 of the* Df *column and of the* Sum of Sq *column are then sequential decreases from fitting, in turn,* vapPress, *then three parallel lines, and then finally three separate lines.*

	Df	Sum of Sq	Mean square	F	Pr(<F)	
vapPress (variable)	1	5.272	5.272	11.3	0.0014	reduction in SS due to fitting one line
Three parallel lines	2	6.544	3.272	7.0	0.0019	additional reduction in SS due to fitting two parallel lines
Three different lines	2	2.126	1.063	2.3	0.1112	additional reduction in SS due to fitting two separate lines
Residuals	61	40.000	0.656			

Recall that CO2level is a factor and vapPress is a variable. Technically, vapPress:CO2level is an interaction. The effect of an interaction between a factor and a variable is to allow different slopes for different levels of the factor.

The analysis of variance table is helpful in allowing a choice between these models. Here it is:[6]

```
Analysis of Variance Table

Model 1: tempDiff ~ 1
Model 2: tempDiff ~ vapPress
Model 3: tempDiff ~ CO2level + vapPress
Model 4: tempDiff ~ CO2level + vapPress + CO2level:vapPress
  Res.Df    RSS Df Sum of Sq      F  Pr(>F)
1     61  40.00
2     60  34.73  1     5.272  11.33  0.0014
3     58  28.18  2     6.544   7.03  0.0019
4     56  26.06  2     2.126   2.28  0.1112
```

This is a sequential analysis of variance table. Thus the quantity in the sum of squares column (Sum of Sq) is the reduction in the residual sum of squares due to the inclusion of that term, given that earlier terms had already been included. The Df (degrees of freedom) column gives the change in the degrees of freedom due to the addition of that term. Table 7.9 explains this in detail.

The analysis of variance table suggests, consistently with what we see in Figure 7.4, that we should choose the parallel line model, shown in panel B of Figure 7.4. The reduction in the mean square from Model 3 (panel B in Figure 7.4) to Model 4 (panel C) in the analysis

---

[6] anova(leaf.lm1, leaf.lm2, leaf.lm3, leaf.lm4)

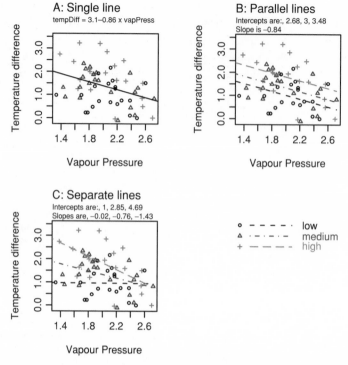

Figure 7.4: A sequence of models fitted to the plot of tempDiff versus vapPress, for low, medium and high levels of CO2level. Panel A relates to Model 2, B to Model 3, and C to Model 4.

of variance table has a *p*-value equal to 0.1112 . The coefficients and standard errors for Model 3 are

```
> summary(leaf.lm2)

Call:
lm(formula = tempDiff ~ CO2level + vapPress,
   data = leaftemp)

Residuals:
      Min        1Q     Median         3Q        Max
-1.696828 -0.542987   0.060763   0.463710   1.358543

Coefficients:
                Estimate Std. Error t value  Pr(>|t|)
(Intercept)        2.685      0.560    4.80  1.16e-05
CO2levelmedium     0.320      0.219    1.46   0.14861
CO2levelhigh       0.793      0.218    3.64   0.00058
vapPress          -0.839      0.261   -3.22   0.00213

Residual standard error: 0.69707 on 58 degrees of freedom
Multiple R-Squared: 0.295,     Adjusted R-squared: 0.259
```

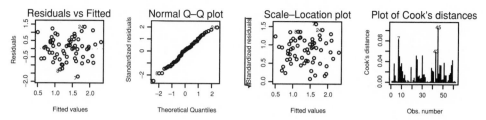

Figure 7.5: Diagnostic plots for the parallel line model of Figure 7.4.

```
F-statistic: 8.106 on 3 and
58 degrees of freedom,          p-value: 0.000135
```

The equations for this parallel line model are spelled out in Figure 7.4B. For the first equation (low CO_2), the constant term is 2.685, for the second equation (medium CO_2) the constant term is $2.685 + 0.320 = 3.005$, while for the third equation the constant term is $2.685 + 0.793 = 3.478$.

In addition, we examine a plot of residuals against fitted values, and a normal probability plot of residuals (Figure 7.5).

These plots seem unexceptional.

7.4* Methods for Passing Smooth Curves through Data

Earlier in this chapter, we used the linear model framework to fit models that have x^2, x^3, etc. terms. Thus, the linear model framework is rich enough to allow for the fitting of polynomial curves to regression data. For a polynomial of degree m, the model matrix must have, in addition to a column of 1s, columns that hold values of x, $x^2, \ldots,$ x^m. Polynomials can be effective when a curve of degree m that is 2 or 3 is appropriate. Polynomial curves where m is much greater than 3 can be problematic. High degree polynomials tend to move up and down between the data values in a snake-like manner. Splines, or piecewise polynomials, which we now consider, are usually preferable to polynomials when a polynomial of degree 3 is not sufficiently flexible to fit the data adequately.

There are several perspectives on spline curves that are helpful to keep in mind:

- A spline curve is formed by joining together two or more polynomial curves, typically cubics, in such a way that there is a smooth transition from one polynomial to the next, i.e., the polynomials have the same slope at the point where they join. Splines are suitable for modeling any smoothly changing response. Here we consider regression splines.
- Once the join points (=knots) have been determined, a spline curve can be fitted by use of an appropriate choice of columns in the model matrix. In technical language, the spline functions have linear bases. There is an elementary explanation of the technical details in Maindonald (1984, Chapter 7).

Table 7.10: *Resistance (ohms), versus apparent juice content. The table shows a selection of the data.*

	Juice (%)	Ohms		Juice (%)	Ohms		Juice (%)	Ohms		Juice (%)	Ohms
1	4	4860	33	20	7500	65	41.5	3350	123	58.5	3650
2	5	5860	34	20.5	8500	66	42.5	2700	124	58.5	3750
3	5.5	6650	35	21.5	5600	67	43	2750	125	58.5	4550
4	7.5	7050	36	21.5	6950	68	43	3150	126	59.5	3300
5	8.5	5960	37	21.5	7200	69	43	3250	127	60	3600
...	...	...	...	...	...	...	...	...	128	9	9850

- There are different possible choices of constraints that must be imposed on the cubic curves at the end points. The different choices of constraints lead to different types of spline and to different sets of basis functions. We work with B-splines.

A modest amount of algebra is required to make the connection between the basis functions and the separate cubic (or other) polynomial curves. It is the basis functions that the R routines use. Do not expect to see the separate polynomial curves pop out of the calculations! Note also that there is a rich variety of different types of smoothing method. The choice of methods that we discuss here is, inevitably, highly selective.

We will take a simple example where there is just one explanatory variable, and try alternative methods, including regression splines, on it.

7.4.1 Scatterplot smoothing – regression splines

We have (in Table 7.10) the apparent juice content and resistance (in ohms) for 128 slabs of fruit (these data relate to Harker and Maindonald, 1994). Figure 7.6 shows four different curves fitted to these data[7]. Figures 7.6A and B show spline curves, the first with one knot and the second with two knots. Figures 7.6C and D show, for comparison, third and fourth degree polynomials. The polynomials, which are inherently less flexible, do quite well here relative to the splines. Also shown are 95% pointwise confidence intervals for the fitted curves.

[7]
```
## Code to obtain Panel A
library(splines)
data(fruitohms)       # From DAAG package
attach(fruitohms)
plot(ohms ~ juice, cex=0.8, xlab="Apparent juice content (%)",
     ylab="Resistance (ohms)")
fruit.lmb4 <- lm(ohms ~ bs(juice,4))
ord <- order(juice)
lines(juice[ord], fitted(fruit.lmb4)[ord], lwd=2)
ci <- predict(fruit.lmb4, interval="confidence")
lines(juice[ord], ci[ord,"lwr"])
lines(juice[ord], ci[ord,"upr"])
## For panels B, C, D replace
## fruit.lmb4 <- lm(ohms ~ bs(juice,4)) with
fruit.lmb5 <- lm(ohms ~ bs(juice,5))    # panel B: bspline, df = 5
fruit.lmp3 <- lm(ohms ~ poly(juice,3))  # panel C: polynomial, df = 3
fruit.lmp4 <- lm(ohms ~ poly(juice,4))  # panel D: polynomial, df = 4
```

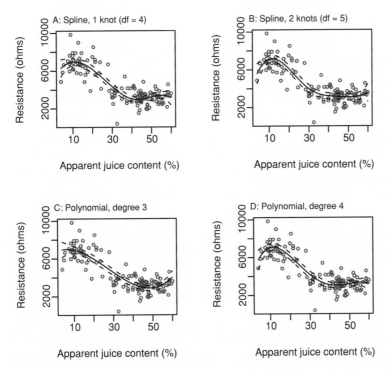

Figure 7.6: Different smooth curves fitted to the data of Table 7.10. The dashed lines show 95% pointwise confidence bounds for the fitted curve.

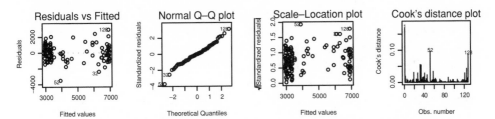

Figure 7.7: Diagnostic plots for the fitted model given in Figure 7.6A.

A problem with all these curves is that they may be unduly influenced by values at the extremes of the range. Diagnostic plots can be used, just as for the models considered in earlier chapters, to highlight points that are associated with large residuals, or that are having a strong influence on the curve. Figure 7.7 shows the default diagnostic plots for the fitted model shown in Figure 7.6A.[8]

Apart from the large residual associated with point 52 (at 32.5% apparent juice content), these plots show nothing of note. The curves have bent to accommodate points near the extremes that might otherwise have appeared as outliers.

---

[8] ```
par(mfrow = c(2,2))
plot(fruit.lmb4)
par(mfrow = c(1,1))
```
```

Here is the summary information:[9]

```
> summary(fruit.lmb4)

Call:
lm(formula = ohms ~ bs(juice, 4))

Residuals:
    Min      1Q  Median      3Q     Max
-3518.2  -618.2   -15.8   515.4  2937.7

Coefficients:
                Estimate Std. Error t value Pr(>|t|)
(Intercept)         6293        457   13.79  < 2e-16
bs(juice, 4)1       2515       1040    2.42  0.01707
bs(juice, 4)2      -5621        652   -8.62  2.7e-14
bs(juice, 4)3      -2471        715   -3.46  0.00076
bs(juice, 4)4      -3120        585   -5.33  4.5e-07

Residual standard error: 957 on 123 degrees of freedom
Multiple R-Squared: 0.743,       Adjusted R-squared: 0.735
F-statistic: 89.1 on 4 and 123 DF,  p-value:     0
```

While it is clear that the coefficients in the spline equation are highly significant, interpreting these coefficients individually is not straightforward and is usually not worth the effort. Attention is best focused on the fitted curve.

The coefficients that are shown are coefficients of the four basis functions. It is best, at this point, to forget the fact that the curve can be constructed by the smooth joining of separate cubic curves. For interpreting the R output, it helps to understand how the curve has been formed as a linear combination of basis functions. We do this by plotting graphs that show the curves for which these are the coefficients. For this, we plot the relevant column of the X-matrix against x, and join up the points, as in Figure 7.8.[10]

Looking back again at the coefficients, basis curve 2 (with a coefficient of -5621) seems somewhat more strongly represented than the other basis curves.

The regression splines that we described above are unusual because they fit easily within a linear model framework, i.e., we can fit them by specifying an appropriate X-matrix. There are a wide variety of other methods, most of which do not fit within the linear model framework.

---

[9] summary(fruit.lmb4)
[10] ## Code to obtain the curve shown in panel A
    plot(juice, model.matrix(fruit.lmb4)[,2], ylab="Column 2 of model matrix",
        xlab="Apparent Juice Content (%)", type="n")
    ord <- order(juice)
    lines(juice[ord], model.matrix(fruit.lmb4)[ord,2])
    ## For panels B, C and D, take columns 3, 4 and 5 respectively of
    ## the model matrix.

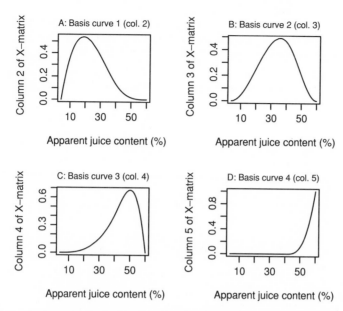

Figure 7.8: Spline basis curves for the B-spline (one knot) fitted in Figure 7.6A. Ordinates of the first basis curve give the entries in the second column of the model matrix, while the three remaining basis curves give entries for columns 3, 4 and 5 respectively.

7.4.2 Other smoothing methods

Lowess curves are a popular alternative to spline curves. Figure 2.6 showed a curve that was fitted to the `fruitohms` data using the lowess smoothing method. The `loess()` function, implemented in the *modreg* package, is an alternative to `lowess()` that is able to handle multi-dimensional smoothing.

The lowess smoothing routine uses a method that is known as *locally weighted regression*. The R implementation uses a robust fitting method, so that the curve is relatively insensitive to large residuals. Special steps are taken to avoid distortions due to end effects. As implemented in R, `lowess()` is not available for use when there are multiple explanatory variables, and there is no mechanism for calculating pointwise confidence bounds.

Kernel smoothing methods further widen the range of possibilities. For example, see the documentation for the function `locpoly()` in the *KernSmooth* package.

There is a useful brief discussion of smoothing methods in Venables and Ripley (2002) and a fuller discussion of kernel smoothing, splines and lowess in Fan and Gijbels (1996). See also Hall (2001).

7.4.3 Generalized additive models

Generalized additive models allow smoothing terms that are more general than splines with fixed knots. Two of the possibilities are:

- Spline terms with optimally located knot points, i.e., the fitting procedure chooses the location of the knots.
- Loess smoothing terms, i.e., equivalent to using the `loess()` function.

Table 7.11: *Average dewpoint* (dewpoint), *for available combinations of monthly averages of maximum temperature* (maxtemp) *and minimum temperature* (mintemp). *The table shows a selection of the data.*

	maxtemp	mintemp	dewpoint			maxtemp	mintemp	dewpoint
1	18	8	7	67	38	26	20
2	18	10	10	68	40	18	5
3	20	6	5	69	40	20	8
4	20	8	7	70	40	22	11
5	20	10	9	71	40	24	14
...	...	...	...	72	40	26	17

7.5 Smoothing Terms in Multiple Linear Models

We now move to models with multiple terms. We will demonstrate the use of spline terms with fixed knots. The use of fixed knots is often adequate for curves that do not bend too sharply. Regression splines may be unable to capture sharp turns in the response curve.

Table 7.11 has data on monthly averages of minimum temperature, maximum temperature and dewpoint. For the background to these data, see Linacre (1992) and Linacre and Geerts (1997). The dewpoint is the maximum temperature at which humidity reaches 100%. Monthly data were obtained for a large number of sites worldwide. For each combination of minimum and maximum temperature the average dewpoint was then determined. Figure 7.9 shows a representation of these data that is given by an additive model with two spline smoothing terms.[11] We can write the model as

$$y = \mu + f_1(x_1) + f_2(x_2) + \varepsilon,$$

where $y = $ dewpoint, $x_1 = $ maxtemp and $x_2 = $ mintemp

Here μ is estimated by the mean of y, so that the estimates of $f_1(x_1)$ and $f_2(x_2)$ give differences from this overall mean. In Figure 7.9, both $f_1(x_1)$ and $f_2(x_2)$ are modeled by spline functions with five degrees of freedom. The left panel is a plot of the estimate of $f_1(x_1)$ against x_1, while the right panel plots $f_2(x_2)$ against x_2.

---

[11]
```
library(splines)
    data(dewpoint)        # DAAG package
    attach(dewpoint)
    ds.lm <- lm(dewpoint ~ bs(maxtemp,5) + bs(mintemp,5), data=dewpoint)
    ds.fit <- predict(ds.lm, type="terms", se=TRUE)
    par(mfrow = c(1,2))
    plot(maxtemp, ds.fit$fit[,1], xlab="Maximum temperature",
         ylab="Change from dewpoint mean", type="n")
    lines(maxtemp, ds.fit$fit[,1])
    lines(maxtemp, ds.fit$fit[,1]-2*ds.fit$se[,1], lty=2)
    lines(maxtemp, ds.fit$fit[,1]+2*ds.fit$se[,1], lty=2)
    plot(mintemp, ds.fit$fit[,2], xlab="Minimum temperature",
         ylab="Change from dewpoint mean", type="n")
    ord <- order(mintemp)
    lines(mintemp[ord], ds.fit$fit[ord,2])
    lines(mintemp[ord], ds.fit$fit[ord,2]-2*ds.fit$se[ord,2], lty=2)
    lines(mintemp[ord], ds.fit$fit[ord,2]+2*ds.fit$se[ord,2], lty=2)
    par(mfrow = c(1,1))
    detach(dewpoint)
```

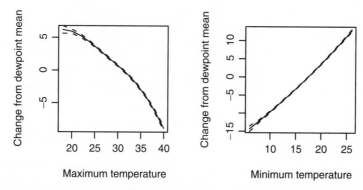

Figure 7.9: Representation of average dewpoint (dewpoint) as the sum of an effect due to maximum temperature (maxtemp), and an effect due to minimum temperature (mintemp). (Data are from Table 7.11.) The dashed lines are 95% pointwise confidence bounds.

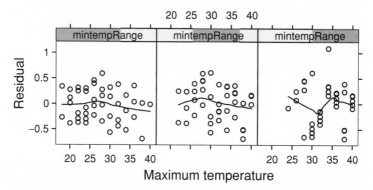

Figure 7.10: Plot of residuals against maximum temperature, for three different ranges of values of minimum temperature. Panel strips are shaded to show the range of values of the conditioning variable.

There is no obvious reason why the additive model should work so well. In general, we might expect an interaction term, i.e., we might expect that $f_1(x_1)$ would be different for different values of x_2, or equivalently that $f_2(x_2)$ would be different for different values of x_1. Even where the effects are not additive, an additive model is often a good starting approximation. We can fit the additive model, and then check whether there are departures from it that require examination of the dependence of y upon x_1 and x_2 jointly.

One check is to take, e.g., for x_2 = mintemp, three perhaps overlapping ranges of values, which we might call "low", "medium" and "high". For this purpose we are then treating mintemp as a *conditioning* variable. We then plot residuals against x_1 = maxtemp for each range of values of x_2, as in Figure 7.10. If there is a pattern in these plots that changes with the range of the conditioning variable, this is an indication that there are non-additive effects that require attention.[12]

[12] library(lattice)
```
mintempRange <- equal.count(dewpoint$mintemp, number=3)
xyplot(residuals(ds.lm) ~ maxtemp | mintempRange,
       data=dewpoint, aspect=1, layout=c(3,1), type=c("p","smooth"),
       xlab="Maximum temperature", ylab="Residual")
```

7.6 Further Reading

There is a brief review of the methodologies we have described, and of extensions to them, in Venables and Ripley (2002). Hastie et al. (2001) discuss substantial extensions of the approaches that we have described. See also Hastie and Tibshirani (1990), and references on the help pages for functions in the *modreg* and *mgcv* packages. Maindonald (1984) has an elementary introduction to simple forms of spline curve. Eubank (1999) gives a comprehensive and readable introduction to the use of splines in nonparametric regression.

References for further reading

Eubank, R.L. 1999. *Nonparametric Regression and Spline Smoothing,* 2nd edn. Marcel Dekker.

Hastie, T.J. and Tibshirani, R.J. 1990. *Generalized Additive Models.* Chapman and Hall.

Hastie, T., Tibshirani, R. and Friedman, J. 2001. *The Elements of Statistical Learning. Data Mining, Inference and Prediction.* Springer-Verlag.

Maindonald, J.H. 1984. *Statistical Computation.* Wiley.

Venables, W.N. and Ripley, B.D. 2002. *Modern Applied Statistics with* S-PLUS, 4th edn. Springer-Verlag.

7.7 Exercises

1. Re-analyze the sugar weight data of Subsection 7.1.1 using `log(weight)` in place of `weight`.

2. Use the method of Section 7.3 to determine, formally, whether there should be different regression lines for the two data frames `elastic1` and `elastic2` from Exercise 1 in Section 5.11.

3. The data frame `toycars` consists of 27 observations on the distance (in meters) traveled by one of three different toy cars on a smooth surface, starting from rest at the top of a 16-inch-long ramp tilted at varying angles (measured in degrees). Because of differing frictional effects for the three different cars, we seek three regression lines relating distance traveled to angle.

    (a) As a first try, fit the model in which the three lines have the same slope but have different intercepts.

    (b) Note the value of R^2 from the summary table. Examine the diagnostic plots carefully. Is there an influential outlier? How should it be treated?

    (c) The physics of the problem actually suggests that the three lines should have the same intercept (very close to 0, in fact), and possibly differing slopes, where the slopes are inversely related to the coefficient of dynamic friction for each car. Fit the model, and note that the value of R^2 is slightly lower than that for the previously fitted model. Examine the diagnostic plots. What has happened to the influential outlier? In fact, we have exhibited an example where taking R^2 too seriously could be somewhat hazardous; in this case, a more carefully thought out model can accommodate all of the data satisfactorily. Maximizing R^2 does not necessarily give the best model!

4. The data frame cuckoos holds data on the lengths and breadths of eggs of cuckoos, found in the nests of six different species of host birds. Fit models for the regression of length on breadth that have:

   A: a single line for all six species.
   B: different parallel lines for the different host species.
   C: separate lines for the separate host species.

   Use the anova() function to print out the sequential analysis of variance table. Which of the three models is preferred? Print out the diagnostic plots for this model. Do they show anything worthy of note? Examine the output coefficients from this model carefully, and decide whether the results seem grouped by host species. How might the results be summarized for reporting purposes?

5. Fit the three models A, B and C from the previous exercise, but now using the robust regression function rlm() from the *MASS* package. Do the diagnostic plots look any different from those from the output from lm()? Is there any substantial change in the regression coefficients?

6. Apply polynomial regression to the seismic timing data in the data frame geophones. Specifically, check the fits of linear, quadratic, cubic and quartic (degree = 4) polynomial estimates of the expected thickness as a function of distance. What do you observe about the fitted quartic curve? Do any of the fitted curves capture the curvature of the data in the region where distance is large?

7. Apply spline regression to the geophones data frame. Specifically, regress thickness against distance, and check the fits of 4-, 5- and 6-degree-of-freedom cases. Which case gives the best fit to the data? How does this fitted curve compare with the polynomial curves obtained in the previous exercise? Calculate pointwise confidence bounds for the 5-degree-of-freedom case.

8. Apply lowess() to the geophones data as in the previous two exercises. You will need to experiment with the f argument, since the default value oversmooths this data. Small values of f (less than .2) give a very rough plot, while larger values give a smoother plot. A value of about .25 seems a good compromise.

9. Check the diagnostic plots for the results of exercise 7 for the 5-degree-of-freedom case. Are there any influential outliers?

10. Continuing to refer to exercise 7, obtain plots of the spline basis curves for the 5-degree-of-freedom case. That is, plot the relevant column of the model matrix against *y*.

11. The ozone data frame holds data, for nine months only, on ozone levels at the Halley Bay station between 1956 and 2000. (See Christie (2000) and Shanklin (2001) for the scientific background. Up to date data are available from the web page http://www.antarctica.ac.uk/met/jds/ozone/.) Replace zeros by missing values. Determine, for each month, the number of missing values. Plot the October levels against Year, and fit a smooth curve. At what point does there seem to be clear evidence of a decline? Plot the data for other months also. Do other months show a similar pattern of decline?

12. The wages1833 data frame holds data on the wages of Lancashire cotton factory workers in 1833. Plot male wages against age and fit a smooth curve. Repeat using the numbers of male workers as weights. Do the two curves seem noticeably different? Repeat the exercise for female workers. [See Boot (1995) for background information on these data.]

13. *Compare the two results

```
seedrates.lm <- lm(grain ~ rate + I(rate^2),
                    data=seedrates)
seedrates.pol <- lm(grain ~ poly(rate,2),
                    data=seedrates)
```

Check that the fitted values and residuals from the two calculations are the same, and that the t-statistic and p-value are the same for the final coefficient, i.e., the same for the coefficient labeled `poly(rate, 2)2` in the polynomial regression as for the coefficient labeled `I(rate^2)` in the regression on `rate` and `rate^2`.

Regress the second column of `model.matrix(seedrates.pol)` on `rate` and `I(rate^2)`, and similarly for the third column of `model.matrix(seedrates.pol)`. Hence express the first and second orthogonal polynomial terms as functions of `rate` and `rate^2`.

8

Logistic Regression and Other Generalized Linear Models

The straight line regression model we considered in Chapter 5 had the form

$$y = \alpha + \beta x + \varepsilon$$

where, if we were especially careful, we would add subscript is to y, x, and ε. We also generalized the model to allow for more than one x-variable. In this chapter, we will resume with models where there is just one x, in order to simplify the discussion. Later, we will add more predictor variables, as required.

We noted that another form of the regression model is

$$\mathrm{E}[y] = \alpha + \beta x$$

where E is *expectation*. It is this way of writing the equation that is the point of departure for our discussion of generalized linear models. These were first introduced in the 1970s. They have been a powerful addition to the armory of tools that are available to the statistical analyst. We will limit attention to a few special cases.

8.1 Generalized Linear Models

Generalized linear models (GLMs) differ in two ways from the models used in earlier chapters. They allow for a more general form of expression for the expectation, and they allow various types of non-normal error terms. Logistic regression models are perhaps the most widely used GLM.

8.1.1 Transformation of the expected value on the left

GLMs allow a transformation $f()$ to the left hand side of the regression equation, i.e., to E[y]. The result specifies a linear relation with x. In other words,

$$f(\mathrm{E}[y]) = \alpha + \beta x$$

where $f()$ is a function, which is usually called the *link* function. In the fitted model, we call $\alpha + \beta x$ the linear predictor, while E[y] is the expected value of the response. The function $f()$ transforms from the scale of the response to the scale of the linear predictor.

Some common examples of link functions are $f(x) = x$, $f(x) = 1/x$, $f(x) = \log(x)$, and $f(x) = \log(x/(1 - x))$. The last is referred to as the logit link and is the link function for logistic regression. Note that these functions are all monotonic, i.e., they increase or (in the case of $1/x$) decrease with increasing values of x.

8.1.2 Noise terms need not be normal

We may write

$$y = E[y] + \varepsilon.$$

Here the elements of y may have a distribution different from the normal distribution. Common distributions are the binomial where y is the number responding out of a given total n, and the Poisson where y is a count.

Even more common may be models where the random component differs from the binomial or Poisson by having a variance that is larger than the mean. The analysis proceeds as though the distribution were binomial or Poisson, but the theoretical binomial or Poisson variance estimates are replaced by a variance that is estimated from the data. Such models are called, respectively, quasi-binomial models and quasi-Poisson models.

8.1.3 Log odds in contingency tables

With proportions that range from less than 0.1 to greater than 0.9, it is not reasonable to expect that the expected proportion will be a linear function of x. A transformation (link function) such as the logit is required. A good way to think about logit models is that they work on a log(odds) scale. If p is a probability (e.g. that horse A will win the race), then the corresponding odds are $p/(1 - p)$, and

$$\log(\text{odds}) = \log(p/(1 - p)) = \log(p) - \log(1 - p).$$

Logistic regression makes it possible to apply a formal modeling approach to analyses where there is a Bernoulli outcome. It works with odds, or more precisely, with log(odds). The logit function $y = \log(p) - \log(1 - p)$ is applied to p. Figure 8.1 shows the logit transformation.[1]

Logistic regression provides a framework for analyzing contingency table data. Let us now recall the fictitious admissions data presented in Table 3.3. The observed proportion of students (male and female) admitted into Engineering is $40/80 = .5$. For Sociology, the admission proportion is $15/60 = .25$. Thus, we have

$$\log(\text{odds}) = \log(.5/.5) = 0 \text{ for Engineering,}$$

$$\log(\text{odds}) = \log(.75/.25) = 1.0986 \text{ for Sociology.}$$

What determines whether a student will be admitted in Engineering? What determines whether a student will be admitted in Sociology? Is age a factor? Logistic regression allows us to model log(odds of admission) as a function of age, or as a function of any other covariate that we may wish to investigate.

---

[1] ## Simplified version of the figure
```
p <- (1:999)/1000
gitp <- log(p/(1 - p))
plot(p, gitp, xlab = "Proportion", ylab = "", type = "l", pch = 1)
```

Table 8.1: *Terminology used for logistic regression (or more generally for generalized linear models), compared with multiple regression terminology.*

Regression	Logistic regression
Degrees of freedom	Degrees of freedom
Sum of squares (SS)	Deviance (D)
Mean sum of squares (divide by degrees of freedom)	Mean deviance (divide by degrees of freedom)
Fit models by minimizing the residual sum of squares.	Fit models by minimizing the deviance.

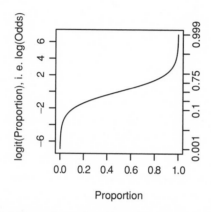

Figure 8.1: The logit or log(odds) transformation. Shown here is a plot of log(odds) versus proportion. Notice how the range is stretched out at both ends.

For such data, we may write

$$\log(\text{odds}) = \text{constant} + \text{effect due to faculty} + \text{effect due to gender}.$$

This now has the form of a two-way analysis of variance model.

8.1.4 Logistic regression with a continuous explanatory variable

The actual fitting of the logistic model is accomplished by minimizing *deviances*. A deviance has a role very similar to a sum of squares in regression (in fact, if the data are normally distributed, the two quantities are equivalent). This aspect of the analogy between regression and logistic regression is furnished by Table 8.1.

Consider the following example. Thirty patients were given an anesthetic agent that was maintained at a pre-determined (alveolar) concentration for 15 minutes before making an incision. It was then noted whether the patient moved, i.e., jerked or twisted. The interest is in estimating how the probability of jerking or twisting varies with increasing concentration of the anesthetic agent.

Table 8.2: *Patients moving (0) and not moving (1), for each of six different alveolar concentrations.*

	Alveolar concentration (conc)					
nomove	0.8	1	1.2	1.4	1.6	2.5
0	6	4	2	2	0	0
1	1	1	4	4	4	2
Total	7	5	6	6	4	2

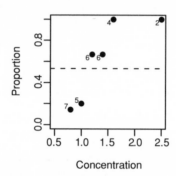

Figure 8.2: Plot, versus concentration, of proportion of patients not moving. The horizontal line is the proportion of moves over the data as a whole.

We take the response as `nomove`, because the proportion then increases with increasing concentration. There is a small number of concentrations; in Table 8.2 we tabulate, for each concentration, the respective numbers with `nomove` equal to 0 (i.e., movement) and `nomove` equal to 1 (i.e., no movement). Figure 8.2 plots the proportions.[2]

We can fit the logit model either directly to the 0/1 data, or to the proportions that we can calculate from the data in Table 8.2. The analysis assumes that individuals respond independently with a probability, estimated from the data, that on a logistic scale is a linear function of the concentration. For any fixed concentration, the assumption is that we have Bernoulli trials, i.e., that individual responses are drawn at random from the same population. Figure 8.3 shows the fitted line.

[2]
```
data(anesthetic)
z <- table(anesthetic$nomove, anesthetic$conc)
tot <- apply(z, 2, sum)          # totals at each concentration
prop <- z[2, ]/(tot)             # proportions at each concentration
oprop <- sum(z[2, ])/sum(tot)    # expected proportion not moving if
                                 # concentration had no effect
conc <- as.numeric(dimnames(z)[[2]])
plot(conc, prop, xlab = "Concentration", ylab = "Proportion", xlim = c(.5,2.5),
     ylim = c(0, 1), pch = 16)
chw <- par()$cxy[1]
text(conc - 0.75 * chw, prop, paste(tot), adj = 1)
abline(h = oprop, lty = 2)
```

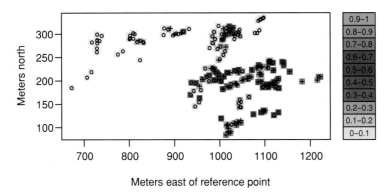

Figure 8.8: Fitted values (model predictions) of the probability of finding a frog are shown on a color density scale. Sites are labeled "o" or "+" according as frogs were not found or were found.

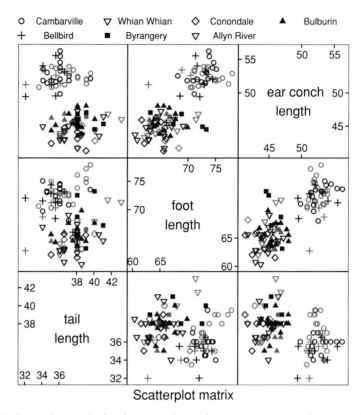

Figure 11.1: Scatterplot matrix for three morphometric measurements on the mountain brushtail possum. Females are in red; males in blue.

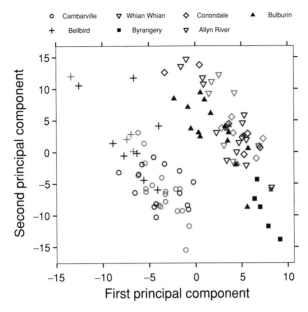

Figure 11.2: Second principal component versus first principal component, for variables in columns 6–14 of the possum data frame. Females are in red; males in blue.

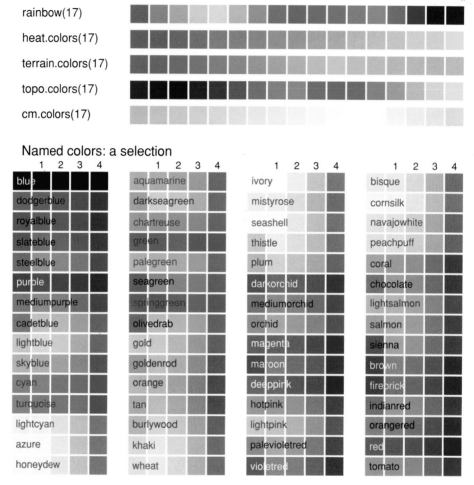

Figure 12.2: A selection of color palettes and named colors that are available in R. For the named colors, we have restricted attention to an incomplete selection of those that come in one of five shades. Thus, in addition to "red", there are "red1", "red2", "red3", and "red4".

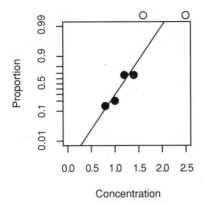

Figure 8.3: Plot, versus concentration, of log(odds) = logit(proportion) of patients not moving. The line gives the estimate of the proportion of moves, based on the fitted logit model.

If individuals respond independently, with the same probability, then we have Bernoulli trials. Justification for assuming the same probability will arise from the way in which individuals are sampled. While individuals will certainly be different in their response, the notion is that, each time a new individual is taken, they are drawn at random from some larger population. Data are in the data set anesthetic. Here is the R output:[3]

```
> anova(anes.logit)

Analysis of Deviance Table

Model: binomial, link: logit

Response: nomove

Terms added sequentially (first to last)


      Df Deviance Resid. Df Resid. Dev
NULL                  29         41.5
conc   1     13.7     28         27.8
> summary(anes.logit)

Call:
glm(formula = nomove ~ conc, family = binomial (link = logit),
    data = anesthetic)

Deviance Residuals:
    Min        1Q    Median       3Q       Max
-1.7666   -0.7441    0.0341   0.6867    2.0689
```

---

[3] anes.logit <- glm(nomove ~ conc, family = binomial(link = logit),
                  data = anesthetic)
  anova(anes.logit)
  summary(anes.logit)

```
Coefficients:
              Estimate Std. Error z value Pr(>|z|)
(Intercept)     -6.47        2.40   -2.69   0.0071
conc             5.57        2.03    2.74   0.0061

(Dispersion parameter for binomial family taken to be 1)

    Null deviance: 41.455  on 29  degrees of freedom
Residual deviance: 27.754  on 28  degrees of freedom
AIC: 31.75

Number of Fisher Scoring iterations: 4
```

Figure 8.3 contains a graphical summary.[4]

Note the two open circles that are located outside the plotting region; these correspond to cases where the observed proportion is 1, giving an infinite logit. With such small sample sizes, it is impossible to do a convincing check of the adequacy of the model.

8.2 Logistic Multiple Regression

We now consider data on the distribution of the Southern Corroboree frog, that occurs in the Snowy Mountains area of New South Wales, Australia (data are from Hunter, 2000). In all, 212 sites were surveyed. In Figure 8.4 filled circles denote sites where the frogs were found; open circles denote the absence of frogs.[5]

The variables in the data set are pres.abs (were frogs found?), easting (reference point), northing (reference point), altitude (in meters), distance (distance in meters to nearest extant population), NoOfPools (number of potential breeding pools), NoOfSites (number of potential breeding sites within a 2 km radius), avrain (mean rainfall for Spring period), meanmin (mean minimum Spring temperature), and meanmax (mean maximum Spring temperature).

As with multiple linear regression, a desirable first step is to check relationships among explanatory variables. Where possible, we will transform so that those relationships are made linear, as far as we can tell from the scatterplot matrix. If we can so transform the variables, this gives us, as we noted earlier in connection with classical multiple

---

[4]
```
pval <- c(0.01, (1:9)/10, 0.99)
gval <- log(pval/(1 - pval))
plot(conc, log(prop/(1 - prop)), xlab = "Concentration", ylab =
    "Proportion", pch = 16, axes=FALSE, ylim = range(gval), xlim = c(0, 2.5))
axis(1); box()
xover <- conc[prop == 1]
yover <- rep(par()$usr[4] + 0.5 * par()$cxy[2], 2)
points(xover, yover, pch = 1, xpd = TRUE)
  # use xpd = TRUE to allow plotting of points outside the plot region
abline(-6.47, 5.57)
axis(2, at = gval, labels = paste(pval))
```
[5]
```
data(frogs)          # DAAG package
plot(northing ~ easting, data=frogs, pch=c(1,16)[frogs$pres.abs+1],
    xlab="Meters east of reference point", ylab="Meters north")
```

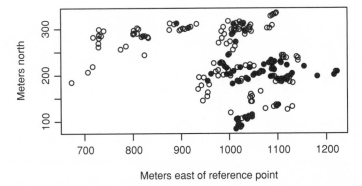

Figure 8.4: Location of sites, relative to reference point, that were examined for frogs. The sites are all in the Snowy Mountains region of New South Wales, Australia. The filled points are for sites where frogs were found.

regression, access to a theory that can be highly helpful in guiding the process of regression modeling.

We wish to explain frog distribution as a function of the other variables. Because we are working within a very restricted geographic area, we do not expect that the distribution will change as a function of latitude and longitude *per se*, so that these will not be used as explanatory variables. Figure 8.5 shows the scatterplot matrix for the remaining explanatory variables.[6]

Notice that the relationships between `altitude`, `avrain`, `meanmin` and `meanmax` are close to linear. For the remaining variables, the distributions are severely skewed, to the extent that it is difficult to tell whether the relationship is linear. For a distance measure, the favored transformation is, depending on the context, the logarithmic transformation. For counts, the first transformation to try is a square root transformation. Figure 8.6 shows density plots for these variables, untransformed, after a square root, and after a logarithmic transformation.[7] For `NoOfSites`, a large number of values of `NoOfSites` are zero. Hence we plot log(`NoOfSites`+1).

For `distance` and `NoOfPools`, we prefer a logarithmic transformation, while `NoOfSites` is best not transformed at all. We now (using Figure 8.7) investigate the scatterplot matrix for the variables that are so transformed.[8]

The smoothing curves for log(`distance`) and log(`NoOfPools`) are perhaps not too seriously non-linear. There appears to be some nonlinearity associated with `NoOfSites`,

---

[6] `pairs(frogs[,4:10], oma=c(2,2,2,2), cex=0.5)`
[7] `## Here is code for the top row of plots`
```
par(mfrow=c(3,3))
for(nam in c("distance","NoOfPools","NoOfSites")){
    y <- frogs[, nam]
    plot(density(y), main="", xlab=nam)
    }
# The other rows can be obtained by replacing y by sqrt(y) or log(y)
# (or log(y+1) for NoOfSites)
par(mfrow=c(1,1))
```
[8] `attach(frogs)`
```
pairs(cbind(altitude, log(distance), log(NoOfPools), NoOfSites),
    panel=panel.smooth, labels=c("altitude","log(distance)",
    "log(NoOfPools)","NoOfSites"))
detach(frogs)
```

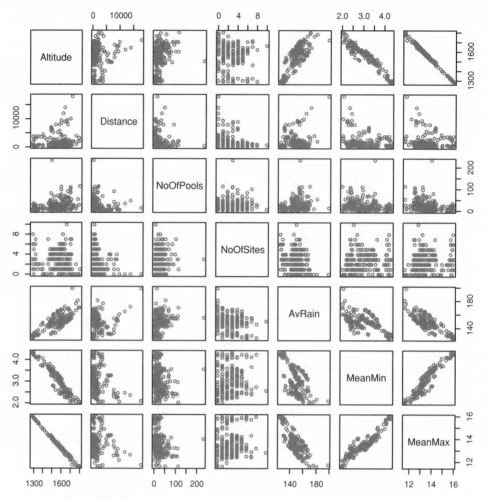

Figure 8.5: Scatterplot matrix for predictor variables in the `frogs` data set.

so that we ought perhaps to consider including `NoOfSites2` (the square of `NoOfSites`) as well as `NoOfSites` in the regression equation. We try the simpler model first:[9]

```
> summary(frogs.glm0)

Call:
glm(formula = pres.abs ~ altitude + log(distance)
  + log(NoOfPools) + NoOfSites + avrain + meanmin
  + meanmax, family = binomial, data = frogs)

Deviance Residuals:
   Min      1Q   Median      3Q      Max
-1.979  -0.719  -0.278   0.796   2.566
```

---

[9] frogs.glm0 <- glm(formula = pres.abs ~ altitude + log(distance) +
                  log(NoOfPools) + NoOfSites + avrain + meanmin + meanmax,
                  family = binomial, data = frogs)
  summary(frogs.glm0)

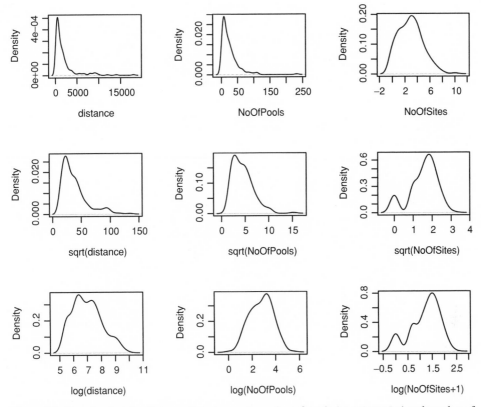

Figure 8.6: Density plots for distance (`distance`), number of pools (`NoOfPools`) and number of sites (`NoOfSites`), before and after transformation.

```
Coefficients:
                Estimate Std. Error z value Pr(>|z|)
(Intercept)     4.10e+01   1.32e+02    0.31  0.75621
altitude       -6.67e-03   3.84e-02   -0.17  0.86225
log(distance)  -7.59e-01   2.54e-01   -2.98  0.00284
log(NoOfPools)  5.73e-01   2.16e-01    2.65  0.00796
NoOfSites      -8.88e-04   1.07e-01   -0.01  0.99339
avrain         -6.81e-03   5.98e-02   -0.11  0.90936
meanmin         5.30e+00   1.53e+00    3.46  0.00054
meanmax        -3.17e+00   4.82e+00   -0.66  0.50985

(Dispersion parameter for binomial family taken to be 1)

    Null deviance: 279.99  on 211  degrees of freedom
Residual deviance: 197.62  on 204  degrees of freedom
AIC: 213.6

Number of Fisher Scoring iterations: 4
```

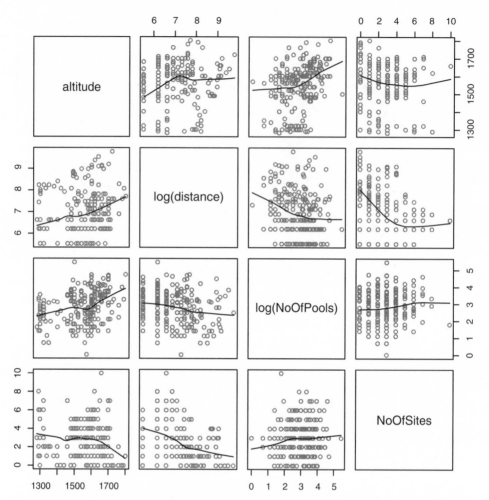

Figure 8.7: Scatterplot matrix for altitude, transformed versions of `distance` and `NoOfPools`, and `NoOfSites`. We are particularly interested in whether the relationship with `altitude` is plausibly linear. If the relationship with `altitude` is linear, then it will be close to linear also with the temperature and rain variables.

Because we have taken no account of spatial clustering, and because there are questions about the adequacy of the asymptotic approximation that is required for their calculation, the *p*-values should be used with caution. We do however note the clear separation of the predictor variables into two groups – in one group the *p*-values are 0.5 or more, while for those in the other group the *p*-values are all very small.

We note that `NoOfSites` has a coefficient that is in the large *p*-value category. Replacing it with `NoOfSites2` makes little difference. We may, without loss of predictive power, be able to omit all variables for which the *p*-values are in the "large" category, i.e., we consider omission of `altitude`, `NoOfSites`, `avrain` and probably also meanmax. Because the first three of these variables are the clearest candidates for omission, we try omitting them first.[10]

```
> summary(frogs.glm)

Call:
glm(formula = pres.abs ~ log(distance) + log(NoOfPools)
    + meanmin + meanmax, family = binomial, data = frogs)

Deviance Residuals:
   Min      1Q   Median      3Q      Max
-1.975  -0.722   -0.278   0.797    2.574

Coefficients:
                Estimate Std. Error z value Pr(>|z|)
(Intercept)       18.526      5.256    3.53  0.00042
log(distance)     -0.755      0.226   -3.34  0.00082
log(NoOfPools)     0.571      0.215    2.66  0.00787
meanmin            5.379      1.190    4.52  6.2e-06
meanmax           -2.382      0.622   -3.83  0.00013

(Dispersion parameter for binomial family taken to be 1)

    Null deviance: 279.99  on 211  degrees of freedom
Residual deviance: 197.66  on 207  degrees of freedom
AIC: 207.7

Number of Fisher Scoring iterations: 4
```

The residual deviance is almost unchanged, as we might have expected. It is interesting that meanmax now joins the group of variables for which *p*-values are small. The coefficients are similar in magnitude to those in the model that used all explanatory variables. Conditioning on the variables that have been left out of the regression equation makes little difference, except in the case of meanmax, to the way that we should interpret the effects of variables that have been included.

Figure 8.8 allows a comparison of model predictions with observed occurrences of frogs, with respect to geographical co-ordinates. Because our modeling has taken no account of spatial correlation, examination of such a plot is more than ordinarily desirable. The main interest of this plot is in points where although frogs were predicted with high probability, none were found; and in points where frogs were not expected, but were found. A further question is whether there are clusters of sites where model predictions are obviously wrong. There is little overt evidence of such clusters.

---

[10] frogs.glm <- glm(formula = pres.abs ~ log(distance) + log(NoOfPools) +
                meanmin + meanmax, family = binomial, data = frogs)
    summary(frogs.glm)

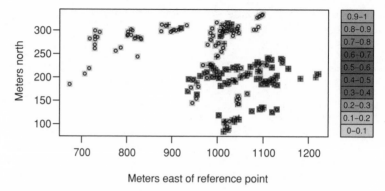

Figure 8.8: Fitted values (model predictions) of the probability of finding a frog are shown on a color density scale. Sites are labeled "o" or "+" according as frogs were not found or were found. For color version, see plate section between pages 200 and 201.

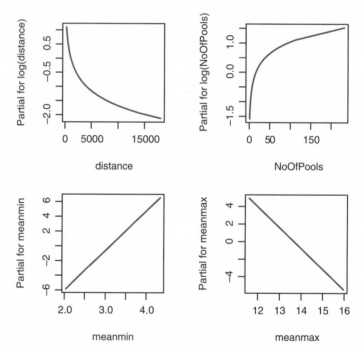

Figure 8.9: Plots showing the contributions of the explanatory variables to the fitted values, on the scale of the linear predictor.

8.2.1 A plot of contributions of explanatory variables

The equation that we have just fitted adds together the effects of `log(distance)`, `log(NoOfPools)`, `meanmin`, and `meanmax`. Figure 8.9 shows the change due to each of these variables in turn, when other terms are held at their means.[11] Note first that the *y*-scale on these plots is on the scale of the linear predictor. As we are using the logit

---

[11] `par(mfrow=c(2,2))`
   `termplot(frogs.glm, data=frogs)`

link, zero corresponds to a probability of 0.5. Probabilities for other values are −4: 0.02, −2: 0.12, 2: 0.88, and 4: 0.98. The fitted line for meanmin indicates that it is making a huge contribution, with the probability of finding frogs increasing substantially as meanmin goes from low to high values. For meanmax, the effect goes in the other direction, i.e., frogs are more likely to be present when meanmax is small. The contributions from log(distance) and log(NoOfPools) are, relatively, much smaller.[12]

8.2.2 Cross-validation estimates of predictive accuracy

Our function cv.binary() calculates cross-validation estimates of predictive accuracy for these models:[13]

```
Fold:  6  4  2  3  10  9  1  5  7  8
Resubstitution estimate of accuracy = 0.797
Cross-validation estimate of accuracy = 0.651
>
Fold:  9  8  2  7  5  1  4  10  3  6
Resubstitution estimate of accuracy = 0.778
Cross-validation estimate of accuracy = 0.703
```

The resubstitution estimate assesses the accuracy of prediction for the data used in deriving the model. Notice that this goes down slightly when we reduce the number of predictor variables. The cross-validation estimate estimates the accuracy that might be expected for a new sample. Notice that this is in both cases much less than the internal estimate.

The cross-validation estimates are highly variable. Four separate runs for the model that included all explanatory variables gave accuracies of 62%, 65%, 68% and 59%. Four separate runs for the reduced model gave 65%, 68%, 74% and 58%. A more careful analysis would try to model this dependence. For an accurate comparison between the two models, we should make the comparison using the same random assignments of observations to folds. We do this by making a random assignment of observations to folds before we call the function, first with frogs.glm0 as the argument and then with frogs.glm as the argument:[14]

```
All: 0.665    Reduced: 0.726
All: 0.693    Reduced: 0.684
All: 0.627    Reduced: 0.708
All: 0.594    Reduced: 0.642
```

---

[12] ## Readers may also wish to try
```
  termplot(frogs.glm, data=frogs, partial.resid=TRUE)
  # Interpretation of such a graph is non-trivial.
```
[13] cv.binary(frogs.glm0)   # All explanatory variables
```
   cv.binary(frogs.glm)     # Reduced set of explanatory variables
```
[14] ## The cross-validated estimate of accuracy is stored in the list
```
  ## element acc.cv, in the output from the function cv.binary(), thus:
  for (j in 1:4){
      randsam <- sample(1:10, 212, replace=TRUE)
      all.acc <- cv.binary(frogs.glm0, rand=randsam)$acc.cv
      reduced.acc <- cv.binary(frogs.glm, rand=randsam)$acc.cv
      cat("\nAll:", round(all.acc,3), "  Reduced:", round(reduced.acc,3))
      }
```

We compare .726 with .665, .684 with .693, .708 with .627, and .642 with .594. We need to repeat the comparison a number of further times in order to be sure that the comparison favors the reduced model.

8.3 Logistic Models for Categorical Data – an Example

The array UCBAdmissions in the *base* package is a $2 \times 2 \times 6$ array: Admit (Admitted/Rejected) $\times$ Gender (Male/Female) $\times$ Dept (A, B, C, D, E, F).

```
data(UCBAdmissions)  # base package
UCB<- data.frame(admit=as.vector(UCBAdmissions[1, , ]),
                 reject=as.vector(UCBAdmissions[2, , ]),
                 Gender=rep(c("Male","Female"), 6),
                 Dept=rep(LETTERS[1:6], rep(2,6)))
UCB$Gender <- relevel(UCB$Gender, ref="Male")
UCB$total <- UCB$admit + UCB$reject
UCB$p <- UCB$admit/UCB$total
```

We use a loglinear model to model the probability of admission of applicants. It is important, for our purposes, to fit Dept before fitting Gender.

```
UCB.glm <- glm(p ~ Dept*Gender, family=binomial,
               data=UCB, weights=total)
anova(UCB.glm, test="Chisq")
```

The output is

```
Analysis of Deviance Table

Model: binomial, link: logit

Response: p

Terms added sequentially (first to last)
```

	Df	Deviance	Resid. Df	Resid. Dev	P(>|Chi|)
NULL			11	877	
Dept	5	855	6	22	1.2e-182
Gender	1	2	5	20	2.2e-01
Dept:Gender	5	20	0	-2.6e-13	1.1e-03

```
>
```

Once there has been allowance for overall departmental differences in admission rate, there is no detectable main effect of Gender. However, the significant interaction term suggests that there are department-specific gender biases, which average out to reduce the main effect of Gender to close to zero.

We now examine the individual coefficients in the model:

```
> summary(UCB.glm)$coef
                   Estimate Std. Error z value Pr(>|z|)
(Intercept)          0.4921     0.0717   6.859 6.94e-12
DeptB                0.0416     0.1132   0.368 7.13e-01
DeptC               -1.0276     0.1355  -7.584 3.34e-14
DeptD               -1.1961     0.1264  -9.462 3.02e-21
DeptE               -1.4491     0.1768  -8.196 2.49e-16
DeptF               -3.2619     0.2312 -14.110 3.31e-45
GenderFemale         1.0521     0.2627   4.005 6.20e-05
DeptB:GenderFemale  -0.8321     0.5104  -1.630 1.03e-01
DeptC:GenderFemale  -1.1770     0.2995  -3.929 8.52e-05
DeptD:GenderFemale  -0.9701     0.3026  -3.206 1.35e-03
DeptE:GenderFemale  -1.2523     0.3303  -3.791 1.50e-04
DeptF:GenderFemale  -0.8632     0.4026  -2.144 3.20e-02
```

The first six coefficients relate to overall admission rates, for males, in the six departments. The strongly significant positive coefficient for `GenderFemale` indicates that log(odds) is increased by 1.05, in department A, for females relative to males. In departments C, D, E and F, the log(odds) is reduced for females, relative to males.

8.4 Poisson and Quasi-Poisson Regression

8.4.1 Data on aberrant crypt foci

We begin with a simple example. The data frame `ACF6` consists of two columns: `count` and `endtime`. The first column contains the counts of simple aberrant crypt foci (ACFs) – these are aberrant aggregations of tube-like structures – in the rectal end of 22 rat colons after administration of a dose of the carcinogen azoxymethane. Each rat was sacrificed after 6, 12 or 18 weeks (recorded in the column `endtime`). For further background information, see McLellan et al. (1991).

The argument is there are a large number of sites where ACFs might occur, in each instance with the same low probability. Because "site" does not have a precise definition, the total number of sites is unknown, but it is clearly large. If we can assume independence between sites, then we might expect a Poisson model, for the total number of ACF sites. The output from fitting the model will include information that indicates whether a Poisson model was, after all, satisfactory. If a Poisson model does not seem satisfactory, then a *quasi-Poisson* model may be a reasonable alternative. A model of a quasi-Poisson type is consistent with clustering in the appearance of ACFs. This might happen because some rats are more prone to ACFs than others, or because ACFs tend to appear in clusters within the same rat.

Figure 8.10 provides plots of the data.[15] We note that the counts increase with time. For fitting a Poisson regression model, we use the `poisson` family in the `glm()` function.

---

[15]
```
data(ACF1)              # DAAG package
  plot(count ~ endtime, data=ACF1, pch=16)
```

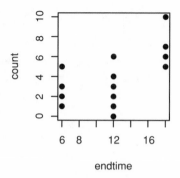

Figure 8.10: Plot of number of simple aberrant crypt foci (count) versus endtime.

Momentarily ignoring the apparent quadratic relation, we fit a simple linear model:[16]

```
> summary(ACF.glm0)

Call:
glm(formula = count ~ endtime, family = poisson, data = ACF1)

Deviance Residuals:
    Min       1Q    Median       3Q       Max
-2.4620   -0.4785   -0.0794   0.3816    2.2633

Coefficients:
             Estimate Std. Error z value Pr(>|z|)
(Intercept)   -0.3215     0.4002   -0.80     0.42
endtime        0.1192     0.0264    4.51  6.4e-06

(Dispersion parameter for poisson family taken to be 1)

    Null deviance: 51.105  on 21  degrees of freedom
Residual deviance: 28.369  on 20  degrees of freedom
AIC: 92.21

Number of Fisher Scoring iterations: 4
```

We see that the relation between count and endtime is highly significant. In order to accommodate the quadratic effect, we need to add an endtime^2 term. The result is[17]

```
. . . .
Coefficients:
             Estimate Std. Error z value Pr(>|z|)
(Intercept)   1.72235    1.09177    1.58    0.115
endtime      -0.26235    0.19950   -1.32    0.188
I(endtime^2)  0.01514    0.00795    1.90    0.057
```

---

[16] ACF.glm0 <- glm(formula = count ~ endtime, family = poisson, data = ACF1)
    summary(ACF.glm0)
[17] ACF.glm <- glm(formula = count ~ endtime + I(endtime^2),
                family = poisson, data = ACF1)
    summary(ACF.glm)

```
(Dispersion parameter for poisson family taken to be 1)

    Null deviance: 51.105  on 21   degrees of freedom
Residual deviance: 24.515  on 19   degrees of freedom
AIC: 90.35
. . . .
```

Observe that the residual mean deviance in $24.515/19 = 1.29$. For a Poisson model this should, unless a substantial proportion of fitted values are small (e.g., less than about 2), be close to 1. A more cautious analysis would use the quasi-Poisson model (specify `family=quasipoisson`) that we demonstrate below for a different set of data. The reader is encouraged to repeat the analysis, using the quasi-Poisson model. This increases standard errors, and reduces t-statistics, by the square root of the estimated *dispersion*. The *dispersion* is the amount by which the variance is increased relative to a Poisson model.

For the present data, the dispersion estimate, obtained by specifying `family=quasipoisson`, is 1.25. Note that it is just a little smaller than the mean residual deviance, which equalled 1.29. Standard errors thus increase by a factor of $\sqrt{1.25} = 1.12$. This is enough that the `endtime^2` term now falls well short of the conventional 5% level of significance. We will comment (Subsection 8.4.3) below on how this estimate is obtained.

8.4.2 Moth habitat example

The `moths` data are from a study of the effect of habitat on the densities of two species of moth. Transects were set out across the search area. Within transects, sections were identified according to habitat type. The end result was that there were

- 41 different lengths (`meters`) of transect ranging from 2 meters to 233 meters,
- grouped into eight habitat types within the search area – Bank, Disturbed, Lowerside, NEsoak, NWsoak, SEsoak, SWsoak, Upperside
- with records taken at 20 different times (morning and afternoon) over 16 weeks.

The variables A and P give the numbers of moths of the two different species. Here is a tabular summary, by habitat:[18]

	Bank	Disturbed	Lowerside	NEsoak	NWsoak	SEsoak	SWsoak	Upperside
	1	7	9	6	3	7	3	5
meters	21	49	191	254	65	193	116	952
A	0	8	41	14	71	37	20	28
P	4	33	17	14	19	6	48	8

An extension of a Poisson regression model (McCullagh and Nelder, 1989, pp. 200–208) provides a way to model such data. For data such as these, there is no reason to accept the Poisson distribution assumption that moths appear independently. The `quasipoisson`

---

[18] `data(moths)`   `# DAAG package`
    `rbind(table(moths[,4]), sapply(split(moths[,-4], moths$habitat), apply, 2, sum))`

family, used in place of the `poisson`, is a better starting point for analysis. We can check the output from the `quasipoisson` model to see whether a Poisson distribution assumption might have been adequate.

For these data, the model takes the form

$$y = \text{habitat effect} + \beta \log(\text{length of section})$$

where $y = \log(\text{expected number of moths})$.

The argument for working with log(expected number of moths), rather than with the expected number, is that effects are likely to be multiplicative. This form of model allows for the possibility that the number of moths per meter might be relatively lower for long than for short sections, or, on the other hand, that the number might be relatively higher for long than for short sections.

Comparisons are made against a reference (or baseline) factor level. The default reference level chosen by R is not always the most useful, so the `relevel` function must be used to change it. In our example, any comparisons with Bank, which is the default reference level, will be inaccurate, as it has only one short section and few moths. Let's proceed up this blind alley to see how it affects the output:[19]

```
> summary(A.glm)


Call:
glm(formula = A ~ log(meters) + factor(habitat),
    family = quasipoisson, data = moths)


Deviance Residuals:
    Min        1Q     Median        3Q       Max
-3.5562   -1.4635   -0.0605    0.9248    3.0821


Coefficients:
```

	Estimate	Std. Error	t value	Pr(>|t|)
(Intercept)	-6.696	23.270	-0.29	0.78
log(meters)	0.129	0.153	0.85	0.40
factor(habitat)Disturbed	6.623	23.274	0.28	0.78
factor(habitat)Lowerside	7.906	23.267	0.34	0.74
factor(habitat)NEsoak	7.084	23.270	0.30	0.76
factor(habitat)NWsoak	9.468	23.266	0.41	0.69
factor(habitat)SEsoak	7.968	23.267	0.34	0.73
factor(habitat)SWsoak	8.137	23.269	0.35	0.73
factor(habitat)Upperside	7.743	23.270	0.33	0.74

```
(Dispersion parameter for quasipoisson family taken to be 2.69)


    Null deviance: 257.108  on 40  degrees of freedom
Residual deviance:  93.995  on 32  degrees of freedom
AIC: NA


Number of Fisher Scoring iterations: 4
```

[19] `A.glm <- glm(formula = A ~ log(meters) + factor(habitat),`
          `family = quasipoisson, data = moths)`
     `summary(A.glm)`

Note that the dispersion parameter, which for a Poisson distribution should equal 1, is 2.69. (There are a variety of possible explanations for the dispersion. One is that there is a cluster effect, with an average cluster size of 2.69. Such clustering does not imply that moths will necessarily be seen together – they may appear at different times and places in the same transect.) Thus, standard errors and *p*-values from a model that assumed Poisson errors would be highly misleading. Use of a quasi-Poisson model, rather than a Poisson model, has increased standard errors by a factor of $\sqrt{2.69} = 1.64$.

The estimates are on a logarithmic scale. Thus, the model predicts $\exp(7.11) = 1224$ times as many moths on Disturbed as on Bank. However, these estimates and associated *z*-statistics are for comparisons with Bank, where the observed number was zero, and all have large standard errors. We cannot, from this output, say anything about comparisons between other types of habitat.

Now consider results from taking Lowerside as the reference:[20]

```
> summary(A.glm)$coef
```

	Estimate	Std. Error	t value	Pr(>|t|)
(Intercept)	1.2098	0.454	2.662	1.20e-02
habitatBank	-7.9057	23.267	-0.340	7.36e-01
habitatDisturbed	-1.2831	0.643	-1.995	5.46e-02
habitatNEsoak	-0.8217	0.536	-1.532	1.35e-01
habitatNWsoak	1.5624	0.334	4.681	5.01e-05
habitatSEsoak	0.0621	0.385	0.161	8.73e-01
habitatSWsoak	0.2314	0.477	0.485	6.31e-01
habitatUpperside	-0.1623	0.583	-0.278	7.83e-01
log(meters)	0.1292	0.153	0.847	4.03e-01

The only change to the output is in coefficients and associated statistical information that are displayed. The model says that

$$\log(\text{expected number of moths}) = 1.21 + 0.13 \log(\text{meters})$$

$$+ \text{habitat (change from Lowerside as reference)},$$

i.e., expected number of moths $= 0.93 \times \text{meters}^{0.13} \times e^{\text{habitat}}$.

The above output indicates that moths are much more abundant in NWsoak than in other habitats. Notice that the coefficient of log(meters) is not statistically significant. This species (A) does not like flying over long open regions. Their preferred habitats were typically at the ends of transects. As a result, the length of the transect made little difference to the number of moths observed.

8.4.3* Residuals, and estimating the dispersion

The R function residuals.glm(), which is called when residuals() is used with glm object as its argument, offers a choice of three different types of residuals – deviance residuals (type="deviance"), Pearson residuals (type="pearson") and working

---

[20] moths$habitat <- relevel(moths$habitat, ref="Lowerside")
   A.glm <- glm(A ~ habitat + log(meters), family=quasipoisson, data=moths)
   summary(A.glm)$coef

Table 8.3: *Data from a randomized trial – assessments of the clarity of the instructions provided to the inhaler.*

	Easy (category 1)	Needed re-reading (category 2)	Not clear (category 3)
Inhaler 1	99	41	2
Inhaler 2	76	55	13

residuals (`type="working"`). For most diagnostic uses, the deviance residuals, which are the default, seem preferable. The R system does not offer deletion residuals, which may be the preferred residual to use in checking for outliers. However, the working residuals may be a reasonable approximation.

Plots of residuals can be hard to interpret. This is an especially serious problem for models that have a binary (0/1) response, where it is essential to use a suitable form of smooth curve as an aid to visual interpretation.

In `quasibinomial` and `quasipoisson` models, a further issue is the estimation of what is called the `dispersion`. The default is to divide the sum of squares of the Pearson residuals by the degrees of freedom of the residual, and use this as the estimate. This estimate is included in the output.

In `lm` models, which are equivalent to GLMs with identity link and normal errors, the dispersion and the residual mean deviance both equal the mean residual sum of squares, and have a chi-squared distribution with degrees of freedom equal to the degrees of freedom of the residual. This works quite well as an approximation for Poisson models whose fitted values are of a reasonable magnitude (e.g., at least 2), and for some binomial models. In other cases, it can be highly unsatisfactory. For models with a binary (0/1) outcome, the approximation is invalid, and gives meaningless results.

The difference in deviance between two nested models has a distribution that is often quite well approximated by a chi-squared distribution. Where there are two such models, with nesting such that the mean difference in deviance (i.e., difference in deviance divided by degrees of freedom) gives a plausible estimate of the dispersion, this is the preferred way to estimate the dispersion. For example, see McCullagh and Nelder (1989).

8.5 Ordinal Regression Models

Here, our intention is to draw attention to an important class of models that further extend the generalized linear model framework. We will demonstrate how logistic and related generalized linear models that assume binomial errors can be used for an initial analysis. The particular form of logistic regression that we will demonstrate is *proportional odds logistic regression*.

Ordinal logistic regression is relevant when there are three or more ordered outcome categories that might, e.g., be (1) complete recovery, (2) continuing illness, (3) death. Here, in Table 8.3, we give an example where patients who were randomly assigned to

Table 8.4: *Odds ratios for two alternative choices of cutpoint in Table 8.3.*

	Odds	
	Easy versus some degree of difficulty	Clear after study versus not clear
Inhaler 1	99/43	140/2
Inhaler 2	76/68	131/13

Table 8.5: *Comparison of log(odds) in Table 8.4.*

	Easy versus some degree of difficulty (1 versus 2 & 3) (cutpoint between 1 & 2)	Clear after study versus not clear (1 & 2 versus 3) (cutpoint between 2 & 3)
Inhaler 1	$\log(99/43) = 0.83$	$\log(140/2) = 4.25$
Inhaler 2	$\log(76/68) = 0.11$	$\log(131/13) = 2.31$
log(Odds ratio)	0.72	1.94

two different inhalers were asked to compare the clarity of leaflet instructions for their inhaler (data, initially published with the permission of 3M Health Care Ltd., are adapted from Ezzet and Whitehead, 1991).

8.5.1 Exploratory analysis

There are two ways to split the outcomes into two categories: we can contrast "easy" with the remaining two responses (some degree of difficulty), or we can contrast the first two categories (clear, perhaps after study) with "not clear". Table 8.4 presents, side by side in parallel columns, odds based on these two splits.

Table 8.5 summarizes the results in terms of log(odds).

Wherever we make the cut, the comparison favors the instructions for inhaler 1. The picture is not as simple as we might have liked. The log(odds ratio), i.e., the difference on the log(odds) scale, may depend on which cutpoint we choose.

8.5.2* Proportional odds logistic regression

The function `polr()` in the MASS package allows the fitting of a formal model. We begin by fitting a separate model for the two rows of data:

```
library(MASS)
inhaler <-  data.frame(freq=c(99,76,41,55,2,13),
                       choice=rep(c("inh1","inh2"), 3),
                       ease=ordered(rep(c("easy","re-read",
                                      "unclear"), rep(2,3))))
```

```
inhaler1.polr <-  polr(ease~1, weights=freq, data=inhaler,
                       subset=inhaler$choice=="inh1")
inhaler2.polr <-  polr(ease~1, weights=freq, data=inhaler,
                       subset=inhaler$choice=="inh2")
```

Notice that the dependent variable specifies the categories, while frequencies are specified as weights. The output is

```
> summary(inhaler1.polr)

Re-fitting to get Hessian

Call:
polr(formula = ease ~ 1, data = inhaler, weights = freq,
  subset = inhaler$choice == "inh1")

No coefficients

Intercepts:
               Value  Std. Error t value
easy|re-read     0.8339 0.1826      4.5659
re-read|unclear 4.2485 0.7121      5.9658

Residual Deviance: 190.3357
AIC: 194.3357
> summary(inhaler2.polr)

Re-fitting to get Hessian

Call:
polr(formula = ease ~ 1, data = inhaler, weights = freq,
  subset = inhaler$choice == "inh2")

No coefficients

Intercepts:
               Value  Std. Error t value
easy|re-read     0.1112 0.1669      0.6663
re-read|unclear 2.3102 0.2908      7.9448

Residual Deviance: 265.5394
AIC: 269.5394
```

For interpreting the output, observe that the intercepts for the model fitted to the first row are $0.8339 = \log(99/43)$ and $4.2485 = \log(140/2)$. For the model fitted to the second row, they are $0.1112 = \log(76/68)$ and $2.3102 = \log(131/13)$.

We now fit the combined model:

```
inhaler.polr <-  polr(ease~choice, weights=freq, data=inhaler)
```

Table 8.6: *The entries are* log*(odds) and odds estimates for the proportional odds logistic regression model that we fitted to the combined data.*

	log(odds). Odds in parentheses	
	Easy versus some degree of difficulty	Clear after study versus not clear
Inhaler 1	0.8627 (2.3696)	3.3527 (28.5798)
Inhaler 2	0.8627 − 0.7903 (1.0751)	3.3527 − 0.7903 (12.9669)

The difference in deviance between the combined model and the two separate models is

```
> deviance(inhaler.polr) - (deviance(inhaler1.polr)
                          + deviance(inhaler2.polr))
[1] 3.42
```

We compare this with the 5% critical value for a chi-squared deviate on 1 degree of freedom – there are 2 parameters for each of the separate models, making a total of 4, compared with 3 degrees of freedom for the combined model. The difference in deviance is just short of significance at the 5% level.

The parameters for the combined model are

```
> summary(inhaler.polr)

Re-fitting to get Hessian

Call:
polr(formula = ease ~ choice, data = inhaler, weights = freq)

Coefficients:
     Value Std. Error    t value
 0.7902877  0.2447535  3.2289125

Intercepts:
                Value   Std. Error t value
easy|re-read     0.8627  0.1811      4.7638
re-read|unclear  3.3527  0.3070     10.9195

Residual Deviance: 459.2912
AIC: 465.2912
```

The value that appears under the heading "Coefficients" is an estimate of the reduction in log(odds) between the first and second rows. Table 8.6 gives the estimates for the combined model: The fitted probabilities for each row can be derived from the fitted log(odds). Thus for inhaler 1, the fitted probability for the `easy` category is $\exp(0.8627)/(1 + \exp(0.8627)) = 0.7032$, while the cumulative fitted probability for `easy` and `re-read` is $\exp(3.3527)/(1 + \exp(3.3527)) = 0.9662$.

8.6 Other Related Models

8.6.1* Loglinear Models

Loglinear models model the frequencies in a multi-way table directly. For the model-fitting process, all margins of the table have the same status. However, one of the margins has a special role for interpretative purposes; it is known as the *dependent* margin. For the `UCBAdmissions` data that we discussed in Section 8.3, the interest was in the variation of admission rate with `Dept` and with `Gender`. A loglinear model, with `Admit` as the dependent margin, offers an alternative way to handle the analysis. Loglinear models are however generally reserved for use when the dependent margin has more than two levels, so that logistic regression is not an alternative.

Examples of the fitting of loglinear models are included with the help page for `loglm()`, in the *MASS* package. To run them, type in

```
library(MASS)
example(loglm)
```

8.6.2 Survival analysis

Survival (or failure) analysis introduces features different from any of those that we have encountered in the regression methods discussed in earlier chapters. It is an elegant methodology that is too little known outside of medicine and industrial reliability testing. It has been widely used for comparing the times of survival of patients suffering a potentially fatal disease who have been subject to different treatments.

Other possibilities are

- the survival of different types of businesses, of different types of light bulb, or of kitchen toasters,
- the failure time distributions of industrial machine components, electronic equipment, automobile components, etc. (failure time analysis, or reliability),
- the waiting time to germination of seeds,
- the waiting time to marriage, or to pregnancy, or to getting one's first job,
- the waiting time to recurrence of an illness.

The outcomes are survival times, but with a twist. The methodology is able to handle data where failure (or another event of interest) has, for a proportion of the subjects, not occurred at the time of termination of the study. It is not necessary to wait till all subjects have died, or all items have failed, before undertaking the analysis! For some subjects, all we have is some final time up to which they had survived. Such objects are said to be right censored. Censoring implies that information about the outcome is incomplete in some respect, but not completely missing. For example, while the exact point of failure of a component may not be known, it may be known that it did not survive more than 720 hours (= 30 days). Thus, for each object we have two pieces of information: a time, and censoring information. Commonly the censoring information indicates either right censoring denoted by a 0, or failure denoted by a 1.

The terms *Failure Time Analysis* and *Reliability* are mostly used in non-medical contexts. Yet another term is *Event History Analysis*. The focus is on time to any event of interest, not necessarily failure.

Many of the same issues arise as in more classical forms of regression analysis. One important set of issues has to do with the diagnostics used to check on assumptions. Here there have been large advances in the past five years. A related set of issues has to do with model choice and variable selection. There are close connections with variable selection in classical regression. Yet another set of issues has to do with the incomplete information that is available when there is censoring.

We plan to include a discussion of survival analysis on our web page. Readers whose curiosity is sufficiently stimulated may wish to run the code that is included in the help pages for functions that are in the *survival* package. Begin by examining the help page for `Surv`, which sets up a survival object.

8.7 Transformations for Count Data

Transformations were at one time commonly used to make count data amenable to analysis using normal theory methods. Generalized linear models have largely removed the need for such approaches. They are, however, still sometimes a useful recourse. Here are some of the possibilities:

- **The square root transformation**: $y = \sqrt{n}$: This may be useful for counts. If counts follow a Poisson distribution, this gives values that have constant variance 0.25.
- **The angular transformation:** $y = \arcsin(\sqrt{p})$, where p is a proportion. If proportions have been generated by a binomial distribution, the angular transformation will "stabilize the variance".

    The R system calculates values in radians. To obtain values that run from 0 to 90, multiply by $180/\pi = 57.3$.
- **The probit or normal equivalent deviate**: Again, this is for use with proportions. For this, take a table of areas under the normal curve. We transform back from the area under the curve to the distance from the mean. It is commonly used in bioassay work. This transformation has a severe effect when p is close to 0 or 1. It moves 0 down to $-\infty$ and 1 up to ∞, i.e., these points cannot be shown on the graph. Some authors add 5 on to the normal deviates to get "probits".
- **The logit**: This function is defined by $\text{logit}(p) = \log(p/(1-p))$: This is a little more severe than the probit transformation. Unless there are enormous amounts of data, it is impossible to distinguish data that require a probit transformation from data that require a logit transformation. Logistic regression uses this transformation on the expected values in the model.
- **The complementary log-log transformation**: This function is defined by $\text{cloglog}(p) = \log(-\log(1-p))$. For p close to 0, this behaves like the logit transformation. For

large values of p, it is a much milder transformation than the probit or logit. It is often the appropriate transformation to use when p is a mortality that increases with time.

Proportions usually require transformation unless the range of values is quite small, between about 0.3 and 0.7.

Note that generalized linear models transform, not the observed proportion, but its expected value as estimated by the model.

8.8 Further Reading

Dobson (2001) is an elementary introduction to generalized linear models. McCullagh and Nelder (1989) is a basic text. Harrell's (2001) comprehensive account of both generalized linear models and survival analysis includes extensive practical advice. Venables and Ripley (2002) give a summary overview of the S-PLUS and R abilities, again covering survival analysis as well as generalized linear models. Collett (2003) is a basic introduction to survival analysis. For recent advances in survival analysis methodology, see Therneau and Grambsh (2001).

There are many ways to extend the models that we have discussed. We have not demonstrated models with links other than the logit. Particularly important, for data with binomial or quasi-binomial errors, is the complementary log-log link; for this specify `family = binomial(link=cloglog)`, replacing `binomial` with `quasibinomial` if that is more appropriate. Another type of extension, not currently handled in R, arises when there is a natural mortality that must be estimated. See, e.g., Finney (1978). Multiple levels of variation, such as we will discuss in Chapter 9, are a further potential complication. Maindonald et al. (2001) present an analysis where all of these issues arose. In that instance, the binomial totals (n) were large enough that it was possible to work with the normal approximation to the binomial.

References for further reading

Collett, D. 2003. *Modelling Survival Data in Medical Research*, 2nd edn. Chapman and Hall.

Dobson, A.J. 2001. *An Introduction to Generalized Linear Models*, 2nd edn. Chapman and Hall.

Finney, D.J. 1978. *Statistical Methods in Bioassay*, 3rd edn. Macmillan.

Harrell, F.E. 2001. *Regression Modelling Strategies, with Applications to Linear Models, Logistic Regression and Survival Analysis*. Springer-Verlag.

Maindonald, J.H., Waddell, B.C. and Petry, R.J. 2001. Apple cultivar effects on codling moth (Lepidoptera: Tortricidae) egg mortality following fumigation with methyl bromide. *Postharvest Biology and Technology* 22: 99–110.

McCullagh, P. and Nelder, J.A. 1989. *Generalized Linear Models*, 2nd edn. Chapman and Hall.

Therneau, T.M. and Grambsch P.M. 2001. *Modeling Survival Data: Extending the Cox Model*. Springer-Verlag.

Venables, W.N. and Ripley, B.D. 2002. *Modern Applied Statistics with* S-PLUS, 4th edn. Springer-Verlag.

8.9 Exercises

1. The following table shows numbers of occasions when inhibition (i.e., no flow of current across a membrane) occurred within 120 s, for different concentrations of the protein peptide-C (data are used with the permission of Claudia Haarmann, who obtained these data in the course of her PhD research). The outcome yes implies that inhibition has occurred.

```
conc 0.1 0.5  1 10 20 30 50 70 80 100 150
no      7   1 10  9  2  9 13  1  1   4   3
yes     0   0  3  4  0  6  7  0  0   1   7
```

Use logistic regression to model the probability of inhibition as a function of protein concentration.

2. In the data set (an artificial one of 3121 patients, that is similar to a subset of the data analyzed in Stiell et al. (2001)) minor.head.injury, obtain a logistic regression model relating clinically.important.brain.injury to other variables. Patients whose risk is sufficiently high will be sent for CT (computed tomography). Using a risk threshold of 0.025 (2.5%), turn the result into a decision rule for use of CT.

3. Consider again the moths data set of Section 8.4.

   (a) What happens to the standard error estimates when the poisson family is used in glm() instead of the quasipoisson family?

   (b) To visualize the data, type in the following:

   ```
   library(lattice)
   avmoths <- aggregate(moths[,2], by=list(moths$habitat),
                        FUN=mean)
   names(avmoths) <- c("habitat", "av")
   dotplot(habitat ~ av, data=avmoths, pch=16, cex=1.5)
   ```

   Reconcile this with the Poisson regression output that used Lowerside as the reference.

   (c) Analyze the P moths, in the same way as the A moths were analyzed. Comment on the effect of transect length.

4.* The factor dead in the data set mifem (*DAAG* package) gives the mortality outcomes (live or dead), for 1295 female subjects who suffered a myocardial infarction. (See Subsection 10.7.1 for further details.) Determine ranges for age and yronset (year of onset), and determine tables of counts for each separate factor. Decide how to handle cases for which the outome, for one or more factors, is not known. Fit a logistic regression model, beginning by comparing the model that includes all two-factor interactions with the model that has main effects only.

9

Multi-level Models, Time Series and Repeated Measures

The multiple levels that are in view are multiple levels in the noise or *error* term. Because *error* is the word that is commonly used in the literature on multi-level models, it is the word that we will use in this chapter. Most of the discussion so far has focused on models where there is a simple error structure. It has been sufficient to specify a distribution for independently and identically distributed errors. In a few instances it has been necessary to give different observations different weights, reflecting different prior variances. Subject to any such adjustment, it has been assumed that errors are independently and identically distributed. (*Note:* Much of the computation in this chapter will require access to the *nlme* package. Make sure that it is installed.)

9.1 Introduction

In multi-level modeling, the random part of the model, i.e., the error, has a structure. Multi-level models are widely useful and important, offering a unified approach to a huge range of problems. They are a large step beyond the "independent errors" models met in a first course in statistics.

Examples that we will use to illustrate the ideas of multi-level models are:

- Data on the attitudes of Australian year 7 students to science. There are schools, classes within schools, and pupils within classes. Different numbers of pupils in the different classes make the design unbalanced.
- A kiwifruit field experiment is arranged in blocks, plots and subplots. For example, there may be three blocks, four plots (one for each treatment) in each block, and four vines in each plot. Treatments are applied to whole plots.
- An experiment was conducted to assess the effect of tinting of car windows on visual performance. There were two levels of variation – within individuals and between individuals.

Prior to the advent of capabilities such as those in the *nlme* package, analysis of variance approaches were available for data that had an appropriate "balance", while various ad hoc approaches were widely used for other data. (The Genstat system (Payne et al. 1993) is distinguished for its coherent approach to the analysis of data from suitably balanced designs. It was perhaps the first of the major systems to implement general methods for the analysis of multi-level models.) The analyses that resulted from these ad hoc approaches were in general less insightful and satisfying than were the more adequate analyses that a multi-level modeling framework allows.

The simplest models have a hierarchical error structure that can be modeled using variance components. The repeated measures models that we will describe are multi-level models where measurements consist of profiles in time or space. Animal or plant growth curves, where each "individual" is measured at several different times, are a common example. Typically, the data exhibit some form of time dependence that the model may need to accommodate. In time series analysis, there is typically a single profile in time, though perhaps for multiple variables.

Important ideas will be

- fixed and random effects,
- variance components, and their connection, in special cases, with expected values of mean squares,
- the specification of mixed models with a simple error structure,
- time series modeling,
- sequential correlation in repeated measures profiles.

9.2 Example – Survey Data, with Clustering

The data are measurements of attitudes to science, from a survey where there were results from 20 classes in 12 private schools and 46 classes in 29 public (i.e. state) schools, all in and around Canberra, Australia. Results are from a total of 1385 year 7 students. The variable like is a summary score based on two of the questions. It is on a scale from 1 (dislike) to 12 (like). The number in each class from whom scores were available ranged from 3 to 50, with a median of 21.5. Figure 9.1 compares results for public schools with those for private schools.[1]

9.2.1 Alternative models

Within any one school, we might have

$$y = \text{class effect} + \text{pupil effect}$$

where y represents the attitude measure.

Within any one school, we might use a one-way analysis of variance to estimate and compare class effects. However, this study has the aim of generalizing beyond the classes in the study to all of some wider population of classes, not just in the one school, but in a wider population of schools from which the schools in the study were drawn. In order to be

---

[1]
```
## First use aggregate() to form means for each class
data(science)          # DAAG package
attach(science)
classmeans <- aggregate(like, by=list(PrivPub, Class), mean)
  # NB: Class identifies classes independently of schools
  #     class identifies classes within schools
names(classmeans) <- c("PrivPub","Class","like")
dim(classmeans)
detach(science)
attach(classmeans)
## Boxplot
boxplot(split(like, PrivPub),
  ylab = "Class average of attitude to science score", boxwex = 0.4)
rug(like[PrivPub == "private"], side = 2)
rug(like[PrivPub == "public"], side = 4)
detach(classmeans)
```

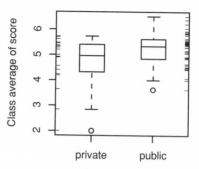

Figure 9.1: Distribution of average scores across classes, compared between public and private schools.

able to generalize in this way, we treat school (`school`), and class (`class`) within school, as *random* effects. We are interested in possible differences between the sexes (`sex`), and between private and public schools (`PrivPub`). The two sexes are not a sample from some larger population of possible sexes (!), nor are the two types of school (for this study at least) a sample from some large population of types of school. Thus we call them *fixed* effects. Our interest is in the specific *fixed* differences between males and females, and between private and public schools.

The preferred approach is a multi-level model analysis. While it is sometimes possible to treat such data using an approximation to the analysis of variance as for a balanced experimental design, it may be hard to know how good the approximation is. We specify sex (`sex`) and school type (`PrivPub`) as fixed effects, while school (`school`) and class (`class`) are specified as random effects. Class is *nested* within school; within each school there are several classes. The model is

$$y = \begin{array}{c} \text{sex effect} \\ \text{(fixed)} \end{array} + \begin{array}{c} \text{type (private or public)} \\ \text{(fixed)} \end{array} + \begin{array}{c} \text{school effect} \\ \text{(random)} \end{array} + \begin{array}{c} \text{class effect} \\ \text{(random)} \end{array} + \begin{array}{c} \text{pupil effect} \\ \text{(random)}. \end{array}$$

Questions we might ask are:

- Are there differences between private and public schools?
- Are there differences between females and males?
- Clearly there are differences among pupils. Are there differences between classes within schools, and between schools, greater than pupil to pupil variation within classes would lead us to expect?

The table of estimates and standard errors is similar to that from an `lm()` (single level) analysis, but with an extra column that is headed DF.[2]

```
> summary(science.lme)$tTable # Print coefficients.
```

	Value	Std.Error	DF	t-value	p-value
(Intercept)	4.722	0.1625	1316	29.07	6.79e-144
sexm	0.182	0.0982	1316	1.86	6.35e-02
PrivPubpublic	0.412	0.1857	39	2.22	3.25e-02

[2] `library(nlme)`
```
  science.lme <- lme(fixed = like ~ sex + PrivPub,
    data = science, random = ~ 1 | school/class, na.action=na.omit)
summary(science.lme)$tTable # Print coefficients.
 # The output in the text uses options (digits=3).
```

Note that we have 1316 degrees of freedom for the comparison between males and females, but only 39 degrees of freedom for the comparison between private and public schools
(there are 12 private schools, and 29 public schools. The degrees of freedom are calculated as
$12 - 1 + 29 - 1 = 39$). The comparison is between different schools of the different types.
On the other hand, schools are made up of classes, each of which includes both males and
females. Thus the between pupils level of variation is relevant for the comparison between
sexes.

There are three variance components:

```
Between schools 0.00105
Between classes 0.318
Between students 3.05
```

These values can be gleaned from `VarCorr(science.lme)`, or alternatively by squar-
ing output from `intervals(science.lme, which="var-cov")`. Note that the
accompanying standard deviations are based on a linear approximation. It is important to
note that the between classes variance component is additional to the variation between
students component in the residual row, i.e. that labeled `Between students`.

This table is interesting in itself. It tells us that differences between classes are greater
than would be expected from differences between students alone. It also suggests we can
do a simpler analysis in which we ignore the effect of schools:[3]

```
> summary(science1.lme)$tTable # Table of coefficients

               Value Std.Error   DF t-value  p-value
(Intercept)    4.722    0.1624 1316   29.07 6.20e-144
sexm           0.182    0.0982 1316    1.86  6.36e-02
PrivPubpublic 0.412    0.1857   64    2.22  3.02e-02
```

The variance components are, to two significant digits, the same as before.[4] Unfortunately
the output is not labeled in a very user-friendly way. In the code given in footnote 4, the
values have been squared, and are variances, not SDs.

```
                 lower   est. upper
sd((Intercept)) 0.190 0.321  0.54
Within Group:    2.83  3.05  3.29
attr(,"label")
[1] "Within-group standard error:"
```

To what extent are differences between classes affecting the attitude to science? The
proportion of variance that is explained by differences between classes is $0.321/(0.321 +
3.05) = 9.5\%$. The main influence comes from outside of the class that the pupil attends, e.g.

[3] `science1.lme <- lme(fixed = like ~ sex + PrivPub, data = science,`
`                    random = ~ 1 | Class, na.action=na.exclude)`
[4] `## The numerical values were extracted thus:`
`intervals(science1.lme, which="var-cov")[[1]]$Class^2`
`intervals(science1.lme, which="var-cov")[[2]]^2`

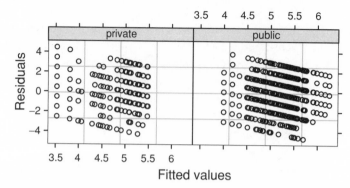

Figure 9.2: Residuals from class means are plotted against class fitted values, separately for private and public schools.

from home, television, friends, inborn tendencies, etc. This statistic has the name *intra-class correlation*.

Do not be tempted to think that, because 0.321 is small relative to the within class component variance of 3.05, it is of little consequence. The variance for the mean of a class that is chosen at random is $0.321 + 3.05/n$. Thus, with a class size of 20, the between class component makes a bigger contribution than the within class component. If all classes were the same size, then the standard error of the difference between class means for public schools and those for private schools would, as there were 20 private schools and 46 public schools, be

$$\sqrt{(0.321 + 3.05/n)\left(\frac{1}{20} + \frac{1}{46}\right)}.$$

From the output table of coefficients and standard errors, we note that the standard error of difference between public and private schools is 0.1857. For this to equal the expression just given, we require $n = 19.1$. Thus the sampling design is roughly equivalent to a balanced design with 19.1 pupils per class.

Figure 9.2 plots standardized residuals from the class means against fitted values.[5] The banded pattern in the residuals makes them hard to interpret. There is, however, no obvious trend or variation in the scatter about the line $y = 0$.

9.2.2 Instructive, though faulty, analyses

Ignoring class as the random effect

It is important that the specification of random effects be correct. It is enlightening to do an analysis that is not quite correct, and investigate the scope that it offers for misinterpretation.

---

[5]
```
library(lattice)
science.lme <- lme(fixed = like ~ sex + PrivPub,
   data = science, random =  ~ 1 | class/school, na.action=na.exclude)
plot(science.lme, resid(.) ~ fitted(.) | PrivPub)
   # Invokes plot.lme(), which uses functions from the lattice package
```

We fit `school`, ignoring `class`, as a random effect. The estimates of the fixed effects change little.[6]

```
> summary(science2.lme)$tTable
```

	Value	Std.Error	DF	t-value	p-value
(Intercept)	4.738	0.163	1341	29.00	5.95e-144
sexm	0.197	0.101	1341	1.96	5.07e-02
PrivPubpublic	0.417	0.185	39	2.25	3.02e-02

The objection to this analysis is that it suggests, wrongly, that the between schools component of variance is substantial. The estimated variance components are[7]

```
Between schools 0.166
Between students 3.219
```

This is misleading. From our earlier investigation, it is clear that the difference is between classes, not between schools!

Ignoring the random structure in the data

Here is the result from a standard regression (linear model) analysis, with `sex` and `PrivPub` as fixed effects:[8]

```
> summary(science.lm)$coef
```

	Estimate	Std. Error	t value	Pr(>|t|)
(Intercept)	4.740	0.0996	47.62	0.000000
sexm	0.151	0.0986	1.53	0.126064
PrivPubpublic	0.395	0.1051	3.76	0.000178

Do not believe this analysis! The SEs are much too small. The contrast between public and private schools is more borderline than this suggests.

9.2.3 Predictive accuracy

The variance of a prediction of the average for a new class of n pupils, sampled in the same way as existing classes, is $0.32 + 3.05/n$. If classes were of equal size, we could derive an equivalent empirical estimate of predictive accuracy by using a resampling method with the class means. With unequal class sizes, use of the class means in this way will be a rough approximation. There were 60 classes. If the training/test set methodology is used, the 60 class means would be divided between a training set and a test set.

---

[6] `science2.lme <- lme(fixed = like ~ sex + PrivPub, data = science,`
    `random = ~ 1 | school, na.action=na.exclude)`
  `summary(science2.lme)$tTable`
[7] `## The numerical values were extracted from`
  `VarCorr(science2.lme)`
[8] `science.lm <- lm(like ~ sex + PrivPub, data=science) # Faulty analysis`
  `summary(science.lm)$coef`

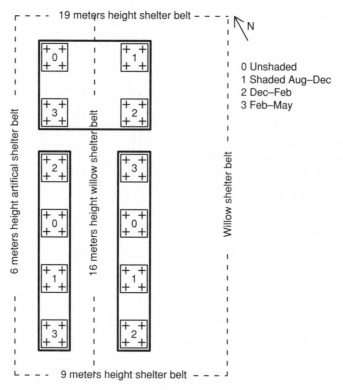

Figure 9.3: The field layout for the kiwifruit shading trial.

An empirical estimate of the within class variation can be derived by applying a re-sampling method (cross-validation, or the bootstrap) to data for each individual class. The variance estimates from the different classes would then be pooled.

The issues here are important. Data do commonly have a hierarchical variance structure comparable with that for the attitudes to science data. The population to which results are to be generalized determines what estimate of predictive accuracy is needed.

9.3 A Multi-level Experimental Design

The data in kiwishade are from a designed experiment that compared different kiwifruit shading treatments. (These data relate to Snelgar et al. (1992). Maindonald (1992) gives the data in Table 9.1, together with a diagram of the field layout that is similar to Figure 9.3. The two papers have different shorthands (e.g. Sept–Nov versus Aug–Dec) for describing the time periods for which the shading was applied.) Figure 9.3 shows the layout. Notice that there are four vines in each plot, and four plots (one for each of four treatments) in each the three blocks. This is a "balanced" design – each plot has the same number of vines, each block has the same number of plots, with each treatment occurring the same number of times.

Note also:

- There are three levels of variation – between vines within plots, between plots within blocks, and between blocks.

Table 9.1: *Data from the kiwifruit shading trial. The level names for the factor* shade *are mnemonics for the time during which shading was applied. Thus* (none) *implies no shading,* Aug2Dec *means "August to December", and similarly for* Dec2Feb *and* Feb2May.

Block	Shade	Vine1	Vine2	Vine3	Vine4
west	none	93.42	100.68	103.49	92.64
west	Aug2Dec	97.42	97.60	101.41	105.77
west	Dec2Feb	84.12	87.06	90.75	86.65
west	Feb2May	90.31	89.06	94.99	87.27
north	none	100.30	106.67	108.02	101.11
north	Aug2Dec	104.74	104.02	102.52	103.10
north	Dec2Feb	94.05	94.76	93.81	92.17
north	Feb2May	91.82	90.29	93.45	92.58
east	none	100.74	98.05	97.00	100.31
east	Aug2Dec	101.89	108.32	103.14	108.87
east	Dec2Feb	94.91	93.94	81.43	85.40
east	Feb2May	96.35	97.15	97.57	92.45

- The experimental unit is a plot. We have four results for each plot.
- Within blocks, treatments were randomly assigned to plots.

The northernmost plots were grouped together because they were similarly affected by shading from the sun in the north. For the remaining two blocks, shelter effects, whether from the west or from the east, were thought more important. Table 9.1 displays the data.

In this example, it is relatively easy to demonstrate the calculation of the analysis of variance sums of squares and mean squares. The variance components can be obtained from these. The principles can then be transferred to more complicated contexts, where the variance components cannot readily be calculated from the mean squares in the analysis of variance table. The model is

$$\text{yield} = \text{overall mean} + \begin{array}{c}\text{block effect}\\\text{(random)}\end{array} + \begin{array}{c}\text{shade effect}\\\text{(fixed)}\end{array} + \begin{array}{c}\text{plot effect}\\\text{(random)}\end{array} + \begin{array}{c}\text{vine effect}\\\text{(random)}\end{array}.$$

We characterize the design thus:

Fixed Effect: shade (treatment).
Random effect: vine (nested) in plot in block, or block/plot/vine.

The block effect is random because we want to be able to generalize to other blocks. Similarly, we want to be able to generalize to other plots and other vines. When horticulturalists apply these treatments in their own orchards, there will be different vines, plots and blocks. A horticulturalist will however reproduce, as far as possible, the same shade treatments as were used in the scientific trial.

Table 9.2: *The analysis of variance information has been structured to reflect the three levels of variation.*

	Df	Sum of Sq	Mean Sq
`block` level	2	172.3	86.17
`block.plot` level			
shade	3	1394.5	464.84
residual	6	125.6	20.93
`block.plot.vines` level	36	438.6	12.18

9.3.1 The anova table

We require an analysis of variance table that structures the sums of squares in a manner that is helpful for extracting the sums of squares and F-statistics that are of interest. For this purpose we need to specify that there are two levels of variation additional to variation between vines – between blocks, and between plots within blocks. We use the `Error()` function for this:[9]

```
> summary(kiwishade.aov)

Error: block
          Df Sum Sq Mean Sq F value Pr(>F)
Residuals  2  172.3    86.2

Error: block:shade
          Df Sum Sq Mean Sq F value Pr(>F)
shade      3   1395     465    22.2 0.0012
Residuals  6    126      21

Error: Within
          Df Sum Sq Mean Sq F value Pr(>F)
Residuals 36    439      12
```

Table 9.2 structures the output, with a view to making it easier to follow.

9.3.2 Expected values of mean squares

The expected values of the mean squares do not appear directly in the model. Instead, the model works with variance components. The expected values of the mean squares are, in suitably balanced designs such as this, linear combinations of the variance components. They are important for investigating the properties of alternative designs. Table 9.3 shows how the variance components combine to give the expected values of mean squares in the analysis of variance table. It imposes an interpretive veneer on the analysis of variance table.

---

[9] `data(kiwishade)      # DAAG package`
  `kiwishade.aov <- aov(yield ~ shade+Error(block/shade), data=kiwishade)`
  `summary(kiwishade.aov)`

Table 9.3: *Mean squares in the analysis of variance table. The final column gives expected values of mean squares, as functions of the variance components.*

	Df	Mean sq	E[Mean sq]
block stratum	2	86.17	$16\sigma_B^2 + 4\sigma_P^2 + \sigma_V^2$
block.plot stratum			
shade	3	464.84	$4\sigma_P^2 + \sigma_V^2$ + treatment component
residual	6	20.93	$4\sigma_P^2 + \sigma_V^2$
block.plot.*Units* stratum	36	12.18	σ_V^2

In this example, we do not need to calculate the variance components. We compare the shade mean square with the residual mean square in the block.plot stratum. The ratio is

$$F\text{-ratio} = \frac{464.84}{20.93} = 22.2, \text{ on 3 and 6 d.f. } (p = 0.0024).$$

Here are the model estimates:[10]

```
none Aug2Dec Dec2Feb Feb2May
100.20  103.23   89.92   92.77
```

The standard error of treatment means is the square root of the residual mean square divided by the sample size: $\sqrt{(20.93/12)} = 1.32$. The sample size is 12 here, since each treatment mean is based on results from 12 vines. The SED for comparing two treatment means is $1.32 \times \sqrt{2} = 1.87$ (recall that we use SED as an abbreviation for "standard error of difference", where the difference that is in view is between two means).

Because of the balance in the data, analysis of variance calculations have so far been able to do everything we needed. In an unbalanced design, it is not usually possible to use an analysis of variance table in this way.

One benefit of determining the variance components, for these data, is that we can then estimate the analysis of variance mean squares for a different experimental design. This can be highly useful in deciding on the design for any further experiment. It is a small step beyond the analysis of variance table to find the variance components, as we will demonstrate below. First, we will examine how the sums of squares in the analysis of variance table are calculated.

Readers who do not at this point wish to study Subsection 9.3.3 in detail may nevertheless find it helpful to examine Figures 9.4 and 9.5, taking on trust the scalings used for the effects that they present. The figures hint at greater variation at the plot level than can be explained by within plot vine to vine variation. The estimate of between plot variance in Table 9.2 was 20.93. While larger than the between vine mean square of 12.18, it is not so much larger that the evidence from Figures 9.4 and 9.5 can be more than suggestive. Variation between treatments does appear much greater than can be explained from variation between plots, and the same is true for variation between blocks.

---

[10] `attach(kiwishade)`
`  sapply(split(yield, shade), mean)`

Table 9.4: *Plot means, and differences of yields for individual vines from the plot mean.*

	Mean	Vine 1	Vine 2	Vine 3	Vine 4
east.Aug2Dec	105.55	3.32	−2.41	2.77	−3.66
east.Dec2Feb	88.92	−3.52	−7.49	5.02	5.99
east.Feb2May	95.88	−3.43	1.69	1.27	0.47
east.none	99.03	1.28	−2.03	−0.98	1.71
north.Aug2Dec	103.60	−0.50	−1.08	0.42	1.14
north.Dec2Feb	93.70	−1.53	0.11	1.06	0.35
north.Feb2May	92.03	0.55	1.42	−1.74	−0.22
north.none	104.03	−2.92	3.99	2.64	−3.73
west.Aug2Dec	100.55	5.22	0.86	−2.95	−3.13
west.Dec2Feb	87.15	−0.50	3.60	−0.09	−3.03
west.Feb2May	90.41	−3.14	4.58	−1.35	−0.10
west.none	97.56	−4.92	5.93	3.12	−4.14

Grand mean = 96.53

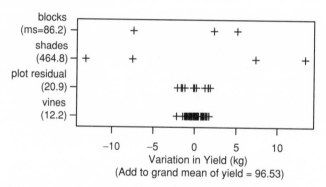

Figure 9.4: Variation at different levels, in the kiwifruit shading trial.

9.3.3* The sums of squares breakdown

This subsection has details of the calculation of the sums of squares and mean squares. It is not essential to what follows, but does provide useful insight.

For each plot, we calculate a mean, and differences from the mean. See Table 9.4.[11] Note that whereas we started with 48 observations we have only 12 means. Now we break the means down into overall mean, plus block effect (the average of differences, for that block, from the overall mean), plus treatment effect (the average of the difference, for that treatment, from the overall mean), plus residual.

[11]
```
kiwi.table <- t(sapply(split(yield, plot), as.vector))
kiwi.means <- sapply(split(yield, plot), mean)
kiwi.means.table <- matrix(rep(kiwi.means,4), nrow=12, ncol=4)
kiwi.summary <- data.frame(kiwi.means, kiwi.table-kiwi.means.table)
names(kiwi.summary)<- c("Mean", "Vine 1", "Vine 2", "Vine 3", "Vine 4")
mean(kiwi.means) # the grand mean (only for balanced design)
detach(kiwishade)
```

Table 9.5: *Each plot mean is expressed as the sum of overall mean, block effect, shade effect, and residual for the* `block.shade` *combination.*

block	shade	Mean	Block effect	Shade effect	block.shade residual
west	none	97.56	−2.618	3.670	−0.027
west	Aug2Dec	100.55	−2.618	6.701	−0.066
west	Dec2Feb	87.15	−2.618	−6.612	−0.158
west	Feb2May	90.41	−2.618	−3.759	0.251
north	none	104.02	1.805	3.670	2.017
north	Aug2Dec	103.60	1.805	6.701	−1.444
north	Dec2Feb	93.70	1.805	−6.612	1.971
north	Feb2May	92.04	1.805	−3.759	−2.545
east	none	99.02	0.812	3.670	−1.990
east	Aug2Dec	105.56	0.812	6.701	1.509
east	Dec2Feb	88.92	0.812	−6.612	−1.813
east	Feb2May	95.88	0.812	−3.759	2.294
			square, add, multiply by 4, divide by 2, to give ms	square, add, multiply by 4, divide by 3, to give ms	square, add, multiply by 4, divide by 6, to give ms

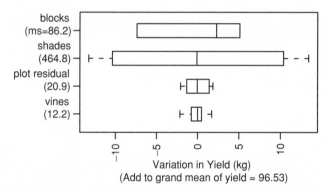

Figure 9.5: Boxplot representation of variation at the different levels, in the kiwifruit shading trial.

Table 9.5 uses the information from Table 9.4 to express each plot mean as the sum of a block effect, a shade effect and a residual for the `block.shade` combination. The notes in the last row of each column show how to determine the mean squares that were given in Table 9.3. Moreover, we can scale the values in the various columns so that their standard deviation is the square root of the error mean square, and plot them. Figure 9.4 plots all this information.

The variation is more easily detected in boxplots of the distributions, as in Figure 9.5. NB. Figures 9.4 and 9.5 plot the same information. Figures 9.4 and 9.5 plot what some authors call *effects*.

We now explain the scaling of effects in Figures 9.4 and 9.5. Consider first the 48 residuals at the vine level. Because 12 degrees of freedom were expended when the 12 plot means were subtracted, the 48 residuals share 36 degrees of freedom and are positively correlated. To enable the residuals to present the appearance of uncorrelated values with the correct variance, we scale the 48 residuals so that the average of their squares is the between vine estimate of variance σ_V^2 – this requires multiplication of each residual by $\sqrt{(48/36)}$.

At the level of plot means, we have 6 degrees of freedom to share between 12 plot effects. In addition, we need to undo the increase in precision that results from taking means of four values. Thus, we multiply each plot effect by $\sqrt{(12/6)} \times \sqrt{(4)}$. If the between plot component of variance is zero, the expected value of the average of the square of these scaled effects will be σ_V^2. If the between plot component of variance is greater than zero, the expected value of the average of these squared effects will be greater than σ_V^2.

In moving from plot effects to treatment effects, we have a factor of $\sqrt{(4/3)}$ that arises from the sharing of 3 degrees of freedom between 4 effects, further multiplied by $\sqrt{(12)}$ because each treatment mean is an average of 12 vines. For block effects, we have a multiplier of $\sqrt{(3/2)}$ that arises from the sharing of 2 degrees of freedom between 3 effects, further multiplied by $\sqrt{(16)}$ because each block mean is an average of 16 vines. Effects that are scaled in this way allow visual comparison, as in Figures 9.4 and 9.5.

9.3.4 The variance components

The mean squares in an analysis of variance table for data from a "balanced" multi-level model can be broken down further, into variance components. The variance components analysis gives more detail about model parameters. Importantly, it provides information that will help design another experiment. Here is the breakdown for the kiwifruit shading data:

- Variation between vines in a plot is made up of one source of variation only. Denote this variance by σ_V^2.
- Variation between vines in different plots is partly a result of variation between vines, and partly a result of additional variation between plots. In fact, if σ_P^2 is the (additional) component of the variation that is due to variation between plots, the expected mean square equals

$$4\sigma_P^2 + \sigma_V^2.$$

(NB: the 4 comes from 4 vines per plot.)
- Variation between treatments is

$$4\sigma_P^2 + \sigma_V^2 + T$$

where T (> 0) is due to variation between treatments.
- Variation between vines in different blocks is partly a result of variation between vines, partly a result of additional variation between plots, and partly a result of additional variation between blocks. If σ_B^2 is the (additional) component of variation that is due to differences between blocks, the expected value of the mean square is

$$16\sigma_B^2 + 4\sigma_P^2 + \sigma_V^2$$

(16 vines per block; 4 vines per plot).

We do not need estimates of the variance components in order to do the analysis of variance. The variance components are helpful when we want to design another experiment. We calculate the estimates thus:

$$\widehat{\sigma}_V^2 = 12.18,$$
$$4\widehat{\sigma}_P^2 + \widehat{\sigma}_V^2 = 20.93, \text{ i.e. } 4\widehat{\sigma}_P^2 + 12.18 = 20.93.$$

This gives the estimate $\widehat{\sigma}_P^2 = 2.19$. We can also estimate $\widehat{\sigma}_B^2 = 4.08$.

We are now in a position to work out how much the precision would change if we had 8 (or, say, 10) vines per plot. With n vines per plot, the variance of the plot mean is

$$(n\widehat{\sigma}_P^2 + \widehat{\sigma}_V^2)/n = \widehat{\sigma}_P^2 + \widehat{\sigma}_V^2/n = 2.19 + 12.18/n.$$

We could also ask how much of the variance, for an individual vine, is explained by vine to vine differences. This depends on how much we stretch the other sources of variation. If the comparison is with vines that may be in different plots, the proportion is $12.18/(12.18 + 2.19)$. If we are talking about different blocks, the proportion is $12.18/(12.18 + 2.19 + 4.08)$.

9.3.5 The mixed model analysis

For a mixed model analysis, we specify that treatment (`shade`) is a fixed effect, that `block` and `plot` are random effects, and that `plot` is nested in `block`. The software works out for itself that the remaining part of the variation is associated with differences between vines. It is not necessary to specify it separately.

For using `lme()`, the command is

```
kiwishade.lme <- lme(fixed = yield ~ shade, random = ~ 1 |
                     block/plot, data=kiwishade)
```

Fixed effects, random effects, and sequential correlation structure

The art in setting up an analysis for these models is in getting the description of the model correct. Specifically it is necessary to

- identify which are fixed and which random effects,
- correctly specify the nesting of the random effects.

In repeated measures designs, it is necessary to specify the pattern of correlation within profiles.

In addition, skill and care may be needed to get output into a form that directly addresses the questions that are of interest. Finally, output must be interpreted. Multi-level analyses often require high levels of professional skill.

Plots of residuals

We urge users to plot and examine the residuals. There are different choices of fitted values, depending on how we account for the random effects, and hence different choices of residuals. By default, fitted values adjust for all except random variation between individual

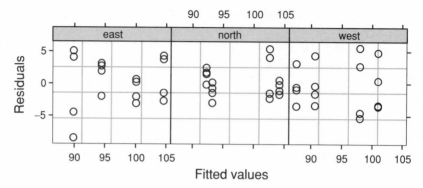

Figure 9.6: Standardized residuals after fitting block and shading effects, plotted against fitted values. There are 12 distinct fitted values, one for each plot.

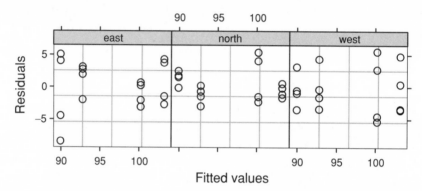

Figure 9.7: Standardized residuals from the treatment means, plotted against the four treatment means. This plot is useful for drawing attention to yields that may be unusually high or low for the specific combination of block and treatment.

vines, i.e., they account for treatment, block and plot effects. For this, set level=2 when calculating fitted values, or the equivalent residuals. Other choices are to calculate fitted values as treatment plus block (level=1) or as treatment effects only (level=0). The block *effects* are differences between fitted values at level 1 and fitted values at level 0, while the plot effects are differences between fitted values at level 2 and fitted values at level 1.

Figure 9.6 fits the block and treatment effects, and takes residuals from those.[12]

In Figure 9.7, residuals from the treatment means are plotted against the treatment means.[13] This allows us to see which blocks are high for a treatment, and which are low.

It will help in interpreting Figure 9.7 to remember that the treatment means are, in order,

```
Dec2Feb Feb2May  none    Aug2Dec
89.92    92.77   100.20  103.23
```

[12]plot(kiwishade.lme, residuals(.) ~ fitted(.))
    # By default, residuals and fitted values are at level 2,
    # i.e., they are adjusted for plot and block effects
[13]## For this, use the level 0 fitted values, i.e. do not adjust for any of
    ## the random effects
  plot(kiwishade.lme, residuals(.) ~ fitted(., level = 0)|block, aspect=1,
    layout=c(3,1))

Thus the `Dec2Feb` and `no shading` treatments may have fared particularly well in the north block.

9.3.6 Predictive accuracy

We have data for one site only. We thus cannot estimate any between site error component. The best we can do is to assess predictive accuracy for new blocks on the current site. Because there are just three blocks, the use of a resampling method to assess accuracy is out of the question!

Experiments on multiple sites will inevitably give results that differ from one site to another. Even after the use of covariates that attempt to adjust for differences in rainfall, temperature, soil type, etc., there will almost inevitably be a site to site component of variation that we will need to estimate. Different results from another experimenter on another site do not at all imply that at least one of the experiments must be flawed. Rather, the implication is that both sets of results require to be incorporated into an analysis that accounts for site to site variation. Better still, plan the experiment from the beginning as a multi-site experiment!

9.3.7 Different sources of variance – complication or focus of interest?

In this discussion, the main interest has been in the parameter estimates. The different sources of variance are a complication. In other applications, the variances are the focus of interest. Many animal and plant breeding trials are of this type. The aim may be to design a breeding program that will lead to an improved variety or breed.

We have used an analysis of data from a field experimental design to demonstrate the calculation and use of components of variance. Other contexts for multi-level models are the analysis of data from designed surveys, and general regression models in which the "error" term is made up of several components. In all these cases, we no longer have independently and identically distributed errors.

9.4 Within and Between Subject Effects – an Example

We consider data (in the data frame `tinting`) from an experiment that aimed to model the effects of the tinting of car windows on visual performance. (For more information, see Burns et al., 1999.) Interest is focused on effects on side window vision, and hence on visual recognition tasks that would be performed when looking through side windows.

The variables are `csoa` (critical stimulus onset asynchrony, i.e., the time in milliseconds required to recognize an alphanumeric target), `it` (inspection time, i.e., the time required for a simple discrimination task) and `age`, while `tint` (three levels) and `target` (two levels) are ordered factors. The variable `sex` is coded `f` for females and `m` for males, while the variable `agegp` is coded `Younger` for young people (all in their 20s) and `Older` for older participants (all in the 70s).

We have two levels of variation – within individuals (who were each tested on each combination of `tint` and `target`), and between individuals. Thus we need to specify id (identifying the individual) as a random effect. Plots of the data make it clear that, to have variances that are approximately homogeneous, we need to work with log(csoa) and log(it). Here, we describe the analysis for log(it).

For conceptual clarity, and in order to keep inference as simple as possible, we limit initial attention to three models:

1. All possible interactions (this is likely to be more complex than we need):

```
data(tinting)              # DAAG package
itstar.lme <- lme(log(it) ~ tint*target*agegp*sex,
              random= ~ 1 | id, data=tinting, method="ML")
```

2. All two-factor interactions (this is a reasonable guess; two-factor interactions may be all we need):

```
it2.lme <- lme(log(it) ~ (tint+target+agegp+sex)^2,
                  random = ~1|id, data=tinting,
                  method="ML")
```

3. Main effects only (this is a very simple model):

```
it1.lme <- lme(log(it)~(tint+target+agegp+sex),
                  random = ~1|id, data=tinting,
                  method= "ML")
```

The reason for specifying method="ML", i.e. use the maximum likelihood estimation criterion, is that we can then do the equivalent of an analysis of variance comparison:

```
anova(itstar.lme, it2.lme, it1.lme)
```

Here is the outcome:

	Model	df	AIC	BIC	logLik	Test	L.Ratio	p-value
itstar.lme	1	26	8.15	91.5	21.93			
it2.lme	2	17	-3.74	50.7	18.87	1 vs 2	6.11	0.7288
it1.lme	3	8	1.14	26.8	7.43	2 vs 3	22.88	0.0065

Notice that df is now used for degrees of freedom, where Df had been used earlier. Such inconsistencies should, in time, be removed.

From this table we see that the p-value for comparing model 1 with model 2 is 0.73, while that for comparing model 2 with model 3 is 0.0065. This suggests that the model that limits attention to two-factor interactions is adequate. (Note also that the AIC statistic favors model 2. The BIC statistic, which is an alternative to AIC, favors the model that has main effects only. The BIC's penalty for model complexity can be unduly severe when the number of residual degrees of freedom is small. See Hastie et al., 2001, p. 208.)

The analysis of variance table indicated that main effects together with two-factor interactions were enough to explain the outcome. Interaction plots, looking at the effects of factors two at a time, are therefore an effective visual summary of the analysis results. Examination of the t-statistics that we give below shows that the highest t-statistics for interaction terms are associated with tint.L:agegpsenior, targethicon:agegpsenior, `tint.L:targethicon` and `tint.L:sexm`. It makes sense to look first at those plots where the interaction effects are clearest, i.e. where the *t*-statistics are largest. The plots may be based on either observed data or fitted values, at the analyst's discretion.[14]

We leave the reader to pursue this investigation, and marry the results with the lattice graphs that we presented in Subsection 2.1.5.

Estimates of model parameters

For exploration of parameter estimates in the model that includes all two-factor interactions, we re-fit the model used for `it2.lme`, but now using `method="REML"` (restricted maximum likelihood estimation). The parameter estimates that come from the REML analysis are in general preferable, because they avoid or reduce the biases of maximum likelihood estimates. (See, e.g., Diggle et al. (2002). The difference from maximum likelihood can however be of little consequence.)

```
it2.reml <- update(it2.lme, method="REML")
summary(it2.reml)$tTable
```

We now examine the estimated effects:

```
> summary(it2.reml)$tTable
```

	Value	Std.Error	DF	t-value	p-value
(Intercept)	3.61907	0.1301	145	27.817	5.30e-60
tint.L	0.16095	0.0442	145	3.638	3.81e-04
tint.Q	0.02096	0.0452	145	0.464	6.44e-01
targethicon	-0.11807	0.0423	145	-2.789	5.99e-03
agegpOlder	0.47121	0.2329	22	2.023	5.54e-02
sexm	0.08213	0.2329	22	0.353	7.28e-01
tint.L:targethicon	-0.09193	0.0461	145	-1.996	4.78e-02
tint.Q:targethicon	-0.00722	0.0482	145	-0.150	8.81e-01
tint.L:agegpOlder	0.13075	0.0492	145	2.658	8.74e-03
tint.Q:agegpOlder	0.06972	0.0520	145	1.341	1.82e-01
tint.L:sexm	-0.09794	0.0492	145	-1.991	4.83e-02
tint.Q:sexm	0.00542	0.0520	145	0.104	9.17e-01
targethicon:agegpOlder	-0.13887	0.0584	145	-2.376	1.88e-02
targethicon:sexm	0.07785	0.0584	145	1.332	1.85e-01
agegpOlder:sexm	0.33164	0.3261	22	1.017	3.20e-01

Because `tint` is an ordered factor, its effect is, for `tint` which has three levels, split up into two parts. The first, which always carries a `.L` (linear) label, checks if there is a linear change across levels. The second part is labeled `.Q` (quadratic), and as `tint` has only three levels, accounts for all the remaining sum of squares that is due to `tint`. This partitioning of the effect of `tint` carries across to interaction terms also. Note that the *t*-statistics are

[14] `## Code that gives the first four such plots, for the observed data`
```
interaction.plot(tinting$tint, tinting$agegp, log(tinting$it))
interaction.plot(tinting$target, tinting$sex, log(tinting$it))
interaction.plot(tinting$tint, tinting$target, log(tinting$it))
interaction.plot(tinting$tint, tinting$sex, log(tinting$it))
```

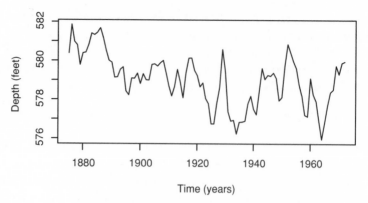

Figure 9.8: A trace plot of annual depth measurements of Lake Huron versus time.

all substantially less than 2.0 in terms that include a `tint.Q` component, suggesting that we could simplify the output by restricting attention to `tint.L` and its interactions.

For these data, the message is in the lattice plots that we presented in Subsection 2.1.5, and in the interaction plots that we suggested above, rather than in this table of parameter estimates.

9.5 Time Series – Some Basic Ideas

Sequences of observations recorded over time are referred to as time series. Many techniques have been developed to deal with the special nature of the dependence that is often associated with such series. We will provide only the most cursory treatment here. The interested reader is directed to any of several excellent books listed in the references section.

9.5.1 Preliminary graphical explorations

The time series object `LakeHuron`, in the *ts* package, gives annual depth measurements at a specific site on Lake Huron. Figure 9.8 is a trace plot, i.e. an unsmoothed scatterplot of the data against time.[15] There is a slight downward trend for the first half of the series. Observe also that, with some notable exceptions, depth measurements that are close together in time are often close together in value. This is a first indication that we are not looking at a series of independent observations. Time series typically exhibit some form of sequential dependence, in which observations that are close together in time are more similar than those that are widely separated. A key challenge for time series analysis is to find good ways to model this dependence.

Lag plots may give a useful visual impression of the dependence. Suppose that our observations are $x_1, x_2, \ldots, x_n$. Then the lag 1 plot plots x_i against x_{i-1} ($i = 2, \ldots, n$), thus:

y-value	x_2	x_3	$\ldots$	x_n
lag 1 (x-axis)	x_1	x_2	$\ldots$	x_{n-1}

[15] `library(ts)`   # required if the ts package is not already loaded
  `data(LakeHuron)`
  `plot(LakeHuron, ylab="depth (in feet)", xlab = "Time (in years)")`

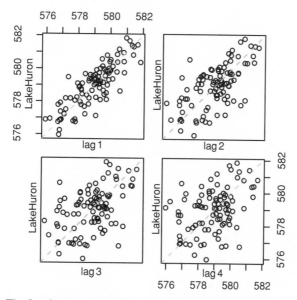

Figure 9.9: The first four lag plots of the annual depth measurements of Lake Huron.

For a lag 2 plot, x_i is plotted against x_{i-2} $(i = 3, \ldots, n)$, and so on for higher lags. Notice that the number of points that are plotted is reduced by the number of lags.

Figure 9.9 shows the first four lag plots for the Lake Huron data.[16] The scatter of points about a straight line in the first lag plot suggests that the dependence between consecutive observations is linear. The increasing scatter observed in the second, third and fourth lag plots indicates that points separated by two or three lags are successively less dependent. Note that the slopes, and therefore the correlations, are all positive.

9.5.2 The autocorrelation function

For each of the lag plots in Figure 9.9, there is a correlation. Technically, such correlations are *autocorrelations* (literally, self-correlations). The autocorrelation function (ACF), which gives the autocorrelations at all possible lags, often gives useful insight. Figure 9.10 plots the first 19 lag autocorrelations.[17] The autocorrelation at lag 0 is included by default; this always takes the value 1, since it represents the correlation between the data and themselves. Of more interest are the autocorrelations at other lags. In particular, as we inferred from the lag plots, the autocorrelation at lag 1 (sometimes called the serial correlation) is fairly large and the lag 2 and lag 3 autocorrelations are successively smaller. The autocorrelations continue to decrease as the lag becomes larger, indicating that there is no linear association among observations separated by lags larger than about 10.

As we will see in the next subsection, autocorrelation in a time series complicates the estimation of such quantities as the standard error of the mean. There must be appropriate modeling of the dependence structure. If there are extensive data, it helps to group the data

---

[16] lag.plot(LakeHuron, lags=3, do.lines=FALSE)
[17] acf(LakeHuron)

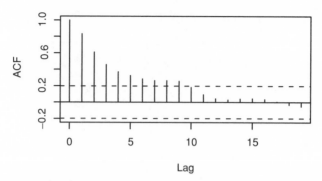

Figure 9.10: A plot of the autocorrelation function for the Lake Huron depth measurements.

in sets of k successive values, and take averages. The serial correlation then reduces to ρ_1/k approximately. See Miller (1986) for further comment.

9.5.3 Autoregressive (AR) models

We have indicated earlier that, whenever possible, the rich regression framework provides additional insights. This remains true when analyzing time series. If we can compute correlations between successive observations, we should be able to compute regressions relating successive observations as well.

The AR(1) model

The autoregressive model of order 1 (AR(1)) for a time series $X_1, X_2, \ldots$, has the recursive formula

$$X_t = \mu + \alpha(X_{t-1} - \mu) + \varepsilon_t, \quad t = 1, 2, \ldots,$$

where μ and α are parameters. Usually, α takes values in the interval $(-1, 1)$ (the so-called stationary case). (If there is a clear trend in the series, the series is not stationary, and the trend should be removed before proceeding. To remove a linear trend, take successive differences.[18]) The error term ε_t is the familiar independent noise term with constant variance σ^2. The distribution of X_0 is assumed fixed and will not be of immediate concern.

For the AR(1) model, the ACF at lag i is α^i, for $i = 1, 2, \ldots$. If $\alpha = .8$, then the observed autocorrelations should be .8, .64, .512, .410, .328, $\ldots$, a geometrically decaying pattern, not too unlike that in Figure 9.10.

To gain some appreciation for the importance of models like the AR(1) model, we consider estimation of the standard error for the estimate of the mean μ. Under the AR(1) model, a large sample approximation to the standard error for the mean is

$$\frac{\sigma}{\sqrt{n}} \frac{1}{(1 - \alpha)}.$$

For a sample of size 100 from an AR(1) model with $\sigma = 1$ and $\alpha = 0.5$, the standard error of the mean is 0.2. This is exactly twice the value that results from the use of the usual

[18] `series.diff <- diff(series)`

$\sigma/\sqrt{n}$ formula. Use of the usual standard error formula will result in misleading and biased confidence intervals for time series where there is substantial autocorrelation.

There are several alternative estimates for the parameter α. One is to use the autocorrelation at lag 1. The maximum likelihood estimator is an alternative. For the Lake Huron data, we find these estimates to be .8319 and .8376, respectively.[19]

The general AR(p) model

We are not restricted to considering regressions of X_t against X_{t-1} only. It is also possible to regress X_t against observations at any previous lag. The autoregressive model of order p (AR(p)) is based on the regression of X_t against $X_{t-1}, X_{t-2}, \ldots, X_{t-p}$:

$$X_t = \mu + \alpha_1(X_{t-1} - \mu) + \cdots + \alpha_p(X_{t-p} - \mu) + \varepsilon_t, \quad t = 1, 2, \ldots,$$

where $\alpha_1, \alpha_2, \ldots, \alpha_p$ are additional parameters that would need to be estimated.

Estimation proceeds in the same way as for the AR(1) model. For example, if we fit an AR(2) model (using maximum likelihood) to the Lake Huron data, we obtain estimates[20] $\widehat{\alpha_1} = 1.044$ and $\widehat{\alpha_2} = -.250$. The question then arises as to how large p should be, i.e., how many AR parameters are required. We will briefly consider two approaches.

The first is based on the Akaike Information Criterion (AIC), which was introduced in Subsection 6.4.2. Use of this criterion, with models fitted using maximum likelihood,[21] suggests an AR(2) model, with the parameter estimates given above.

A plot of the *partial autocorrelation* function allows a more informal approach. A partial autocorrelation at a particular lag measures the strength of linear correlation between observations separated by that lag, after adjusting for correlations between observations separated by fewer lags. The partial autocorrelation at lag 4, for example, is the estimate of α_4 for an AR(4) model. Figure 9.11 displays a plot of the partial autocorrelations for the Lake Huron data.[22] This plot indicates, for example, that, after taking into account observations at the previous two lags, the correlation between observations separated by three or more lags is quite small. Thus, an AR(2) model seems appropriate, in agreement with the result obtained using AIC.

9.5.4 Autoregressive moving average (ARMA) models – theory*

In a moving average (MA) process of order q, the *error* term is the sum of an *innovation* ϵ_t that is specific to that term, and a linear combination of earlier *innovations* $\epsilon_{t-1}, \epsilon_{t-2}, \ldots,$

---

[19] `LH.yw <- ar(x = LakeHuron, order.max = 1, method = "yw")`
```
                          # autocorrelation estimate
                          # order.max = 1 for the AR(1) model
    LH.yw$ar              # autocorrelation estimate of alpha
    LH.mle <- ar(x = LakeHuron, order.max = 1, method = "mle")
                          # maximum likelihood estimate
    LH.mle$ar             # maximum likelihood estimate of alpha
    ## Estimates of the series mean and of sigma^2 can be obtained by typing
    LH.mle$x.mean         # estimated series mean
    LH.mle$var.pred       # estimated innovation variance
```
[20] `LH.mle2 <- ar(LakeHuron, order.max=2, method="mle")`
[21] `ar(LakeHuron, method="mle")`   `# AIC is used by default if`
```
                          # order.max is not specified
```
[22] `acf(LakeHuron, type="partial")`

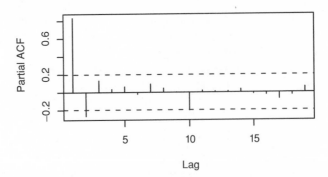

Figure 9.11: Partial autocorrelation function for the Lake Huron depth measurements.

ϵ_{t-q}. The equation is

$$X_t = \mu_t + \epsilon_t + b_1\epsilon_{t-1} + \cdots + b_q\epsilon_{t-q} \tag{9.1}$$

where $\epsilon_1, \epsilon_2, \ldots, \epsilon_q$ are independent random normal variables, all with mean 0. The autocorrelation between terms that are more than q time units apart is zero. Moving average terms can be helpful for modeling autocorrelation functions that are spiky, or that have a sharp cutoff.

An autoregressive moving average (ARMA) model is an extension of an autoregressive model to include "moving average" terms. Autoregressive integrated moving average (ARIMA) models, which we will not discuss further here, allow even greater flexibility. (The model (AR or MA or ARMA) is applied, not to the series itself, but to a *differenced* series, where the differencing process may be repeated more than once.) There are three parameters: the order p of the autoregressive component, the order d of differencing (in our example 0) and the order q of the moving average component. For details, see `help(arima)`.

Figure 9.12 shows the autocorrelation functions for simulations of several different moving average models. Notice that

- with $b_1 = 0.5$ ($q = 1$) there is a spike at lag 1,
- with $b_1 = b_2 = 0, b_3 = 0.5$ ($q = 3$) there is a spike at lag 3,
- with $b_1 = b_2 = 0, b_3 = 0.5, b_4 = 0, b_5 = 0.5$ ($q = 5$) there are spikes at lags 2, 3 and 5,
- with $b_i = 0.5$ ($i = 1, \ldots, 5$), there are spikes at the first five lags.

9.6* Regression modeling with moving average errors – an example

The Southern Oscillation Index (SOI) is the difference in barometric pressure at sea level between Tahiti and Darwin. Annual SOI and Australian rainfall data, for the years 1900–2001, are in the data frame `bomsoi`. (See Nicholls et al. (1996) for background. The data are taken from the Australian Bureau of Meteorology web pages http://www.bom.gov.au/climate/change/rain02.txt and http://www.bom.gov.au/climate/current/soihtm1.shtml.) To what extent is the SOI useful for predicting Australian annual average rainfall?

It will turn out that we can use a moving average model for the error structure in a regression model. The function `gls()` in the *nlme* package will fit such regression models. We will, in addition, need functions from the *ts* time series package.

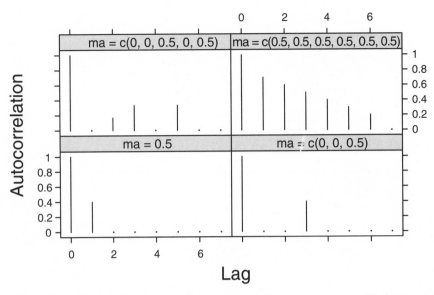

Figure 9.12: Theoretical autocorrelation functions for a moving average process, with the parameters shown.

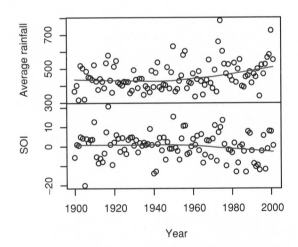

Figure 9.13: Plots of `avrain` and `SOI` against time, for the Southern Oscillation Index (`bomsoi`) data.

Figure 9.13 plots the two series. The code is

```
data(bomsoi)            # DAAG package
plot(ts(bomsoi[, 15:14], start=1900),
     panel=function(y,...)panel.smooth(1900:2001, y,...))
```

The saucer-shaped trend for the rainfall series combines with the inverted saucer-shaped trend for the Southern Oscillation Index data to give a small negative correlation between the trend curves for the two series. We will demonstrate this relationship shortly.

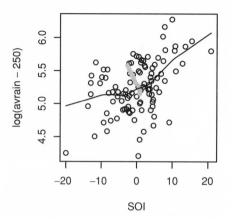

Figure 9.14: Plot of `log(avrain - 250)` against `SOI`, with fitted smooth curve. Superimposed on the plot is a thick gray curve that shows the relationship between the trend curve for `log(avrain - 250)` and the trend curve for `SOI`.

Note also that the rainfall data seem positively skewed. Experimentation with transformations of the form $\log(\text{avrain} - a)$, for different values of a, suggests that $\log(\text{avrain} - 250)$ will be satisfactory.[23]

Figure 9.14 shows a plot of the data, with fitted smooth curve. Superimposed on the plot is a thick gray curve that shows the relationship between the two trend curves. It is interesting and instructive that these two curves show quite different relationships. The code is

```
plot(bomsoi$SOI, log(bomsoi$avrain - 250), xlab = "SOI",
     ylab = "log(avrain - 250)")
lines(lowess(bomsoi$SOI)$y, lowess(log(bomsoi$avrain - 250))$y,
      lwd=2)
  # NB: separate lowess fits against time
lines(lowess(bomsoi$SOI, log(bomsoi$avrain - 250)))
```

In order to obtain data that are free of any effect from the separate trends over time, i.e., free of any effect from the heavy curve in Figure 9.14, we need to remove the trends from the separate series, and work with the detrended series.

```
detsoi <- data.frame(
  detSOI = bomsoi[, "SOI"] - lowess(bomsoi[, "SOI"])$y,
  detrain = log(bomsoi$avrain - 250) -
               lowess(log(bomsoi$avrain - 250))$y)
row.names(detsoi) <- paste(1900:2001)
```

---

[23] ## Check by comparing the normal probability plots for different a, e.g.
```
 par(mfrow = c(2,3))
 for (a in c(50, 100, 150, 200, 250, 300))
 qqnorm(log(bomsoi[, "avrain"] - a))
   # a = 250 leads to a nearly linear plot
 par(mfrow = c(1,1))
```

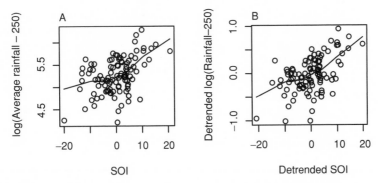

Figure 9.15: Plot of (A) log(`avrain` − 250) against SOI, and (B) `detrain` against `detSOI`. Plot B is for the detrended series.

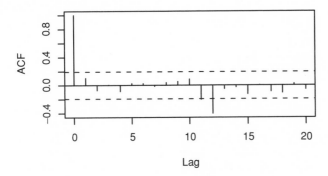

Figure 9.16: Plot of autocorrelation function for the residuals from the regression of `detrain` on `detSOI`.

Figure 9.15A plots `log(avrain - 250)` against `SOI`, while Figure 9.15B gives the equivalent plot for the detrended series.[24]

The analysis that follows will model the relationship that seems apparent in Figure 9.15B. The time series structure of the data will almost inevitably lead to correlated errors. It is necessary to model this structure, if we are to obtain realistic standard errors for model parameters and make accurate inferences.

Our first step however is to fit a linear curve as for uncorrelated errors; we then use the residuals to get an initial estimate of the error structure. Figure 9.16 shows the autocorrelation structure of the residuals. There is a noticeable negative autocorrelation at lag 12. The code is

```
acf(resid(lm(detrain ~ detSOI, data = detsoi)))
```

In view of our earlier discussion of MA processes (Subsection 9.5.4), a moving average series with $q = 12$ appears a good choice for modeling the error term. The function `gls()`

[24]`par(mfrow=c(1,2))`
```
plot(log(avrain-250) ~ SOI, data = bomsoi, ylab = "log(Average rainfall - 250)")
lines(lowess(bomsoi$SOI, log(bomsoi$avrain-250)))
plot(detrain ~ detSOI, data = detsoi,
  xlab="Detrended SOI", ylab = "Detrended log(Rainfall-250)")
lines(lowess(detsoi$detrain ~ detsoi$detSOI))
par(mfrow = c(1,1))
```

in the *nlme* package allows the fitting of regression type models with correlated error terms. The syntax is similar to that for `lme()`, but without a *random* parameter. The code is

```
soi.gls <- gls(detrain ~ detSOI, data = detsoi, correlation
                = corARMA(q=12))
```

The output is

```
> summary(soi.gls)
Generalized least squares fit by REML
  Model: detrain ~ detSOI
  Data: detsoi
   AIC BIC logLik
   70.4 109  -20.2

Correlation Structure: ARMA(0,12)
 Formula: ~1
 Parameter estimate(s):
 Theta1   Theta2   Theta3   Theta4   Theta5   Theta6   Theta7
-0.0630  -0.0970   0.0161  -0.1360  -0.1057  -0.1250  -0.0286
 Theta8   Theta9  Theta10  Theta11  Theta12
-0.0309   0.1095   0.2357  -0.2823  -0.4754

Coefficients:
               Value Std.Error  t-value p-value
(Intercept) -0.0181   0.00614    -2.94  0.0041
detSOI       0.0338   0.00329    10.27  <.0001

 Correlation:
       (Intr)
detSOI -0.059

Standardized residuals:
   Min      Q1     Med      Q3     Max
-3.120  -0.593   0.126   0.833   2.405

Residual standard error: 0.313
Degrees of freedom: 102 total; 100 residual
```

In summary, there are two phenomena that seem relatively unconnected. The net effect of the different trend curves over the past 102 years for rainfall and SOI is a negative correlation; a downward trend in the SOI over the past half-century has been matched by an upward trend of very nearly constant relative magnitude in rainfall. Values of rainfall and SOI vary widely about the individual trend lines. Whenever the SOI for any individual year is above the trend line, the Australian average rainfall in that year tends to be above the trend line for rainfall. Note however that departures from the trend for `SOI` explain only 29% of the variance in departures from the trend for `log(avrain - 250)`. Additionally, rainfall is an average taken over the whole of Australia. The relationship is likely to be stronger for rainfall in north-eastern regions, and weaker in the south.

Formal significance tests

A first step might be to use `Box.test()` to compare a straight line model with uncorrelated errors against the alternative of autocorrelation coefficients up to some specified lag. In the following code we specify a lag of 20, arguing that a meaningful autocorrelation at any higher lag than this is not plausible, for a series of this length:

```
> Box.test(resid(lm(detrain ~ detSOI, data = detsoi)),
          type="Ljung-Box", lag=20)

Box-Ljung test

data: resid(lm(detrain ~ detSOI, data = detsoi))
X-squared = 33.058, df =20, p-value = 0.03325
```

The t-values that are given in the table of coefficients can be an inaccurate approximation. For a more reliable check on model coefficients, the model should be fitted with and without the term that is of interest, in both cases specifying `method = "ML"`, in place of the default `method = "REML"`. The `anova()` function then allows a comparison of the loglikelihoods. Here then is a check on the coefficients in the model that we have just fitted.

```
soi1ML.gls <- update(soi.gls, method = "ML")
soi0ML.gls <- update(soi.gls, detrain ~ 1, method = "ML")
anova(soi1ML.gls, soi0ML.gls)
```

The output is

```
            Model df   AIC   BIC logLik   Test L.Ratio p-value
soi1ML.gls      1 15  52.4  91.8  -11.2
soi0ML.gls      2 14 113.1 149.8  -42.5 1 vs 2    62.7  <.0001
```

Figure 9.15 suggested that a straight line was adequate as a model for the relationship. We may however wish to carry out a formal check on whether a squared (`detrain^2`) term is justified. The code is

```
soi2ML.gls <- update(soi.gls, detrain ~ detSOI + detSOI^2,
                    method = "ML")
anova(soi2ML.gls, soi1ML.gls)
```

We may also wish to check whether the order q of the moving average error term is about right.

```
## Compare models with q = 11, q = 12 and q = 13
## NB Comparison of loglikelihoods
soi11.gls <- update(soi.gls, correlation=corARMA(q=11))
soi13.gls <- update(soi.gls, correlation=corARMA(q=13))
anova(soi11.gls, soi.gls, soi13.gls)
```

In fitting `soi.gls`, the `method` parameter was left at its default, which is `"REML"`. This gives a valid comparison providing that, as here, the two models have the same fixed effects. The output is

```
> anova(soi11.gls, soi.gls, soi13.gls)
          Model df  AIC BIC logLik    Test L.Ratio p-value
soi11.gls     1 14 81.6 118  -26.8
soi.gls       2 15 70.4 110  -20.2 1 vs 2   13.22  0.0003
soi13.gls     3 16 71.6 113  -19.8 2 vs 3    0.86  0.3542
```

Other checks and computations

We give code for various follow-up computations, leaving the reader to work through them.

```
acf(resid(soi.gls))                  # (Correlated) residuals
acf(resid(soi.gls, type="normalized"))      # Uncorrelated
## Examine normality of estimates of uncorrelated "residuals"
qqnorm(resid(soi.gls, type="normalized"))  # Examine normality
## Now extract the moving average parameters, and
## plot the theoretical autocorrelation function
## that they imply.
beta <- summary(soi.gls$modelStruct)$corStruct
plot(ARMAacf(ma=beta,lag.max=20), type="h")
## Next plot several simulated autocorrelation functions
## We plot autocorrelation functions as though they were
## time series!
plot.ts(ts(cbind(
  "Series 1" = acf(arima.sim(list(ma=beta), n=83), plot=F,
                  lag.max = 20)$acf,
  "Series 2" = acf(arima.sim(list(ma=beta), n=83), plot=F,
                  lag.max = 20)$acf,
  "Series 3" = acf(arima.sim(list(ma=beta), n=83), plot=F,
       lag.max = 20)$acf), start=0), type="h", main = "",
       xlab = "Lag")
## Show confidence bounds for the MA parameters
intervals(soi.gls)
```

9.7 Repeated Measures in Time – Notes on the Methodology

Whenever we make repeated measurements on a treatment unit we are, technically, working with repeated measures. In this sense, both the kiwifruit shading data and the window tinting data sets were examples of repeated measures data sets. Here, our interest is in generalizing the multi-level modeling approach to handle models for longitudinal data, i.e., data where the measurements were repeated at different times. We comment on the principles involved.

In the kiwifruit shading experiment, we gathered data from all vines in a plot at the one time. In principle, we might have taken data from different vines at different points in time. For each plot, there would be data at each of four time points.

There is a close link between a wide class of repeated measures models and time series models. In the time series context, there is usually just one realization of the series, which may however be observed at a large number of time points. In the repeated measures context, there may be a large number of realizations of a series that is typically quite short.

Perhaps the simplest case is where there is no apparent trend with time. Thus consider data from a clinical trial of a drug (progabide) used to control epileptic fits. (For an analysis of data from the study to which this example refers, see Thall and Vail, 1990.) The analysis assumes that patients were randomly assigned to the two treatments – placebo and progabide. After an eight-week run-in period, data were collected, both for the placebo group and for the progabide group, in each of four successive two-week periods. The outcome variable was the number of epileptic fits over that time.

One way to do the analysis is to work with the total number of fits over the four weeks, perhaps adjusted by subtracting the baseline value. It is possible that we might obtain extra sensitivity by taking account of the correlation structure between the four sets of fortnightly results, taking a weighted sum rather than a mean.

Suppose there is a trend with time. Then taking a mean over all times will not usually make sense. In all, we have the following possibilities.

1. There is no trend with time.
2. The pattern with time may follow a simple form, e.g., a line or a quadratic curve.
3. A general form of smooth curve, e.g., a curve fitted using splines, may be required to account for the pattern of change with time.

We may have any of these possibilities both for the overall pattern of change and for the pattern of difference between one profile and another.

9.7.1 The theory of repeated measures modeling

For the moment, profiles (or subjects) are assumed independent. The analysis must allow for dependencies between the results of any one subject at different times. For a balanced design, we will assume n subjects ($i = 1, 2, \ldots, n$) and p times ($j = 1, 2, \ldots, p$), though perhaps with missing responses (gaps) for some subjects at some times. The plot of response versus time for any one subject is that subject's "profile".

A key idea is that there are (at least) two levels of variability – between subjects and within subjects. In addition, there is a measurement error. Repeating the same measurement on the same subject at the same time will not give exactly the same result. The between subjects component of variation is never observable separately from sources of variation that operate "within subjects". In any data that we collect, measurements are always affected by "within subjects" variability, plus measurement error. Thus the simplest model that is commonly used has a between subjects variance component denoted by ν^2, while there is a within subjects variance at any individual time point that is denoted by σ^2. The measurement error may be bundled in as part of σ^2. The variance of the response for one subject at a particular time point is $\nu^2 + \sigma^2$.

In the special case just considered, the variance of the difference between two time points for one subject is $2\sigma^2$. Comparisons "within subjects" are more accurate than comparisons "between subjects".

9.7.2 Correlation structure

The time dependence of the data has implications for the correlation structure. The simple model just described takes no account of this structure. Points that are close together in time are in some sense more closely connected than points that are widely separated in time. Often, this is reflected in a correlation between time points that decreases as the time separation increases. The variance for differences between times typically increases as points move further apart in time.

We have seen that correlation structure is also a key issue in time series analysis. A limitation, relative to repeated measures, is that in time series analysis the structure must typically be estimated from just one series, by assuming that the series is in some sense part of a repeating pattern. In repeated measures there may be many realizations, allowing a relatively accurate estimate of the correlation structure. By contrast with time series, the shortness of the series has no effect on our ability to estimate the correlation structure. Multiple realizations are preferable to a single long series.

While we are typically better placed than in time series analysis to estimate the correlation structure there is, for most of the inferences that we commonly wish to make, less need to know the correlation structure. Typically our interest is in the consistency of patterns between individuals. For example we may want to know: "Do patients on treatment A improve at a greater rate than patients on treatment B?"

For further discussion of repeated measures modeling, see Diggle at al. (2002) and Pinheiro and Bates (2000). The Pinheiro and Bates book is based around the S-PLUS version of the *nlme* package.

9.7.3 Different approaches to repeated measures analysis

Traditionally, repeated measures models have been analyzed in many different ways. Here is a summary of methods that have been used:

- A simple way to analyze repeated measures data is often to form one or more summary statistics for each subject, and then use these summary statistics for further analysis.
- When the variance is the same at all times and the correlation between results is the same for all pairs of times, data can in principle be analyzed using an analysis of variance model. This allows for two components of variance, (1) between subjects and (2) between times. An implication of this model is that the variance of the difference is the same for all pairs of time points. The assumption that the correlation is the same for all pairs of times is, in general, optimistic. It effectively ignores the time series structure.
- Various adjustments adapt the analysis of variance approach to allow for the possibility or likelihood that the variance of time differences are not all equal. These should be avoided now that there are good alternatives to the analysis of variance approach.
- Multivariate comparisons accommodate all possible patterns of correlations between time points. This approach accommodates the time series structure, but does not take advantage of it to simplify the parameterization of the correlation structure.
- Repeated measures models aim to reflect the time series structure, in the fixed effects, in the random effects, and in the correlation structure. This modeling approach often allows insights that are hard to gain from approaches that ignore or do not take advantage of the time series structure.

9.8 Further Notes on Multi-level Modeling

9.8.1 An historical perspective on multi-level models

Multi-level models are not new. The inventor of the analysis of variance was R.A. Fisher. Although he would not have described it that way, many of the analysis of variance calculations that he demonstrated were analyses of specific forms of multi-level model. Data have a structure that is imposed by the experimental design. The particular characteristic of the experimental design models that Fisher worked with was that the analysis could be handled using analysis of variance methods. The variance estimates that are needed for different comparisons may be taken from different lines of the analysis of variance table. This circumvents the need to estimate the variances of the random effects that appear in a fully general analysis.

Until the modern computing era, it was hard or impossible to analyze data from multi-level models that were not suited to an analysis of variance approach. Data whose structure did not fit the analysis of variance framework required some form of approximate analysis.

Statistical analysts who used Fisher's experimental designs and methods of analysis followed Fisher's rules for the calculations. Each different design had its own recipe. After working through a few such analyses, some of those who followed Fisher's methods began to feel that they understood the rationale fairly well at an intuitive level. Books appeared that gave instructions on how to do the analyses. The most comprehensive of these is Cochran and Cox (1957).

Regression models are another starting point for talking about multi-level models. Both the fixed effects part of the model and the random effects part of the model have a structure. We have moved beyond the models with a single error term that have been the stock in trade of courses on regression modeling. Even now, most regression texts give no recognition, or scant recognition, to structure in the random part of the model. Yet data, as often as not, do have structure – students within classes within institutions, nursing staff within hospitals within regions, managers within local organizations within regional groupings, and so on.

As we have noted above, models have not always been written down. Once theoretical statisticians did start to write down models there was a preoccupation with models that had a single error term that described the random part of the model. Theoretical development, where the discussion centered around models, was disconnected from the practical analysis of experimental designs, where most analysts were content to follow Cochran and Cox and avoid any formal mathematical description of the models that were used.

Modern computer software makes it possible to specify directly the fixed and random terms in the model. It works just about equally well with balanced and unbalanced designs. We do not have to look up Cochran and Cox to find how to do an analysis, and we can analyze data that have not come from Cochran and Cox style experimental designs! There are still advantages, for suitably balanced designs, in using programs that take advantage of the balance to simplify and structure the output.

Modern software makes it straightforward, having identified what seems a reasonable model for observational data, to estimate parameters for the model. Except in very simple cases, it is hard to be sure that, for observational data, estimates of model parameters are meaningful. The power of modern software can become a trap, perhaps leading to inadequate care in the design of data collection in the expectation that computer software will take care of any problems. The result may be data whose results are hard to interpret or cannot be interpreted at all, or that make poor use of resources.

9.8.2 Meta-analysis

Meta-analysis is a name for analyses that bring together into a single analysis framework data from, e.g., multiple agricultural trial sites, or multiple clinical trial centers, or multiple psychological laboratories. We will use "location" as a general term for "site" or "center", or "laboratory", etc. If treatment or other effects can be shown to be consistent relative to location to location variation, the result can be compelling for practical application. Multi-level modeling makes it possible to do analyses that take proper account of site to site or center to center variation. Indeed, in such areas of investigation, it is unsatisfactory to give an assessment of treatment or other effects that fails to take account of site to site variation. In such research, multi-level modeling, and various extensions of multi-level modeling, provide the appropriate framework for planning and for handling the analysis.

Meta-analysis is uncontroversial when data come from a carefully planned multi-location trial. More controversial is the bringing together into one analysis of data from quite separate investigations. There may however be little choice; the alternative is an informal and often unconvincing qualitative evaluation of the total body of evidence. Clearly such analyses challenge the critical acumen of the analyst. A wide range of methodologies have been developed to handle the problems that may arise. Gaver et al. (1992) is a useful summary.

9.9 Further Reading

Fisher (1935) is a non-mathematical account that takes the reader step by step through the analysis of important types of experimental design. It is useful background for reading more modern accounts. Williams et al. (2002) is similarly example-based, with an emphasis on tree breeding. See also Cox (1958). Cox and Reid (2000) is an authoritative up to date account of the area. It has a more practical focus than its title might seem to imply.

On multi-level and repeated measures models see Snijders and Bosker (1999), Diggle et al. (2002), Goldstein (1995), Pinheiro and Bates (2000) and Venables and Ripley (2002).

Talbot (1984) is an interesting example of the use of multi-level modeling, with important agricultural and economic implications. It summarizes a large amount of information that is of importance to farmers, on yields for many different crops in the UK, including assessments both of center to center and of year to year variation.

The relevant chapters in Payne et al. (1997), while directed to users of the Genstat system, have helpful commentary on the use of the methodology and on the interpretation of results. Pinheiro and Bates (2000) describes the use of the *nlme* package for handling multi-level analyses.

Chatfield (2003) is a relatively elementary introduction to time series. Brockwell and Davis (2002) is an excellent and practically oriented introduction, with more mathematical detail than Chatfield. Diggle (1990) offers a more advanced practically oriented approach.

On meta-analysis see Chalmers and Altman (1995) and Gaver at al. (1992).

References for further reading

Analysis of variance with multiple error terms

Cochran, W.G. and Cox, G.M. 1957. *Experimental Designs*, 2nd edn. Wiley.

Cox, D.R. 1958. *Planning of Experiments.* Wiley.

Cox, D.R. and Reid, N. 2000. *Theory of the Design of Experiments.* Chapman and Hall.

Fisher, R.A. 1935 (7th edn. 1960). *The Design of Experiments.* Oliver and Boyd.

Payne, R.W., Lane, P.W., Digby, P.G.N., Harding, S.A., Leech, P.K., Morgan, G.W., Todd, A.D., Thompson, R., Tunnicliffe Wilson, G., Welham, S.J. and White, R.P. 1997. *Genstat 5 Release 3 Reference Manual.* Oxford University Press.

Williams, E.R., Matheson, A.C. and Harwood, C.E. 2002. *Experimental Design and Analysis for Use in Tree Improvement*, revised edn. CSIRO Information Services, Melbourne.

Multi-level models and repeated measures

Diggle, P.J., Heagerty, P.J., Liang, K.-Y. and Zeger, S.L. 2002. *Analysis of Longitudinal Data*, 2nd edn. Clarendon Press.

Goldstein, H. 1995. *Multi-level Statistical Models.* Arnold.

Payne, R.W., Lane, P.W., Digby, P.G.N., Harding, S.A., Leech, P.K., Morgan, G.W., Todd, A.D., Thompson, R., Tunnicliffe Wilson, G., Welham, S.J. and White, R.P. 1997. *Genstat 5 Release 3 Reference Manual.* Oxford University Press.

Pinheiro, J.C. and Bates, D.M. 2000. *Mixed Effects Models in* S *and* S-PLUS. Springer-Verlag.

Snijders, T.A.B. and Bosker, R.J. 1999. *Multilevel Analysis. An Introduction to Basic and Advanced Multilevel Modelling.* Sage.

Talbot, M. 1984. Yield variability of crop varieties in the UK. *Journal of the Agricultural Society of Cambridge* 102: 315–321.

Venables, W.N. and Ripley, B.D. 2002. *Modern Applied Statistics with* S, 4th edn. Springer-Verlag.

Meta-analysis

Chalmers, I. and Altman, D.G. 1995. *Systematic Reviews.* BMJ Publishing Group.

Gaver, D.P., Draper, D.P., Goel, K.P., Greenhouse, J.B., Hedges, L.V., Morris, C.N. and Waternaux, C. 1992. *Combining Information: Statistical Issues and Opportunities for Research.* National Research Council, National Academy Press.

Time series

Brockwell, P. and Davis, R.A. 2002. *Time Series: Theory and Methods.* 2nd edn, Springer-Verlag.

Chatfield, C. 2003. *The Analysis of Time Series: an Introduction.* 6th edn, Chapman and Hall/CRC.

Diggle, P. 1990. *Time Series: a Biostatistical Introduction.* Clarendon Press.

9.10 Exercises

1. Repeat the calculations of Subsection 9.3.5, but omitting results from two vines at random. Here is code that will handle the calculation:

```
n.omit <- 2
take <- rep(T, 48)
take[sample(1:48,2)] <- F
kiwishade.lme <- lme(yield ~ shade, random = ~1 |
                     block/plot, data = kiwishade,subset=take)
VarCorr(kiwishade.lme)[4, 1]  # Plot component of variance
VarCorr(kiwishade.lme)[5, 1]  # Vine component of variance
```

Repeat this calculation five times, for each of n.omit = 2, 4, 6, 8, 10, 12 and 14. Plot (i) the plot component of variance and (ii) the vine component of variance, against number of points omitted. Based on these results, for what value of n.omit does the loss of vines begin to compromise results? Which of the two components of variance estimates is more damaged by the loss of observations? Comment on why this is to be expected.

2. Repeat the previous exercise, but now omitting 1, 2, 3, 4 complete plots at random.

3. A time series of length 100 is obtained from an AR(1) model with $\sigma = 1$ and $\alpha = -.5$. What is the standard error of the mean? If the usual $\sigma/\sqrt{n}$ formula were used in constructing a confidence interval for the mean, with σ defined as in Section 9.5.3, would it be too narrow or too wide?

4. Use the ar function to fit the second order autoregressive model to the Lake Huron time series.

5. The data set Gun (*nlme* package) reports on the numbers of rounds fired per minute, by each of nine teams of gunners, each tested twice using each of two methods. In the nine teams, three were made of men with slight build, three with average, and three with heavy build. Is there a detectable difference, in number of rounds fired, between build type or between firing methods? For improving the precision of results, which would be better – to double the number of teams, or to double the number of occasions (from 2 to 4) on which each team tests each method?

6. *The data set ergoStool (*nlme* package) has data on the amount of effort needed to get up from a stool, for each of nine individuals who each tried four different types of stool. Analyze the data both using aov() and using lme(), and reconcile the two sets of output. Was there any clear winner among the types of stool, if the aim is to keep effort to a minimum?

7. *In the data set MathAchieve (*nlme* package), the factors Minority (levels yes and no) and sex, and the variable SES (socio-economic status) are clearly fixed effects. Discuss how the decision whether to treat School as a fixed or as a random effect might depend on the purpose of the study? Carry out an analysis that treats School as a random effect. Are differences between schools greater than can be explained by within school variation?

8. The function Box.test() (in the *ts* package) may be used to compare the straight line model with uncorrelated errors that was fitted in Section 9.6, against the alternative of autocorrelation at some lag greater than zero. Try, e.g.

```
Box.test(resid(lm(detrain ~ detSOI, data=detsoi)),
         type="Ljung-Box", lag=20)
```

It is necessary to guess at the highest possible lag at which an autocorrelation is likely. The number should not be too large, so that the flow-on from autocorrelation at lower lags is still evident. A common, albeit arbitrary, choice is 20, as here. Try the test with lag set to values of 1 (the default), 5, 20, 25 and 30. Comment on the different results.

10

Tree-based Classification and Regression

Tree-based methods, or *decision tree* methods, may be used for two broad types of problem –
classification, and regression. These methods are derived from an earlier methodology that
had the name "automatic interaction detection". They may be appropriate when there are
extensive data, and there is uncertainty about the form in which explanatory variables ought
to enter into the model. They may be useful for initial data exploration. Tree-based methods
have been especially popular among those who call themselves data miners.

Tree-based regression and classification methodology is radically different from the methods discussed thus far in this monograph. The theory that underlies those methods has limited
relevance to tree-based methods. The tree-based methodology is relatively easy to use and
can be applied to a wide class of problems. It is at the same time insensitive to the nuances
of particular problems to which it may be applied. It is not a precision tool! It is not the
right tool for every problem.

The methodology makes weak assumptions about the form of the regression model. It
makes limited use of the ordering of values of continuous or ordinal explanatory variables.
In small data sets, it is unlikely to reveal data structure. Its strength is that, in large data
sets, it has the potential to reflect relatively complex forms of structure, of a kind that may
be hard to detect with conventional regression modeling. It is well-suited for use with data
sets that are so large that we can afford to ignore information.

There has been extensive work on alternatives to regression tree modeling that build in
less structure than classical regression modeling, but more structure than classification and
regression trees. Note also that the inferences that we describe assume that the random part
of the model is independent between observations.

Note

This chapter relies on the *rpart* package to handle the calculations. Before starting, ensure
that it is installed. Entry of `library(rpart)` will then load it.

10.1 The Uses of Tree-based Methods

10.1.1 Problems for which tree-based regression may be used

We will consider four types of problem:

* regression with a continuous outcome variable,
* binary regression,

- ordered classification problems where there are more than two outcomes,
- unordered classification problems.

There are other possibilities, including tree-based survival analysis, which the *rpart* package implements.

We will use simple regression examples with a continuous outcome variable to illustrate the methodology. However, the main focus will be on classification with two outcomes, i.e., to problems that can be treated as binary regression problems. For such binary outcomes, the use of a classification tree is a competitor for a logistic or related regression. Unordered classification with multiple outcomes is a simple extension, which we do not however have room to include in this chapter.

Several simple toy data sets, i.e., data sets that have been constructed to demonstrate how the methodology works, will be used as a basis for initial discussion. Also used mainly for illustrating the methodology is a data set that gives mileage versus weight, for cars described in US April 1990 *Consumer Reports*. Data sets that demonstrate realistic applications of the methodology are:

- A data set on email spam. For simplicity, our discussion will limit attention to 6 out of the 57 explanatory variables on which there is information. The choice is based on the first author's intuition, educated by his own experience of spam mail!
- A data set, with 11 explanatory variables, that examines the mortality of hospitalized female heart attack patients.

10.1.2 Tree-based regression versus parametric approaches

Tree-based approaches can be used in tandem with parametric approaches, in ways that combine the strengths of the different approaches. Thus, tree-based regression may suggest interaction terms that ought to appear in a parametric model. We might also apply tree-based regression analysis to residuals from conventional parametric modeling, in order to check whether there is residual structure that the parametric model has not captured. Another variation is to apply tree-based regression to fitted values of a parametric model, in order to cast predictions in the form of a decision tree.

In small data sets, it may be necessary to use parametric models that build in quite strong assumptions, e.g., no interactions, and the assumption of linear effects for the main explanatory variables, in order to gain adequate predictive power. Results are, as a consequence, likely to contain some bias, but this is often a price worth paying. In larger data sets, we can make weaker assumptions, e.g., allow simple forms of curve rather than lines and include interactions. Tree-based regression may be helpful for data sets that are so large that very limited assumptions will allow useful inference. Even for such data sets, however, interesting structure may be missed because assumptions have been too weak. Also, even given its limited assumptions, it does not make optimal use of the data. While each local split is optimal, the overall tree may be far from optimal.

Note again the point made earlier, that tree-based regression may be a useful adjunct to parametric regression. Exploration of a new data set with tree-based regression or classification may be a good way to gain a quick handle on which variables have major effects on the outcome.

10.1.3 Summary of pluses and minuses

Here is a more detailed comparison that matches tree-based regression against the use of linear models and generalized additive models. Strengths of tree-based regression include:

- Results are invariant to a monotone re-expression of explanatory variables.
- The methodology is readily adapted to handle missing values, without omission of complete observations.
- Tree-based regression is adept at capturing non-additive behaviour. Interactions are automatically included.
- It handles regression, and in addition unordered and ordered classification.
- Results are in an immediately useful form for classification or diagnosis.

Weaknesses include:

- The overall tree may not be optimal. The methodology assures only that each split will be optimal.
- Continuous predictor variables are treated, inefficiently, as discrete categories.
- Assumptions of monotonicity or continuity across category boundaries are lost.
- Low order interaction effects do not take precedence over higher order interactions.
- Limited notions of what to look for may result in failure to find useful structure.
- It may obscure insights that are obvious from parametric modeling, e.g., a steadily increasing risk of cancer with increasing exposure to a carcinogen.
- Large trees make poor intuitive sense; their predictions must be used as black boxes.

10.2 Detecting Email Spam – an Example

The originator of these data, George Forman, of Hewlett–Packard Laboratories, collected 4601 email items, of which 1813 items were identified as spam (these data are available from the University of California at Irvine Repository of Machine Learning Databases and Domain Theories. The address is http://www.ics.uci.edu/~mlearn/MLRepository.html).

Here, for simplicity, we will work with 6 of the 57 explanatory variables on which Forman collected information. In Figure 10.1 (untransformed data in the upper series of panels; log transformed data in the lower series of panels) we show boxplots for 6 of the variables. In the lower series of panels, we added 0.5 before taking logarithms.[1]

The explanatory variables are

- `crl.tot`, total length of words that are in capitals,
- `dollar`, the frequency of the $ symbol, as a percentage of all characters,
- `bang`, the frequency of the ! symbol, as a percentage of all characters,
- `money`, frequency of the word "money", as a percentage of all words,
- `n000`, frequency of the text string "000", as a percentage of all words,
- `make`, frequency of the word "make", as a percentage of all words.

The outcome is the variable `yesno`, which is n for non-spam and y for spam.

---

[1] ## To get another sample and to obtain one of the plots, type in:
```
data(spam7)        # DAAG package
spam.sample <- spam7[sample(seq(1,4601),500,replace=FALSE), ]
boxplot(split(spam.sample$crl.tot, spam.sample$yesno))
```

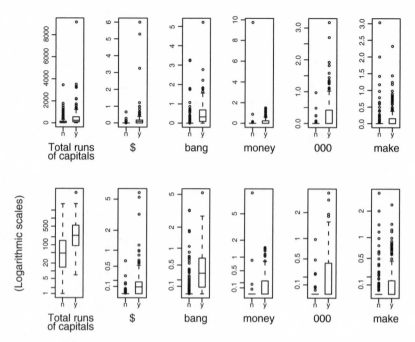

Figure 10.1: Boxplots for the 6 selected explanatory variables, in a random sample of 500 out of 4601 rows (email messages) in the SPAM database. (Presenting data for a sub-sample makes it easier to see the pattern of outliers.) The upper series of panels are for untransformed data, while in the lower series of panels 0.5 was added to each variable prior to logarithmic transformation.

If we were to use logistic regression or another parametric technique with these data, use of transformations, for the final five variables, more severe than the logarithmic transformation used in the lower series of panels, would be essential. For tree-based regression, this is not necessary. This is fortunate, because it is not straightforward to find transformations that work well with these data.

Now consider a tree-based regression, using the same 6 variables as predictors. Except for the setting method="class" that ensures that we get a classification tree, we use default settings of control parameters.[2] Figure 10.2 shows the tree.

Reading the tree is done as follows. If the condition that is specified at a node is satisfied, then we take the branch to the left. Thus, dollar ($) < 0.0555 and bang (!) < 0.0915 leads to the prediction that the email is not spam.

Table 10.1 displays information on the predictive accuracy of the tree:[3] It gives two types of error rate:

(1) The final row of the column headed rel error gives the relative error rate for predictions for the data that generated the tree: 35%. To determine the absolute error

---

[2] `library(rpart)`
`spam.rpart <- rpart(formula = yesno ~ crl.tot + dollar + bang +`
`                    money + n000 + make,  method="class", data=spam7)`
`plot(spam.rpart)`
`text(spam.rpart)`
[3] `options(digits=5)`
`printcp(spam.rpart)`

Table 10.1: *The final row gives information on the performance of the decision tree in Figure 10.2. Earlier rows show the performances of trees with fewer splits.*

```
> printcp(spam.rpart)

Classification tree:
rpart(formula = yesno ~ crl.tot + dollar + bang + money +
    n000 + make, method="class", data = spam7)

Variables actually used in tree construction:
[1] bang     crl.tot dollar

Root node error: 1813/4601 = 0.394

n= 4601

      CP nsplit rel error xerror    xstd
1 0.4766      0     1.000  1.000 0.0183
2 0.0756      1     0.523  0.557 0.0155
3 0.0116      3     0.372  0.386 0.0134
4 0.0105      4     0.361  0.378 0.0133
5 0.0100      5     0.350  0.377 0.0133
```

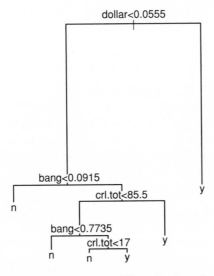

Figure 10.2: Output from the function `rpart()`, with `method="class"`, for classification with the email spam data set, using the 6 explanatory variables noted in the text. At each split, observations for which the condition is satisfied take the branch to the left.

rate, multiply this by the root node error, giving an absolute error rate of $0.394 \times 0.35 =$ 0.138, or 13.8%. This is often called the *resubstitution* error rate. For prediction of the error rate in a new sample, it is optimistic.

(2) The column headed `xerror` presents the more useful measure of performance. The x in `xerror` is an abbreviation for cross-validated. The cross-validated error rate is $0.394 \times 0.377 = 0.149$, or 14.9%.

The cross-validated error rate estimates the expected error rate for use of the prediction tree with new data that are sampled in the same way as the data used to derive Figure 10.2. Examination of the cross-validated error rate suggests that four splits may be marginally better than three splits. It is unclear whether more than four splits would lead to a tree with greater predictive power. The use of five or more splits has not, at this point, been investigated.

10.2.1 Choosing the number of splits

In classical regression, the inclusion of too many explanatory variables may lead to a loss of predictive power, relative to a more parsimonious model. With tree-based regression, the more immediate issue is the number of splits, rather than the number of explanatory variables. The use of cross-validation error estimates to choose the size of the tree protects against choosing too many splits. For this problem, it turns out to be advantageous to include all 6 explanatory variables in the tree-based regression. Readers who wish to pursue investigation of these data are encouraged to carry through the style of analysis that we describe with the complete data, available from the internet web site given at the beginning of this section. We will return to this example later, still working with just 6 explanatory variables.

The next section will explain the terminology and the methodology for forming trees. From there, the discussion will move to showing the use of cross-validation for assessing predictive accuracy and for choosing the optimal size of tree. Key concerns are to find an optimum tree size, and to get an unbiased assessment of predictive accuracy.

10.3 Terminology and Methodology

The simple tree in Figure 10.3 illustrates basic nomenclature and labeling conventions.[4] In Figure 10.3, there is a single factor (called `Criterion`) with five levels. By default, the function `text.rpart()` labels the levels a, b, ... , in order. Each *split* carries a label that gives the decision rule that was used in making the split, e.g., `Criterion=ac`, etc. The label gives the factor levels, or more generally the range of variable values, that lead to choosing the left branch. We have arranged, for this trivial example, that the outcome value is 1 for the level a, 2 for level b, etc.

---

[4]
```
## Code to plot such a tree is
 Criterion <- factor(paste("Leaf", 1:5))
 Node <- c(1,3,2,4,5)
 demo.df <- data.frame(Criterion = Criterion, Node = Node)
 demo.rpart <- rpart(Node ~ Criterion, data = demo.df,
                     control = list(minsplit = 2, minbucket = 1))
 plot(demo.rpart, uniform=TRUE)
 text(demo.rpart)
```

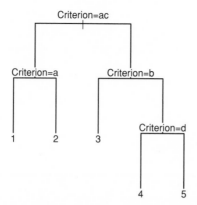

Figure 10.3: Tree labeling. For this illustrative tree, the only term used to determine the splitting is a factor with the name "Criterion". The *split* labels give the factor levels (always coded a, b, . . .) that lead to choosing the left branch. More generally, different variable or factor names may appear at different nodes. Terminal nodes (*leaves*) are numbered 1, . . . , 5.

There are nine *nodes* in total, made up of four splits and five terminal nodes or *leaves*. In this example, the leaves are labeled 1, 2, . . . , 5. The number of splits is always one less than the number of leaves. Output from the *rpart* package uses the number of splits as a measure of the size of the tree.

10.3.1 Choosing the split – regression trees

We first describe the anova splitting rule. To ensure use of the anova spliting rule, call `rpart()` with `method="anova"`. This splitting rule is the default when the outcome variable is a continuous or ordinal variable. The anova splitting rule minimizes the "residual sum of squares", calculated in a manner that we will describe below.

Let $\mu_{[j]}$ be the mean for the cell to which y_j is currently assigned. Then the residual sum of squares is

$$D = \sum_j (y_j - \mu_{[j]})^2.$$

Calculations proceed as in forward stepwise regression. At each step, the split is chosen so as to give the maximum reduction in D. Observe that D is the sum of the "within cells" sums of squares.

Prior to the first split, the deviance is

$$D = \sum_j (y_j - \bar{y})^2.$$

The split will partition the set of subscripts j into two subsets – a set of j_1s (write $\{j_1\}$) and a set of j_2s (write $\{j_2\}$). For any such partition,

$$D = \sum_j (y_j - \bar{y})^2 = \sum_{j_1} (y_{j_1} - \bar{y}_1)^2 + \sum_{j_2} (y_{j_2} - \bar{y}_2)^2 + n_1(\bar{y}_1 - \bar{y})^2 + n_2(\bar{y}_2 - \bar{y})^2$$

Table 10.2: *Toy example, used
to illustrate the methodology
used in generating a tree.*

fac	x	y
left	1	1
right	1	2
left	2	3
right	2	4
left	3	5
right	3	6

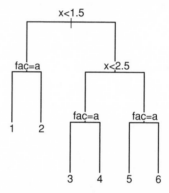

Figure 10.4: This illustrative tree is generated from one explanatory factor (`fac`) and one explanatory variable (x). For factors, the levels that lead to a branch to the left are given. The levels are labeled a, b, ... , rather than with the actual level names. For variables, the range of values is given that leads to taking the left branch.

where the first two terms comprise the new "within group" sum of squares, and the final two terms make up the "between group" sum of squares. If the values y_j are ordered, then the partition between $\{j_1\}$ and $\{j_2\}$ must respect this ordering, i.e., the partitioning is defined by a cutpoint. If the outcome variable is an unordered factor, i.e., if the values are unordered, then every possible split into $\{j_1\}$ and $\{j_2\}$ must in principle be considered.

The split is chosen to make the sum of the first two terms as small as possible, and the sum of the final two terms as large as possible. The approach is equivalent to maximizing the between groups sum of squares, as in a one-way analysis of variance.

For later splits, each of the current cells is considered as a candidate for splitting, and the split is chosen that gives the maximum reduction in the residual sum of squares.

10.3.2 Within and between sums of squares

Typically there will be several explanatory variables and/or factors. Figure 10.4 uses the data in Table 10.2. There are one factor (with levels "left" and "right") and one variable. Observe

Table 10.3: *Comparison of candidate splits at the* root, *i.e., at the first split.*

	Overall mean	Cell means	"Within" sum of squares	"Between" sum of squares
fac=="a" (y==1,3,5) versus fac=="b" (y==2,4,6)	3.5	3, 4	$8 + 8 = 16$	$3 \times (3 - 3.5)^2 + 3 \times (4 - 3.5)^2 = 1.5$
x==1 (y==1,2) versus x==2 or 3 (y==3,4,5,6)	3.5	1.5, 4.5	$0.5 + 5.0 = 5.5$	$2 \times (1.5 - 3.5)^2 + 4 \times (4.5 - 3.5)^2 = 12$
x==1 or 2 (y=1,2,3,4) versus x==3 (y==5, 6)	3.5	2.5, 5.5	$5.0 + 0.5 = 5.5$	$4 \times (2.5 - 3.5)^2 + 2 \times (5.5 - 3.5)^2 = 12$

that, here, $y_j = j$ ($j = 1, 2, \ldots, 6$). Table 10.3 shows the sums of squares calculations for the candidate splits at the root.

The split between fac=="a" and fac=="b" gives a between sum of squares $= 1.5$, and a "within" sum of squares $= 16$. Either of the splits on x (between $x = 1$ and $x > 1$, or between $x < 3$ and $x = 3$) give a within sum of squares equal to 5.5. As the split on x leads to the smaller "within" sum of squares, it is the first of the splits on x that is chosen. (The second of the splits on x might equally well be taken. The tie must be resolved somehow!)

The algorithm then looks at each of the subcells in turn, and looks at options for splits within each subcell. The split chosen is the one that gives the largest "between" sum of squares, for the two new subcells that are formed.

10.3.3 Choosing the split – classification trees

If the dependent variable is a factor the default is to fit a classification tree, with the factor levels specifying the classes. To specifically request a classification tree, include the parameter setting method="class" in the rpart() call.

As we saw, the split that minimizes the residual sum of squares is the usual choice for regression trees, but there are alternatives. For classification trees, there is more variety among software programs in the measure that is used. The *rpart* documentation and output use the generic term *error* for whatever measure is used.

The classes (indexed by k) are the categories of the classification. Then n_{ik} is the number of observations at the ith leaf who are assigned to class k. The n_{ik} are used to estimate the proportions p_{ik}. Each leaf becomes, in turn, a candidate to be the node for a new split.

For the split at the ith node of a classification tree, *rpart* uses a modified version of the Gini index

$$\sum_{j \neq k} p_{ij} p_{ik} = 1 - \sum_{k} p_{ik}^2$$

as its default measure of "error", or "impurity". The deviance $D_i = \sum_{\text{classes } k} n_{ik} \log(p_{ik})$ is available in *rpart* as an alternative. The deviance differs from the entropy, another measure that is sometimes used, by a constant factor, and thus would give the same tree if the same stopping rule were used. For the two-class problem (a binary classification), the Gini index and the deviance will almost always choose the same split.

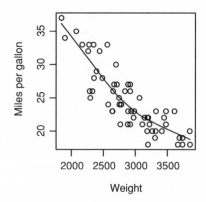

Figure 10.5: `Mileage` versus `Weight`, for cars described in US April 1990 *Consumer Reports*.

The splitting rule, if specified, is set by specifying, e.g., `parms=list(split=gini)`. A classification tree is the default when the outcome variable is a factor.

10.3.4 The mechanics of tree-based regression – a trivial example

As well as demonstrating how tree-based regression works, this example is designed to highlight differences between tree-based regression and other approaches. The data are from the US *Consumer Reports* for April 1990. It would be silly to make serious use of tree-based regression for these data. Nonetheless useful insights may be gained by comparing the predictions from tree-based regression with the predictions from another approach. The `loess` routine from the R *modreg* package gave the smooth curve in Figure 10.5.[5] To fit a regression tree to the car mileage data that are displayed in Figure 10.5, we specify the outcome variable (`Mileage`) and the predictor variable (`Weight`).[6]

Figure 10.6 shows the fitted regression decision tree.

Observe how the predictions of Figure 10.6 compare with the predicted values from the use of `loess` (Figure 10.5). Notice how the tree-based regression has given several anomalous predictions. Later splits have relied on information that is too local to improve predictive ability. Figure 10.7 is the plot that results when the split criteria are left at their defaults.[7] We obtain much coarser prediction.

Table 10.5 compares the predictions for this less ambitious tree with the predictions from the loess regression. Beyond a certain point, adding additional leaves reduces genuine

---

[5] `library(modreg)`
`data(car.test.frame)  # example data set in the rpart package`
`car.lo <- loess(Mileage ~ Weight, car.test.frame)`
`plot(car.lo, xlab="Weight", ylab="Miles per gallon")`
`lines(seq(1850,3850), predict(car.lo, data.frame(Weight=seq(1850,3850))))`
[6] `car.tree <- rpart(Mileage ~ Weight, data=car.test.frame,`
`                   control = list(minsplit = 10, minbucket = 5, cp = 0.0001))`
`plot(car.tree, uniform = TRUE)`
`text(car.tree, digits = 3, use.n = TRUE)`
[7] `car.tree <- rpart(Mileage ~ Weight, data = car.test.frame)`
`plot(car.tree, uniform = FALSE)`
`text(car.tree, digits = 3, use.n = TRUE)`

Table 10.4: *Predictions from the regression tree of Figure 10.6. For comparison, we give the range of predictions from the smooth curve that we fitted in Figure 10.5.*

Range of `Weight`	Predicted `Mileage`	Range of predictions (Figure 10.5)
1845.0 – 2280	34	36.4 – > 30.3
> 2280.0 – 2567.5	28.9	30.3 – > 26.9
> 2567.5 – 2747.5	25.6	26.9 – > 24.9
> 2747.5 – 2882.5	23.3	24.9 – > 23.9
> 2882.5 – 3087.5	24.1	23.9 – > 22.1
> 3087.5 – 3322.5	20.5	22.1 – > 20.7
> 3322.5 – 3637.5	22	20.7 – > 19.7
> 3637.5 – 3855	18.7	19.7 – > 18.9

Table 10.5: *Predictions from the regression tree of Figure 10.6.*

Range of weights	Predicted mileage	Range of loess predictions
– 2567.5	30.9	36.4 – > 26.9
> 2567.5 – 2747.5	25.6	26.9 – > 24.9
> 2747.5 – 3087.5	23.8	24.9 – > 22.1
> 3087.5	20.4	22.1 – > 18.9

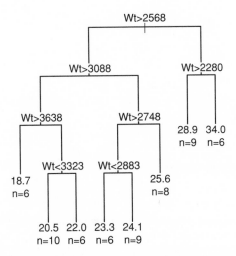

Figure 10.6: Tree-based model for predicting `Mileage` given `Weight`, for cars described in US April 1990 *Consumer Reports*. For present illustrative purposes, split criteria have been changed from the *rpart* default, to increase the number of splits. Notice that for this plot we have used uniform vertical spacing between levels of the tree.

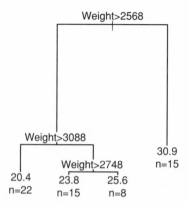

Figure 10.7: Tree-based model for the data of Figure 10.6, but with split criteria set to the *rpart* default, and with vertical spacing now set to reflect the change in residual sum of squares. In Figure 10.6, such non-uniform spacing would have given splits that were bunched up at the lower nodes.

predictive power, even though the fit to the data used to develop the predictive model must continue to improve.

10.4 Assessments of Predictive Accuracy

Realistic estimates of the predictive power of a model must take account of the extent to which the model has been selected from a wide range of candidate models. Two main approaches have been used for determining unbiased assessments of predictive accuracy. These are cross-validation, and the use of training and test sets.

10.4.1 Cross-validation

For cross-validation, the data are split into k subsets, where typically $3 \leq k \leq 10$. Each of the k subsets of the data is left out in turn, the regression is run for the remaining data, and the results used to predict the outcome for the subset that has been left out. As pointed out in our earlier discussion in Subsection 5.5.1, there are k different folds, i.e., k different divisions of the data between *training* set and *test* set.

The prediction error is then calculated. In a regression model, prediction error is usually taken as the squared difference between observed and predicted. In a classification model, prediction error is usually determined by counting 1 for a misclassification and 0 for a correct classification.

The crucial feature of cross-validation is that each prediction is independent of the data to which it is applied. As a consequence, cross-validation gives an unbiased estimate of predictive power. An estimate of average "error" is found by summing up the measure of "error" over all observations and dividing by the number of observations. Once predictions are available in this way for each of the subsets, the average error is taken as (total error)/(total number of observations). The number of splits, here 10, is under the control of the user.

The cross-validation methodology thus allows valid assessments of the predictive power of tree-based models, while using all the data for predictive purposes. Just as importantly, it yields a decision on the optimal size of tree. We will demonstrate its use in *rpart* to assess the extent to which a tree that has already been constructed should be pruned. We prune, from an over-fitted tree, to determine the tree that has the greatest predictive power, or equivalently the smallest cross-validation error.

Cross-validation is built into the *rpart* calculations. It gives a formal way to get a handle on how predictive power changes with tree size, while using all existing data. Cross-validation is computationally intensive.

10.4.2 The training/test set methodology

This methodology is suitable when there are tens of thousands of observations, at the level of variation that is relevant for the use of any model predictions. Thus, 60% of the data might be chosen to form the training set. The remaining 40% of the data will then be used to test the performance of the model.

Some methods require three sets of data. A second set of data – the holdout set – is required for making checks on predictive accuracy that are required as calculations proceed. Finally, results are tested on a third set – the test set. Testing performance under conditions of actual use, which may be different from those that generated the initial data, may require a validation set.

In k-fold cross-validation, there are k different divisions of the data between training and test set. Each of the k subsets becomes, in turn, a test set.

10.4.3 Predicting the future

The choice of test set should, in principle, relate to the way that model predictions will be used. If there is no grouping within the data, then random division between test and training set is appropriate. Consider, however, a case where data are from a (preferably random) sample of schools in New South Wales (Australia's most populous state), and will be applied to all schools in the state. The training set should then consist of pupils from a randomly chosen subset of schools, with pupils from remaining schools making up the test set. Cross-validation should, similarly, divide schools into the chosen number k of splits.

10.5 A Strategy for Choosing the Optimal Tree

We will begin by defining a measure that balances lack of fit against tree complexity.

10.5.1 Cost–complexity pruning

Define the size of the tree to be the number of splits. Let D be the measure of lack of fit ("error"), such as the residual sum of squares, which is summed over all nodes. Set

$$D_\alpha = D + \mathrm{cp} \times \mathrm{size},$$

where cp is a mnemonic for *complexity parameter*. We write cp rather than c_p, because this is what appears in the *rpart* output.

Then we will take the tree that minimizes D_α to be optimal for the given value of cp. The value of D decreases as we increase the size, while cp × size increases. Clearly the optimal tree size increases as cp decreases. It may be shown that, as we increase cp, we can move from one optimum (for one value of cp) to the next (for some larger value of cp) by a sequence of 0 or more snips of the current tree. Each choice of cp thus determines an optimal tree size. Splitting continues until the tree size is reached that is optimal for this value of cp.

It follows that, given any tree, we can thus identify a sequence of prunings from the constructed tree back to the root, such that

- at each pruning the complexity reduces,
- each tree has the smallest number of nodes possible for that complexity, given the previous tree.

A graphical presentation of cost–complexity pruning information can be useful as an informal mechanism for gaining a rough assessment of optimal tree size. A more precise method is to choose, from among the trees in the sequence that is identified, the tree that has near optimal expected predictive power on a new data set. Either predictive power may be estimated directly on a new data set, or it may be a cross-validation estimate that uses the existing data. The rpart() function is set up to do cross-validation.

We emphasize that the resubstitution error rate, because it is calculated from the data used to determine the tree, can never increase as tree size increases. It thus gives an optimistic assessment of relative error, and is of no use in deciding tree size.

Measures of lack of fit

An obvious choice to measure the lack of fit D of a tree is the measure that was used in forming the tree – the residual sum of squares, the Gini index, or the deviance. For classification trees, there is another choice – fraction or percent misclassified. As noted above, we use the term *error* for the particular measure with which we have chosen to work.

10.5.2 Prediction error versus tree size

In Figure 10.8, we return to the car mileage data of Figure 10.5 and Table 10.5. Figures 10.8A and 10.8B plot, against tree size, different assessments of the relative error. As we have a continuous outcome variable (Mileage), the relative error is the sum of squares of residual divided by the total sum of squares.

Figure 10.8A plots the resubstitution assessment of relative error, which shows the performance of the tree-based prediction on the data used to form the tree. Figure 10.8B shows estimates from three cross-validation runs, with the data split into $k = 10$ subsets (or *folds*) at each run. The plot gives an indication of the variability that can be expected, for these data, from one cross-validation run to another.

One way to choose the optimal tree is to look, in Figure 10.8, for the smallest tree that has near minimum cross-validated relative error. In Figure 10.8, all three runs suggest that the optimal tree size is 3. The point at which the cross-validation estimate of error starts to increase will differ somewhat between runs. For this reason, several cross-validation runs are recommended.

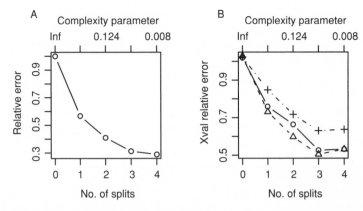

Figure 10.8: While the relative error (A) must inevitably decrease as the number of splits increases, the cross-validation relative error (X-val relative error, shown in B) is likely, once the number of splits is large enough, to increase. Results from three cross-validation runs are shown. Plots are for the car mileage data of Figure 10.5 and Table 10.5.

As noted earlier, cross-validation estimates are a realistic estimate of the performance of the tree on new data, provided these are sampled in the same way as the available data.

10.6 Detecting Email Spam – the Optimal Tree

In Figure 10.2, where the control parameter cp had its default value of 0.01, splitting did not continue long enough for the cross-validated relative error to reach a minimum. In order to find the minimum, we now repeat the calculation, this time with cp = 0.001.[8] The choice of cp = 0.001 in place of cp = 0.01 was, when we first did the calculation, a guess. It turns out to carry the splitting just far enough that the cross-validated relative error starts to increase.

An alternative to choosing the tree that minimizes the cross-validated relative error is to use the one-standard-error rule that we will explain below. The output is available both in graphical and in tabular form. Figure 10.9 shows the information in graphical form,[9] while Table 10.6 shows the information in tabular form.[10]

Choice of the tree that minimizes the cross-validated error leads to nsplit=17 with xerror=0.33. Again, we note that different runs of the cross-validation routine will give slightly different results.

10.6.1 The one-standard-deviation rule

The function rpart() calculates, for each tree, both the cross-validated estimate of "error" and a standard deviation for that error. Where the interest is in which splits are likely to be meaningful, users are advised to choose the smallest tree whose error is less than

$$\text{minimum error} + 1 \text{ standard deviation.}$$

[8] spam7a.rpart <- rpart(formula = yesno ~ crl.tot + dollar + bang + money +
                         n000 + make, method="class", data = spam7, cp = 0.001)
[9] plotcp(spam7a.rpart)
[10] printcp(spam7a.rpart)

Table 10.6: *Printed summary of the cross-validation information, for the tree that was presented in Figure 10.9.*

```
> printcp(spam7a.rpart)

Classification tree:
rpart(formula = yesno ~ crl.tot + dollar + bang + money +
    n000 + make, data = spam7, method = "class",
    cp = 0.001)

Variables actually used in tree construction:
[1] bang     crl.tot dollar   money     n000

Root node error: 1813/4601 = 0.394

n= 4601

          CP nsplit rel error xerror    xstd
1  0.47656      0     1.000  1.000 0.0183
2  0.07557      1     0.523  0.540 0.0153
3  0.01158      3     0.372  0.386 0.0134
4  0.01048      4     0.361  0.376 0.0133
5  0.00634      5     0.350  0.368 0.0132
6  0.00552     10     0.317  0.368 0.0132
7  0.00441     11     0.311  0.358 0.0130
8  0.00386     12     0.307  0.346 0.0128
9  0.00276     16     0.291  0.336 0.0127
10 0.00221     17     0.288  0.330 0.0126
11 0.00193     18     0.286  0.334 0.0127
12 0.00165     20     0.282  0.333 0.0126
13 0.00100     25     0.274  0.330 0.0126
```

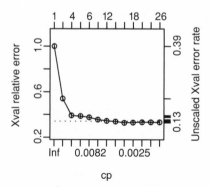

Figure 10.9: Change in cross-validated error as the size of tree changes, for prediction of email spam. The labeling on the left axis is from the *rpart* plotcp() function, while the labeling on the right vertical axis is our addition.

Figure 10.9 has a dotted horizontal line that shows where this error level is attained. The rule is conservative if the interest is in choosing the optimal predictive tree, i.e., the predictive power will on average be slightly less than the optimum. Use of this rule suggests taking nsplit=16. (From Table 10.6, minimum + standard error = 0.330 + 0.013 = 0.343. The smallest tree whose xerror is less than or equal to this has nsplit = 16.) Figure 10.10 plots this tree.[11]

For classification trees, using number misclassified, the standard deviation is, assuming independent errors, approximately equal to the square root of the number misclassified.

10.7 Interpretation and Presentation of the *rpart* Output

We now provide additional information on printed and graphical output. The data are for the mortality of female heart attack patients. There is information on 1295 subjects.

10.7.1 Data for female heart attack patients

A summary of the female heart attack data follows.[12]

```
> summary(mifem)

  outcome          age            yronset        premi      smstat
  live:974    Min.    :35.0   Min.    :85.0   y :311     c :390
  dead:321    1st Qu.:57.0   1st Qu.:87.0   n :928     x :280
              Median :63.0   Median :89.0   nk: 56     n :522
              Mean    :60.9   Mean    :88.8              nk:103
              3rd Qu.:66.0   3rd Qu.:91.0
              Max.    :69.0   Max.    :93.0
  diabetes highbp    hichol    angina     stroke
  y :248   y :813   y :452   y :472    y :  153
  n :978   n :406   n :655   n :724    n :1063
  nk: 69   nk: 76   nk:188   nk: 99    nk:  79

Notes:

premi = previous myocardial infarction event
For smstat, c = current   x = ex-smoker   n = non-smoker
        nk = not known
```

(Technically, these are patients who have suffered a *myocardial infarction*. Data are from the Newcastle (Australia) center of the Monica project; see the web site http://www.ktl.fi/monicaindex.html.)

---

[11] ## Use of prune.rpart() with cp between 0.00276 and 0.00386 will prune
  ## back to nsplit=16. Specify:
  spam7b.rpart <- prune(spam7a.rpart, cp=0.003)
  plot(spam7b.rpart, uniform=TRUE)
  text(spam7b.rpart, cex=0.75)
[12] data(mifem)   # DAAG package

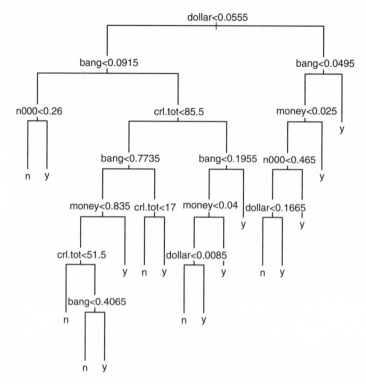

Figure 10.10: Decision tree for a tree size of 16 in Figure 10.9.

In order to fit the tree, we specify

```
mifem.rpart <- rpart(outcome ~ ., method="class",
                     data = mifem, cp = 0.0025)
```

The dot (.) on the right hand side of the formula indicates that all available explanatory variables are included. A choice of $cp = 0.002$ continues splitting to the point where the cross-validated relative error has started to increase. Figure 10.11 shows the change in cross-validated error rate as a function of tree size,[13] while Table 10.7 shows the same information in printed form.[14] Notice the increase in the cross-validated error rate when there are more than two splits.

10.7.2 Printed Information on Each Split

We will prune the tree back to `nsplit=1` and examine the split in more detail. We obtain the following output by setting `cp =0.006`, i.e., between the cp values for no split and one split:[15]

```
> print(mifemb.rpart)
n= 1295
```

---

[13] `plotcp(mifem.rpart)`
[14] `printcp(mifem.rpart)`
[15] `mifemb.rpart <- prune(mifem.rpart, cp=0.006)`
    `print(mifemb.rpart)`

Table 10.7: *Cross-validated error rate for different sizes of tree, for the female heart attack data.*

```
> printcp(mifem.rpart)

Classification tree:
rpart(formula = outcome ~ ., data = mifem, method = "class",
    cp = 0.0025)

Variables actually used in tree construction:
[1] age       angina   diabetes hichol   premi     smstat
    stroke    yronset

Root node error: 321/1295 = 0.248

n= 1295

        CP nsplit rel error xerror    xstd
1 0.20249      0     1.000  1.000 0.0484
2 0.00561      1     0.798  0.832 0.0454
3 0.00467     13     0.717  0.844 0.0456
4 0.00312     17     0.698  0.838 0.0455
5 0.00250     18     0.695  0.850 0.0457
```

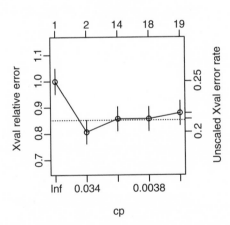

Figure 10.11: Change in cross-validated error as the size of tree changes, for the female heart attack data.

```
node), split, n, loss, yval, (yprob)
      * denotes terminal node

 1) root 1295 321 live ( 0.75 0.25 )
   2) angina:y, n 1196 239 live ( 0.80 0.20 ) *
   3) angina:nk 99  17 dead ( 0.17 0.83 ) *
```

Predictions are:[16]

- At the first split (the *root*) the prediction is `live` (probability 0.752), with the 321 who are `dead` misclassified. Here the `loss` (number misclassified) is 321.
- Take the left branch from node 1 if the person's angina status is `y` or `n`, i.e. if it is known (1196 persons). The prediction is `live` (probability 0.800), with the 239 who die misclassified.
- Take the right branch from node 1 if the angina status is unknown (99 persons). The prediction is `dead`, probability 0.828), with the 17 who are `live` misclassified.

The function `summary.rpart()` gives information on alternative splits.

10.8 Additional Notes

Models with a complex error structure

The methods that we have described are not well suited for use with data where the noise component has a structure such as we described in Chapter 9. If there is clustering in the data, i.e., there are groups of observations where the random component is correlated, then the cross-validation approach that we described is likely to choose an overly complex model. If the size of such groups varies widely, predictions may be strongly biased.

Pruning as variable selection

Venables and Ripley (2002) suggest that pruning might be regarded as a method of variable selection. This suggests using an AIC-like criterion as the criterion for optimality. We refer the reader to their discussion for details.

Other types of tree

The *rpart* package allows two other types of tree. The `poisson` splitting method adapts *rpart* models to event rate data. See Therneau and Atkinson (1997, pp. 35–41). The `survival` splitting method is intended for use with survival data, where each subject has either 0 or 1 events. (The underlying model is a special case of the Poisson event rate model.) See again Therneau and Atkinson (1997, pp. 41–46).

Factors as predictors

There are $2^{k-1} - 1$ ways to divide k factor levels into two groups, each with at least one of the levels. If a factor has a large number of levels (more than perhaps eight or ten), calculations readily become out of hand, and take an eternity to run. A way around this is to replace the k-level factor by k (or possibly $k - 1$) dummy variables. Splits then take place on these dummy variables one at a time. This leads to a different calculation that (except when $k = 2$) will run more quickly.

---

[16] `printcp(mifemb.rpart)`

Instead of obtaining optimality over all splits of factor levels into two groups, we obtain optimality at each split on a dummy variable given previous splits. Note that a sequence of splits that are (locally) optimal does not lead to a sequence that is, necessarily, optimal overall. Also, while there will be sequences of splits on the dummy variables that are equivalent to a single split of factor levels into two groups, it is unlikely that splitting on the dummy variables will find such a sequence.

The trade-off for this weakening of the optimality criterion is that a calculation that uses an inordinate amount of time, or will not run at all, may now run quite easily. Especially if there is a large number of observations, a sequence of splits on the dummy variables may do as well or just about as well as splits on the factor. A serious disadvantage is that a much more complicated tree results.

An ensemble of trees

There are a variety of approaches that build multiple trees, and then somehow aggregate the results. These approaches can for some data sets give a substantial improvement in predictive accuracy, by comparison with the *rpart* methodology. The *randomForest* package implements a "bagging" approach – a number of trees are independently constructed using bootstrap samples, i.e., using random with replacement sampling. In addition *randomForest* introduces a random element into the choice of predictors on which to split at each node. The prediction is determined by a simple majority vote across the multiple trees.

In "boosting" successive trees give extra weight to points that were incorrectly predicted by earlier trees, with a weighted vote finally taken to get a consensus prediction.

10.9 Further Reading

Venables and Ripley (2002) is the reference of choice for a terse discussion. See also Ripley (1996) and Chambers and Hastie (1992). Hastie et al. (2001) is a comprehensive treatment. They discuss new research directions, including such extensions of the methodology as bagging and boosting.

Therneau and Atkinson (pp. 50–52, 1997) give comparisons between *rpart* methodology and other software. Lim and Loh (2000) investigated 10 decision tree implementations, not however including *rpart*. Several of these implementations offer a choice of algorithms.

References for further reading

Chambers, J.M. and Hastie, T.J. 1992. *Statistical Models in* S. Wadsworth and Brooks-Cole Advanced Books and Software.

Hastie, T., Tibshirani, R. and Friedman, J. 2001. *The Elements of Statistical Learning. Data Mining, Inference and Prediction.* Springer-Verlag.

Lim, T.-S. and Loh, W.-Y. 2000. A comparison of prediction accuracy, complexity, and training time of thirty-three old and new classification algorithms. *Machine Learning* 40: 203–228.

Ripley, B.D. 1996. *Pattern Recognition and Neural Networks.* Cambridge University Press.

Therneau, T.M. and Atkinson, E.J. 1997. *An Introduction to Recursive Partitioning Using the RPART Routines.* Technical Report Series No. 61, Department of Health Science Research, Mayo Clinic, Rochester, MN, 2000. Available from http://www.mayo.edu/hsr/techrpt.html

Venables, W.N. and Ripley, B.D. 2002. *Modern Applied Statistics with* S-PLUS, 4th edn. Springer-Verlag.

10.10 Exercises

1. Refer to the `head.injury` data frame.

    (a) Use the default setting in `rpart()` to obtain a tree-based model for predicting occurrence of clinically important brain injury, given the other variables.
    (b) How many splits gives the minimum cross-validation error?
    (c) Prune the tree using the one-standard-error rule.

2. The data set `mifem` is part of the larger data set in the data frame `monica` that we have included in our *DAAG* package. Use tree-based regression to predict mortality in this larger data set. What is the most immediately striking feature of your analysis? Should this be a surprise?

3. Use tree-based regression to predict `re78` in the data frame `nsw74psid1` that is in our *DAAG* package. Compare the predictions with the multiple regression predictions in Chapter 6.

4. Copy down the email spam data set from the web site given at the start of Section 10.2. Carry out a tree-based regression using all 57 available explanatory variables. Determine the change in the cross-validation estimate of predictive accuracy.

11

Multivariate Data Exploration and Discrimination

In our discussion so far, we have made heavy use of exploratory methods that look both at variables individually and at variables in pairs, as a preliminary to regression modeling. For the latter purpose, scatterplot matrices have been an important tool. In this chapter the focus is more on the pattern presented by multiple variables. While the methodology has applications in a regression context, this is not its primary focus.

Principal component analysis (PCA) is a useful tool for the data exploration that we have in mind. It offers "views" of the data that may be insightful, and suggests the views that it may be worthwhile to examine first. It replaces the variables with which we started by new *principal component* variables, called *principal components*. The analysis places them in order, using as an ordering ("importance") criterion the amounts that the individual principal components contribute to the sum of the variances of the original variables. The most insightful plots are often, but by no means inevitably, those that involve the first few principal components.

As noted, the methods have application in regression and related analyses, in problems where it is helpful to reduce the number of candidate explanatory variables. We may, e.g., replace a large number of candidate explanatory variables by the first few principal components, hoping that they will adequately summarize the information in the candidate explanatory variables. If we are fortunate, it may happen that simple modifications of the components give new variables that are readily interpretable. In the analysis of morphometric data, the first component is often an overall measure of size.

Discriminant analysis methods are another important class of methods. In discrimination, observations belong to one of several groups. The aim is to find a rule, based on values of explanatory variables, that will, as far as possible, assign observations to their correct groups. This rule may then be used to classify new observations whose group may be unknown or unclear. An obvious way to evaluate classification rules is to examine their *predictive accuracy*, i.e., the accuracy with which they can be expected to assign new observations.

Discriminant analysis methodology is clearly an important methodology in its own right. In addition it is sometimes used, in a manner akin to one of the uses of PCA, to reduce the data to a small number of "discriminant components", i.e., linear combinations of the candidate variables, that for discrimination purposes sum up the information in those variables. Plots of these components can be useful for data exploration. In some applications, the components may become covariates in a regression analysis, rather than finding use for discrimination *per se*.

11.1 Multivariate Exploratory Data Analysis

Preliminary exploration of the data is as important for multivariate analysis as for classical regression. Faced with a new set of data, what helpful forms of preliminary graphical exploration are available? Here we illustrate a few of the possibilities.

The data set `possum` was described in Chapter 2. The interest is in finding the morphometric characteristics, if any, that distinguish possums at the different sites. For simplicity, we will limit attention to a subset of the morphometric variables.

11.1.1 Scatterplot matrices

It is good practice to begin by examining relevant scatterplot matrices. This may draw attention to gross errors in the data. A plot in which the sites and/or the sexes are identified will draw attention to any very strong structure in the data. For example, one site may be quite different from the others, for some or all of the variables.

There are too many measurements to put them all into one scatterplot matrix. We will therefore look at a selection of variables. We invite the reader to examine the scatterplot matrix for variables that we omit.

The choice of variables in Figure 11.1 anticipates results of later analysis, showing the variables that discriminate the populations most effectively.[1] Already we notice two clusters of values, with some of the six sites more strongly represented in one cluster than in the other.

11.1.2 Principal components analysis

Principal components analysis can be a useful exploratory tool for multivariate data. As noted in the introductory comment at the beginning of the chapter, the idea is to replace the original variables by a small number of "principal components" – linear combinations of the initial variables, that together may explain most of the variation in the data. A useful starting point for thinking about principal component analysis may be to imagine a two-dimensional scatterplot of data that has, roughly, the shape of an ellipse. Then the first principal component coincides with the longer axis of the ellipse. For example, if we had just the two variables `footlgth` (foot length) and `earconch` (ear conch length) from the `possum` data set, then the first principal component would have the

---

[1]
```
## The use of different colors for different sexes, and different
## symbols for different sites, makes for intricate code.
library(lattice)
data(possum)                                    # DAAG package
colr <- c("red", "blue")
pchr <- c(1,3,6,0,5,6,17)
ss <- expand.grid(site=1:7, sex=1:2)            # Site varies fastest
ss$sexsite <- paste(ss$sex, ss$site, sep="-")   # Site varies fastest
sexsite <- paste(possum$sex, possum$site, sep="-")
splom(~ possum[, c(9:11)], panel = panel.superpose,
      groups = sexsite, col = colr[ss$sex], pch = pchr[ss$site],
      varnames=c("tail\nlength","foot\nlength","ear conch\nlength"),
      key = list(points = list(pch=pchr),
                 text=list(c("Cambarville","Bellbird","Whian Whian","Byrangery",
                           "Conondale ","Allyn River", "Bulburin")),
                 columns=4))
```

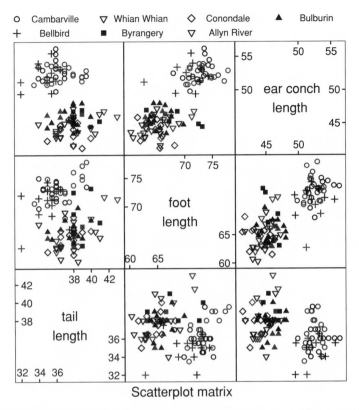

Scatterplot matrix

Figure 11.1: Scatterplot matrix for three morphometric measurements on the mountain brushtail possum. Females are in red; males in blue. For color version, see plate section between pages 200 and 201.

direction of a line from the bottom left to the top right of the scatterplot in the first row and second column of Figure 11.1. (We might equally well examine the plot in the second row and third column, where the *x*- and *y*-axes are interchanged.) In three dimensions we might think about a scatter of data that has the shape of an ellipsoidal object, e.g., the shape of a mango or marrow or other similarly shaped fruit or vegetable, but perhaps flattened on one side. The first principal component is the longest axis. If the first principal component explains most of the variation, then we have a long thin object.

Clearly the scaling of variables is important. It makes sense, in some circumstances, to work with the data as they stand, and in others to standardize values of each variable by dividing by the standard deviation. For morphometric (shape and size) data such as in the `possum` data frame, use of the logarithms of variables has the benefit that it places all variables on a scale on which equal distances correspond to equal relative changes. Columns 6 to 14 in the `possum` data frame are all essentially linear measurements. For the `possum` data, taking logarithms makes very little difference to the appearance of the plots. This is because the ratio of largest to smallest value is relatively small, never more than 1.6.[2] Thus

---

[2] ## Determine the ratios of largest to smallest values
```
sapply(possum[,6:14], function(x)max(x, na.rm=TRUE)/min(x, na.rm=TRUE))
```

for simplicity, in our discussion, we will not use logarithmically transformed variables, instead leaving examination of the effect of a logarithmic transformation as an exercise for the reader.

In order to make the coefficients in the linear combinations unique, there must be a constraint on their magnitudes. The `princomp()` function makes the squares of the coefficients sum to one. Technically, the first principal component is then the component that explains the greatest part of the variation. The second principal component is the component that, among linear combinations of the variables that are uncorrelated with the first principal component, explains the greatest part of the remaining variation, and so on.

Different pairs of principal components allow different two-dimensional views of the data. Consistently with the notion that the components that account for the largest part of the variation may be the most "important", it is usual to begin by plotting the second principal component against the first principal component. Or, we might use a scatter-plot matrix form of presentation to examine all the pairwise plots for the first three or four principal components. In the absence of other clues on what views of the data may be useful to examine, such principal components plots are a reasonable starting point.

The commands we use to do the analysis are

```
library(mva)              # Load multivariate analysis package
possum.prc <- princomp(na.omit(possum[,6:14]))
```

Notice the use of the `na.omit()` function to remove rows that contain missing values, prior to the principal components analysis. Note again that, for present tutorial purposes, we have chosen to work with the variables as they stand. Figure 11.2 plots the second principal component against the first, for variables 6 to 14 in the possum data.[3]

Here are further details of the principal components output.[4] We have edited the output so that the "importance of components" information is printed to two decimal places only:

```
Importance of components:
                           Comp.1    2    3    4    5    6    7    8    9
Standard   deviation       6.80  5.03 2.67 2.16 1.74 1.60 1.29 1.11 0.92
Proportion of Variance     0.50  0.27 0.08 0.05 0.03 0.03 0.02 0.01 0.01
Cumulative Proportion      0.50  0.77 0.85 0.90 0.93 0.96 0.98 0.99 1.00
```

---

[3] `here<- complete.cases(possum[,6:14])`
```
  colr <- c("red", "blue")
  pchr <- c(1,3,6,0,5,6,17)
  ss <- expand.grid(site=1:7, sex=1:2)               # Site varies fastest
  ss$sexsite <- paste(ss$sex, ss$site, sep="-")   # Site varies fastest
  sexsite <- paste(possum$sex, possum$site, sep="-")[here]
  xyplot(possum.prc$scores[,1] ~ possum.prc$scores[,2], panel = panel.superpose,
         groups = sexsite, col = colr[ss$sex], pch = pchr[ss$site],
         xlab="2nd Principal Component", ylab="1st Principal Component",
         key =list(points = list(pch=pchr),
                     text=list(c("Cambarville", "Bellbird", "Whian Whian", "Byrangery",
                               "Conondale  ", "Allyn River", "Bulburin" )),
                        columns=4))
```
[4] `summary(possum.prc, loadings=TRUE, digits=2)`

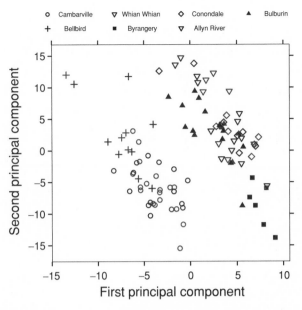

Figure 11.2: Second principal component versus first principal component, for variables in columns 6–14 of the possum data frame. Females are in red; males in blue. For color version, see plate section between pages 200 and 201.

```
Loadings:
          Comp.1 Comp.2 Comp.3 Comp.4 Comp.5 Comp.6 Comp.7 Comp.8
          Comp.9
hdlngth   -0.41   0.28   0.34  -0.19   0.70  -0.28         -0.18
skullw    -0.30   0.27   0.54  -0.34  -0.52   0.28   0.26   0.11
totlngth  -0.52   0.31  -0.65  -0.16          0.23  -0.15   0.34
taill             0.25  -0.35         -0.19          0.44  -0.75   0.11
footlgth  -0.51  -0.47                -0.34  -0.63
earconch  -0.31  -0.65                 0.25   0.58   0.21  -0.17
eye                                                 0.19   0.24   0.94
chest     -0.22          0.17   0.17  -0.18   0.19  -0.76  -0.40   0.27
belly     -0.25   0.18   0.13   0.89          0.10   0.24   0.14
```

Most of the variation is in the first three or four principal components. Notice that component 5 explains only 3% of the variance, while later components explain even less. We expect that most of the variation in these later components will represent noise in the data.

The loadings are the multiples of the original variables that are used in forming the principal components. Notice that the first component is pretty much a size component. The negative signs are an artifact of the computations; these could just as well have been all positive. By default, the printout shows blanks for coefficients that are less than 0.1.

11.2 Discriminant Analysis

We want a rule that will allow us to predict the group to which a new data value will belong. Using language that has been popular in the machine learning literature, our interest is

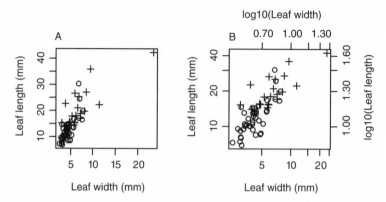

Figure 11.3: Leaf length versus leaf width (A) with untransformed scales; and (B) using logarithmic scales. The symbols are o = plagiotropic, and + = orthotropic.

in supervised classification. For example, we may wish to predict, based on prognostic measurements and outcome information for previous patients, which future patients will remain free of disease symptoms for 12 months or more. We demonstrate two methods – logistic discrimination and classical Fisherian discriminant analysis.

The methods that we discussed in earlier chapters for assessing predictive accuracy apply here also. Note in particular cross-validation, and the training/test set methodology. We emphasize once again that *resubstitution* accuracy, i.e., prediction accuracy on the data used to derive the model, must inevitably improve as the prediction model becomes more complex. For the cross-validation estimate of predictive accuracy, and for predictive accuracy on any test set, we will typically, at some point, start to see a reduction in accuracy. Cross-validation is an attractive approach when data are scarce – no more than a few hundred or a few thousand observations at the level that matters for making predictions.

11.2.1 Example – plant architecture

Our first example is from data on plant architecture (data relate to King and Maindonald (1999). There is a discussion of plant architectures, with diagrams that help make clear the differences between the different architectures, in King, 1998). Orthotropic species have steeply angled branches, with leaves coming off on all sides. Plagiotropic species have spreading branches (a few intermediates and a third uncommon branching pattern are excluded). The interest was in comparing the leaf dimensions of species with the two different architectures. Is leaf shape a good discriminator? The interest is not in prediction *per se*, but in the extent to which leaf shape discriminates between the two architectures.

We examine data from a north Queensland site, one of the six sites that provided data. Figure 11.3 is a plot of the data. A logarithmic scale is clearly preferable, as in Figure 11.3B.

With two explanatory variables, there is not much point in doing principal components analysis. A plot of the second versus first principal components scores from the log transformed data would look like Figure 11.3B, but rotated so that the major axis of variation (running from the lower left to the upper right of Figure 11.3B) is horizontal. We could draw a line by hand through these data at right angles to this major axis, from the lower right to the upper left, that would discriminate orthotropic from plagiotropic species. However, we need a more objective way to discriminate the two groups.

We demonstrate two methods: classical discriminant analysis, and discrimination using logistic regression.

11.2.2 Classical Fisherian discriminant analysis

Classical discriminant analysis is equivalent, when there are two groups, to a regression analysis in which the dependent variable is 0 for the plagiotropic species, and 1 for the orthotropic species. Let π_0 and π_1 be the respective prior probabilities of orthotropic and plagiotropic architectures. Then the discriminant analysis yields a linear function $f(x, y)$ of $x = \log(\text{leaf width})$ and $y = \log(\text{leaf length})$ such that when

$$f(x, y) < -\log\left(\frac{\pi_1}{\pi_0}\right),$$

the prediction is that the plant will be plagiotropic, while otherwise the plant is predicted to be orthotropic. The function $f(x, y)$ has the form $a(x - \bar{x}) + b(y - \bar{y})$ and is called the discriminant function. There is one value for each observation. The values of the discriminant function are known as scores. The constants a and b must be estimated from the data.

Rather than using regression, we use the specifically designed lda() function from the Venables and Ripley *MASS* package. We first extract predictions, and then examine the discriminant function. The code is

```
library(MASS)
data(leafshape17)      # DAAG package
leaf17.lda <- lda(arch ~ logwid+loglen, data=leafshape17)
leaf17.hat <- predict(leaf17.lda)
leaf17.lda
```

Here is the information that we obtain from the analysis:

```
> leaf17.lda
Call:
lda.formula(arch ~ logwid + loglen, data = leafshape17)

Prior probabilities of groups:
      0     1
0.672 0.328

Group means:
   logwid loglen
0    1.43   2.46
1    1.87   2.99
```

```
Coefficients of linear discriminants:
          LD1
logwid 0.156
loglen 3.066
```

Notice that the prior probabilities favor the group that has the zero code, which is the larger group. This may not always be appropriate. The prior probabilities should reflect the proportions in the population to which the discriminant function will be applied. The coefficient estimates are a = 0.156 and b = 3.066.

The object leaf17.hat is a list with the elements class (the classification: 0 or 1), posterior (a matrix where each row has the respective posterior probabilities for groups 0 and 1) and x (the vector of scores).

Assessments of predictive accuracy

We note two sorts of assessments of predictive accuracy. Training set predictions use the data that generated the discriminant function. This leads to a *resubstitution* error rate. The main alternative that we consider is cross-validation. This estimates predictive accuracy on a new set of data, sampled in the same way as the existing training data.

Consider first training set or resubstitution predictive accuracy:

```
> table(leafshape17$arch, leaf17.hat$class)
     0  1
  0 38  3
  1  8 12
```

The predictive accuracy is $(38 + 12)/(38 + 12 + 3 + 8) = 82.0\%$. We can find this as follows:

```
> tab <- table(leafshape17$arch, leaf17.hat$class)
> sum(tab[row(tab)==col(tab)])/sum(tab)
[1] 0.8197
```

One way to do cross-validation is to leave observations out one at a time, then using the remaining data to make a prediction for the one observation that was omitted. To determine such a *leave-one-out* cross-validation estimate of predictive accuracy, specify

```
> leaf17cv.lda <- lda(arch ~ logwid+loglen, data=leafshape17, CV=T)
> tab <- table(leafshape17$arch, leaf17cv.lda$class)
> tab

     0  1
  0 37  4
  1  8 12
> sum(tab[row(tab)==col(tab)])/sum(tab)
[1] 0.8033
```

In this example, it turns out that the two assessments give similar predictive errors. The cross-validation estimate will often be similar to the resubstitution error rate if the error rate is low and there are few explanatory variables.

11.2.3 Logistic discriminant analysis

We have the choice of using logistic regression, or of using the function `multinom()` in the Venables and Ripley *nnet* (neural net) package. Here we use logistic regression.

Binary logistic discrimination works with a model for the probability, e.g., that a plant will be orthotropic. To fix attention on this specific example, the model specifies log(odds orthotropic) as a linear function of the explanatory variables for the classification. We get the function from

```
> leaf17.glm <- glm(arch ~ logwid + loglen, family=binomial,
                     data=leafshape17)
> options(digits=3)
> summary(leaf17.glm)$coef
            Estimate Std. Error z value Pr(>|z|)
(Intercept)  -15.286       4.09  -3.735 0.000188
logwid         0.185       1.57   0.118 0.905911
loglen         5.268       1.95   2.704 0.006856
```

Thus we have

$$\log(\text{odds orthotropic}) = -15.3 + 0.185 \log(\text{width}) + 5.628 \log(\text{length}).$$

Our function `cv.binary()` returns two estimates, the internal (or training set) estimate, and a cross-validation estimate of external predictive accuracy.[5] The resubstitution estimate is

```
    0  1
0  37  4
1   7 13
```

while the cross-validation estimate is

```
    0  1
0  37  4
1   9 11
```

The leave-one-out cross-validation error rate is, again, slightly higher than the resubstitution assessment. The resubstitution estimate can be grossly optimistic, and should not be quoted.

---

[5] `leaf17.one <- cv.binary(leaf17.glm)`
`  table(leafshape17$arch, round(leaf17.one$internal))   # Resubstitution`
`  table(leafshape17$arch, round(leaf17.one$cv))         # Cross-validation`

11.2.4 An example with more than two groups

We present discriminant analysis calculations for the `possum` data frame. We will use the same nine variables as before.[6] The output is

```
> possum.lda
```

```
Call:
lda.formula(site ~ hdlngth + skullw + totlngth + taill + footlgth +
    earconch + eye + chest + belly, data = na.omit(possum))
```

```
Prior probabilities of groups:
    1     2     3     4     5     6     7
0.320 0.117 0.068 0.068 0.126 0.126 0.175
```

```
Group means:
   hdlngth skullw totlngth taill footlgth earconch   eye chest belly
1     93.7   57.2     89.7  36.4     73.0     52.6  15.0  27.9  33.3
2     89.9   55.1     81.7  34.7     70.8     52.1  14.4  26.3  31.2
3     94.6   58.9     88.1  37.2     66.6     45.3  16.1  27.6  34.9
4     97.6   61.7     92.2  39.7     68.9     45.8  15.5  29.6  34.6
5     92.2   56.2     86.9  37.7     64.7     43.9  15.4  26.7  32.0
6     89.2   54.2     84.5  37.7     63.1     44.0  15.3  25.2  31.5
7     92.6   57.2     85.7  37.7     65.7     45.9  14.5  26.1  31.9
```

```
Coefficients of linear discriminants:
                LD1      LD2      LD3      LD4      LD5      LD6
hdlngth     0.15053   0.0583  -0.2557  -0.0124   0.0819  -0.1871
skullw      0.02653   0.0399  -0.2498   0.1245   0.1365   0.1438
totlngth   -0.10696   0.2792   0.3052  -0.1849   0.1390  -0.0892
taill       0.45063  -0.0896  -0.4458  -0.1730  -0.3242   0.4951
footlgth   -0.30190  -0.0360  -0.0391   0.0756  -0.1191  -0.1261
earconch   -0.58627  -0.0436  -0.0731  -0.0882   0.0638   0.2802
eye         0.05614   0.0892   0.7845   0.4644  -0.2848   0.2972
chest      -0.09062   0.1042   0.0498   0.1275  -0.6475  -0.0787
belly      -0.00997  -0.0517   0.0936   0.1672   0.2903   0.1939
```

```
Proportion of trace:
   LD1      LD2      LD3      LD4      LD5      LD6
0.8927  0.0557  0.0365  0.0082  0.0047  0.0022
```

Figure 11.4 shows the scatterplot matrix for the first three discriminant scores.

---

[6] `library(MASS)`
```
possum.lda <- lda(site ~ hdlngth+skullw+totlngth+ taill+footlgth+
                  earconch+eye+chest+belly, data=na.omit(possum))
  # na.omit() omits any rows that have one or more missing values
options(digits=4)
possum.lda$svd   # Examine the singular values
plot(possum.lda, dimen=3)
  # Scatterplot matrix - scores on 1st 3 canonical variates (Figure 11.4)
possum.lda
```

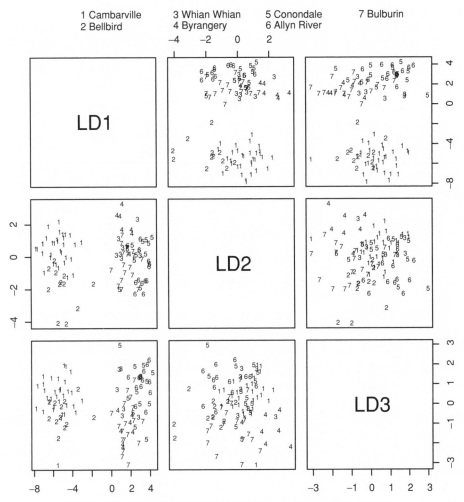

Figure 11.4: Scatterplot matrix for the first three canonical variates.

The successive linear combinations account for reducing parts of the separation between groups. The "proportion of trace" is a measure of the amount. For these data, the first linear discriminant does most of the discriminating. The variables that seem important are `hdlngth`, and `taill`, contrasted with `totlngth` and `footlgth`.

We invite the reader to repeat the analysis with the parameter setting `CV=TRUE` in the call to `lda()`, in order to obtain results from leave-one-out cross-validation.

11.3 Principal Component Scores in Regression

The data set `socsupport` has the following columns:

1. gender: male or female
2. age: 18-20, 21-24, 25-30, 31-40, 40+
3. country: Australia, other

```
4.  marital: married, single, other
5.  livewith: alone, friends, parents, partner, residences, other
6.  employment: full-time, part-time, govt assistance, parental
    support, other
7.  firstyr: first year, other
8.  enrolment: full-time, part-time, blank
9   10. emotional, emotionalsat: availability of emotional support,
    and associated satisfaction (5 questions each)
11  12. tangible, tangiblesat: availability of tangible support
    and associated satisfaction (4 questions each)
13  14. affect, affectsat: availability of affectionate support
    sources and associated satisfaction (3 questions each)
15  16. psi: psisat: availability of positive social interaction
    and associated satisfaction (3 questions each)
17. esupport: extent of emotional support sources (4 questions)
18. psupport: extent of practical support sources (4 questions)
19. socsupport: extent of social support sources  (4 questions)
20. BDI: Score on the Beck depression index (total over
    21 questions)
```

The Beck depression index (BDI) is a standard psychological measure of depression (see for example Streiner and Norman, 1995). The data are from individuals who were generally normal and healthy. One interest was in studying how the support measures (columns 9–19 in the data frame) may affect BDI, and in what bearing the information in columns 1–8 may have. Pairwise correlations between the 11 measures range from 0.28 to 0.85. In the regression of BDI on all of the variables 9–19, nothing appears significant, though the F-statistic makes it clear that, overall, there is a statistically detectable effect. It is not possible to disentangle the effects of these various explanatory variables. Attempts to take account of variables 1–8 will only make matters worse. Variable selection has the difficulties that we noted in Chapter 6. In addition, any attempt to interpret individual regression coefficients focuses attention on specific variables, where a careful account will acknowledge that we observe their combined effect.

We therefore prefer, prior to any use of regression methods, to try to reduce the 11 variables to some smaller number of variables that together account for the major part of the variation. We will use principal components methodology for this purpose. A complication is that the number of questions whose scores were added varied, ranging from 3 to 5. This makes it more than usually desirable to base the principal components calculation on the correlation matrix.

Here now is a summary of the steps that we followed:

1. Following a principal components calculation, we obtained scores for the first six principal components.
2. The six sets of scores were then used as six explanatory variables, in a regression analysis that had BDI as the response variable. The first run of this regression calculation identified an outlier, which was then omitted and the regression calculation repeated.
3. The regression output suggested that only the first of the variables used for the regression, i.e., only the principal component scores for the first principal component, contributed to the explanation of BDI. Compare $p = 0.000\,07$ for scores on the first component with p-values for later sets of scores, all of which have $p > 0.05$.

4. It was then of interest to examine the loadings of the first principal component, to see which of the initial social support variables was involved.

Code to do the initial analysis, then presenting a scatterplot matrix of the scores on the first three principal components, is in the footnote.[7]

The name given with the point that we have identified as an outlier is "36", which is the row name in the initial file. We omit this point and repeat the calculation.[8] The output from summary() is

```
> summary(ss.pr)   # Examine the contribution of the components

Importance of components:
                        Comp.1 Comp.2 Comp.3  Comp.4   Comp.5   Comp.6
Standard deviation      2.3937 1.2190 1.1366 0.84477  0.75447  0.69536
Proportion of Variance  0.5209 0.1351 0.1174 0.06488  0.05175  0.04396
Cumulative Proportion   0.5209 0.6560 0.7734 0.83833  0.89007  0.93403
                        Comp.7  Comp.8  Comp.9  Comp.10   Comp.11
Standard deviation      0.49726 0.45610 0.35952 0.295553 0.231892
Proportion of Variance  0.02248 0.01891 0.01175 0.007941 0.004889
Cumulative Proportion   0.95651 0.97542 0.98717 0.995111 1.000000
```

We now regress BDI on the first six principal components. Because the successive columns of scores are uncorrelated, the coefficients are independent. At the same time, additional terms that contribute little except noise will make little difference to the residual mean square, and hence to the standard errors. Thus, there is no reason for great economy in the number of terms that we choose for initial examination.[9] The coefficients in the regression output are

```
> summary(ss.lm)$coef

                          Estimate Std. Error t value   Pr(>|t|)
(Intercept)                10.4607     0.8934 11.7088  0.0000000
ss.pr$scores[, 1:6]Comp.1   1.3113     0.3732  3.5134  0.0007229
ss.pr$scores[, 1:6]Comp.2  -0.3959     0.7329 -0.5402  0.5905258
ss.pr$scores[, 1:6]Comp.3  -0.6036     0.7860 -0.7679  0.4447445
ss.pr$scores[, 1:6]Comp.4  -1.4248     1.0576 -1.3473  0.1816102
ss.pr$scores[, 1:6]Comp.5   2.1459     1.1841  1.8122  0.0736221
ss.pr$scores[, 1:6]Comp.6  -1.2882     1.2848 -1.0027  0.3189666
```

---

[7] data(socsupport)   # DAAG package
ss.pr1 <- princomp(as.matrix(na.omit(socsupport[, 9:19])), cor=TRUE)
sort(-ss.pr1$scores[,1])   # Minus the largest value appears first
pairs(ss.pr1$scores[,1:3])
Alternative to pairs(), using the lattice function splom()
splom(~ss.pr1$scores[,1:3])
[8] not.na <- complete.cases(socsupport[,9:19])
not.na[36] <- FALSE
ss.pr <- princomp(as.matrix(socsupport[not.na, 9:19]), cor=TRUE)
summary(ss.pr)   # Examine the contribution of the components
[9] ss.lm <- lm(BDI[not.na] ~ ss.pr$scores[, 1:6], data=socsupport)
summary(ss.lm)$coef

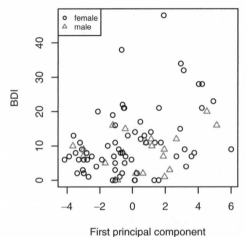

Figure 11.5: Plot of BDI against scores on the first principal component.

Components other than the first do not make an evident contribution to prediction of BDI. We now examine the loadings for the first component:

```
> ss.pr$loadings[,1]
```

emotional	emotionalsat	tangible	tangiblesat	affect
-0.3201	-0.2979	-0.2467	-0.2887	-0.3069
affectsat	psi	psisat	esupport	psupport
-0.2883	-0.3628	-0.3322	-0.2886	-0.2846
socsupport				
-0.2846				

The negative signs are incidental. This first component is pretty much an average of the 11 measures. A further step is then to plot BDI against the scores on the first principal component, using different colors and/or different symbols for females and males. This should be repeated for each of the other seven factors represented by columns 1–8 of the data frame socsupport. Figure 11.5 does this for the factor gender.[10]

Two points seem anomalous, with BDI indices that are high given their scores on the first principal component. Both are females. We leave it as an exercise for the reader to re-calculate the principal components with these points omitted, and repeat the regression.

Regression on principal component scores has made it possible to identify a clear effect from the social support variables. Because we have regressed on the principal components, it is not possible to ascribe these effects, with any confidence, to individual variables. The attempt to ascribe effects to individual social support variables, independently of other support variables, may anyway be misguided. It is unlikely to reflect the reality of the way that social support variables exercise their effects.

---

[10] attach(socsupport)
```
plot(BDI[not.na] ~ ss.pr$scores[ ,1], col=as.numeric(gender[not.na]),
    pch=as.numeric(gender[not.na]), xlab ="1st principal component", ylab="BDI")
topleft < - par()$usr[c(1,4)]
legend(topleft[1], topleft[2], col=1:2, pch=1:2, legend=levels(gender))
detach(socsupport)
```

11.4* Propensity Scores in Regression Comparisons – Labor Training Data

We consider, as in the study of the labor training program data in Section 6.5, the comparison between a treatment group and a non-experimental control group. The methodology is capable of wider extension, e.g., to more than two groups. A key idea is that of propensity score.

A propensity is a measure, determined by covariate values, of the probability that an observation will fall in the treatment rather than in the control group. Various forms of discriminant analysis may be used to determine scores. The propensity score is intended to account for between group differences that are not due to the effect under investigation. If values of this variable for the different groups do not overlap, any inference about differences between groups will be hazardous. If there is substantial overlap, then comparison of observations within the approximate region of overlap may be reasonable, but using the propensity score to adjust for differences that remain. A benefit of this methodology is that we can use standard checks to investigate whether the propensity score effect is plausibly linear – we have just one covariate to investigate, rather than the difficult and often impossible task of checking the linearity of the effects of several covariates.

The discussion sheds light on the obstacles to reaching secure conclusions from observational data.

The training program effect on post-intervention earnings

We return to the data (Lalonde et al. 1986; Dehejia and Wahba 1999) examined in Section 6.5, where we used regression methods to investigate the effect of a US labor training program on post-intervention earnings. The methodology that we now present is in the spirit of Dehejia and Wahba (1999). However, where they were interested in reproducing the results from the experimental comparison, our interest is in deriving a model that accurately characterizes the treatment effect. We work with the `nsw74psidA` subset of the `nsw74psid1` data set that we identified in Section 6.5. This was chosen so that values of the variables `re74` and `re75` had similar ranges in both the control and the treatment group. Our aim is to find a single score that accounts for differences between the control and treatment groups.

We use logistic regression to determine propensity scores. We hope to find a single score that characterizes differences between control and treatment observations. We do this for observations that we already know to be broadly similar, which is why we work with the subset that was identified in Section 6.5. (If the groups differ too much, data values that are in some sense extreme may seriously distort the scores that are given by logistic regression.) We use the values of the linear predictor from the logistic regression as scores. These scores measure the "propensity" for membership in the treatment rather than in the control group.[11] Figure 11.6 shows the distributions of the scores, separately for the control and for the treatment group.

We limit the comparison to the range of values of the scores where the ratios of the numbers of treatment and control are, as estimated by the density curve, between 0.05 and

---

[11]
```
data(nsw74psidA)      # DAAG package
discA.glm <- glm(formula = trt ~ age + educ + black + hisp + marr +
                 nodeg + re74 + re75, family = binomial, data = nsw74psidA)
A.scores <- predict(discA.glm)
```

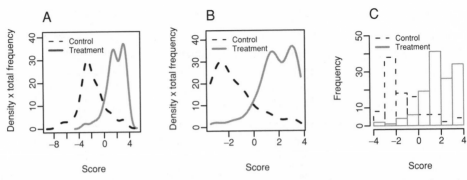

Figure 11.6: Density plots of scores (predicted values on the scale of the linear predictor from the object `discA.glm`), separately for control and treatment groups. Panels B and C are for a restricted range of values of the score.

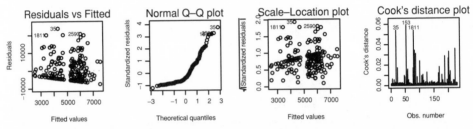

Figure 11.7: Diagnostic plots for the model (`A.lm`) that uses the scores as the only covariate.

20. Obviously this decision on what is a reasonable region of overlap is somewhat arbitrary. We should check whether another restriction, e.g., to ratios between 0.1 and 10, makes much difference. The relevant ranges are those of Figure 11.6B and C. Figure 11.6C shows the actual numbers in the various class intervals of the histogram. We now use the scores as a single covariate to account for differences between treatment and control. Here then is the analysis:

```
> overlap <- (A.scores > -3.5) & (A.scores < 3.8)
> A.lm <- lm(re78 ~ trt + A.scores, data=nsw74psidA,
            subset = overlap)
> summary(A.lm)$coef
            Estimate Std. Error t value  Pr(>|t|)
(Intercept)   3588.0      739.8   4.850 2.286e-06
trt           2890.2     1204.6   2.399 1.723e-02
A.scores      -333.4      270.7  -1.232 2.194e-01
```

Finally (in Figure 11.7), we examine the diagnostic plots. The residuals are clearly not from a normal distribution. However, the data set is so large that we should be able to rely on Central Limit Theorem effects to yield sensible parameter estimates and associated standard errors. The exercises canvass several possible refinements of the analysis.

Further comments

A key requirement, for this methodology, is that the data must include information on all relevant covariates. In general, propensity score methods require extensive data, unless the data sets are extremely well matched. Adequate numbers of observations must remain after omission of observations that lie outside the region of overlap of the scores.

The data that we have analyzed are part of a much larger collection that are available from http://www.columbia.edu/~rd247/nswdata.html

11.5 Further Reading

There is a large literature on the spectrum of methodologies we have discussed, and a large and growing range of methodologies that are available in R and in other software. See Venables and Ripley (2002) for a terse summary overview, aimed at practical data analysts. Krzanowski (2000) is a comprehensive and accessible overview of the methodology of multivariate analysis. Streiner and Norman (1995) discuss important issues that relate to the collection and analysis of multivariate data in medicine, in the health social sciences, and in psychology. On the use of propensity scores, see Rosenbaum and Rubin (1983) and Rosenbaum (1999).

Machine learning and data mining specialists have brought to these problems a research perspective that has been different from that of academic statisticians. Recently, there has been extensive interchange between the two streams of methodological development. Ripley (1996) and Hastie et al. (2001) are important contributions to the ongoing dialogue.

Principal components methodology, though often effective, is a relatively crude tool for the exploration of multi-dimensional data. The XGobi data visualization system (available at no charge from http://www.research.att.com/areas/stat/xgobi/) greatly extends the range of possibilities, offering its users a range of three-dimensional dynamical graphical views. We strongly recommend this highly professional system. The GGobi system (from http://www.ggobi.org) is an adaptation of XGobi that runs on Microsoft Windows systems. A further step is to install the R *xgobi* package and run XGobi from R. There is, at least for Microsoft Windows, a similar R interface for GGobi.

References for further reading

Hastie, T., Tibshirani, R. and Friedman, J. 2001. *The Elements of Statistical Learning. Data Mining, Inference and Prediction.* Springer-Verlag.
Krzanowski, W.J. 2000. *Principles of Multivariate Analysis*, revised edn. *A User's Perspective*. Clarendon Press.
Ripley, B.D. 1996. *Pattern Recognition and Neural Networks.* Cambridge University Press.
Rosenbaum, P. and Rubin, D. 1983. The central role of the propensity score in observational studies for causal effects. *Biometrika* 70: 41–55.
Streiner, D.L. and Norman, G.R. 1995. *Health Measurement Scales. A Practical Guide to Their Development and Use,* 2nd edn. Oxford University Press.
Venables, W.N. and Ripley, B.D. 2002. *Modern Applied Statistics with* S, 4th edn. Springer-Verlag.

11.6 Exercises

1. Carry out the principal components analysis of Subsection 11.1.2, separately for males and females. For each of the first and second principal, plot the loadings for females against the loadings for all data combined, and similarly for males. Are there any striking differences?

2. In the discriminant analysis for the `possum` data (Subsection 11.2.4), determine, for each site, the means of the scores on the first and second discriminant functions. Plot the means for the second discriminant function against the means for the first discriminant function. Identify the means with the names of the sites.

3. The data frame `possumsites` (*DAAG* package) holds latitudes, longitudes, and altitudes, for the seven sites. The following code, which assumes that the *oz* package is installed, locates the sites on a map that shows the eastern Australian coastline and nearby state boundaries.

```
library(DAAG); library(oz)
data(possumsites)
oz(sections=c(3:5, 11:16))
attach(possumsites)
points(latitude, longitude)
chw <- par()$cxy[1]
chh <- par()$cxy[2]
posval <- c(2, 4, 2, 2, 4, 2, 2)
text(latitude+(3-posval)*chw/4, longitude,
     row.names(possumsites), pos=posval)
```

Do the site means that were calculated in Exercise 2 relate in any obvious way to geographical position.

4. Use `merge()` to merge the altitude information from `possumsites` into the `possum` data frame. The variable `altitude` has the same value for all observations within a `site`. What are the implications for any attempt to use `altitude` as an explanatory variable in the linear discriminant function calculations of Subsection 11.2.1?

5. Create a version of Figure 11.3B that shows the discriminant line. In the example of Subsection 11.2.1, investigate whether use of `logpet`, in addition to `logwid` and `loglen`, improves discrimination.

6.* The data set `leafshape` has three leaf measurements – `bladelen` (blade length), `bladewid` (blade width), and `petiole` (petiole length). These are available for each of two plant architectures, in each of six locations. (The data set `leafshape17` that we encountered in Subsection 11.2.1 is a subset of the data set `leafshape`.) Use logistic regression to develop an equation for predicting architecture, given leaf dimensions and location. Compare the alternatives: (i) different discriminant functions for different locations; (ii) the same coefficients for the leaf shape variables, but different intercepts for different locations; (iii) the same coefficients for the leaf shape variables, with an intercept that is a linear function of latitude; (iv) the same equation for all locations. Interpret the equation that is finally chosen as discriminant function.

7. Repeat the principal components calculation omitting the points that appear as outliers in Figure 11.5, and redo the regression calculation. What differences are apparent, in loadings for the first two principal components and/or in the regression results?

to include data where the ratio of treatment to control numbers, as estimated from the density curve, lies between 0.025 and 40. Narrow it to include data where the ratio is restricted to lie between 0.1 and 10. Compare the three sets of results that are now available, and comment on any differences that you notice.

[Use the function `overlap.density()` in our *DAAG* package to estimate the overlap regions that are needed for this exercise.]

9. In the data sets `nsw74psidA`, a substantial proportion of the individuals, both in the treatment group and in the control group, had no earnings in 1978. Thus we create a variable `nonzero78` that is 0 for individuals with no 1978 earnings, and 1 for individuals with non-zero earnings in 1978. Carry out a logistic regression of `nonzero78` on treatment group as a factor and `A.scores` as covariate.

10. This is a continuation of Exercise 9. Fit a regression model, with `log(re78)` as the dependent variable, to observations for which `nonzero78 > 0`. Combine this model with the model of Exercise 9, and obtain predicted values and SEs for predictions from the combined model.

The R System – Additional Topics

12.1 Graphs in R

Much of the information in this section is intended for reference. We note that R's 26 plotting symbols can be supplemented with ASCII characters in the range 32–255. We note functions for plotting various shapes. We comment on R's color choices.

Plotting characters

Setting pch to one of the numbers 0, 1, . . . , 25 gives one of 25 different plotting symbols. The numbers and symbols are shown on the top two lines of Figure 12.1. In addition pch may be set to any value in the range 32–255. Figure 12.1 shows the characters that appear when pch is set to a number in this range. An alternative is to use a quoted string, e.g., pch="a". The full range of characters is given for the font setting font=1. Notice that, for plotting, the characters are centered vertically about their mid-position.

The following statements give a cut-down version of Figure 12.1:

```
plot(c(0, 31), c(1, 22), type="n", ylab="", axes=FALSE)
points(0:25, rep(21,26), pch=0:25)
text(0:25, rep(22,26), paste(0:25), srt=90, cex=0.65)
for(i in 1:8){
    par(font=i)
    ypos <- 21-i
    points(0:31, rep(ypos,32), pch=64:95)
    axis(2, at=ypos, labels=paste("font =",i), las=1)
    }
par(font=1)
for(i in 1:6){
    lines(c(0, 31), c(i, i), lty=i)
    axis(2, at=i, labels=paste("lty =",i), las=1)
    }
```

Symbols

The function symbols() offers a choice of circles, squares, rectangles, stars, thermometers and boxplots. The parameter fg (foreground) specifies the color of the symbol outline, while bg (background) specifies the fill color. See help(symbols) for details.

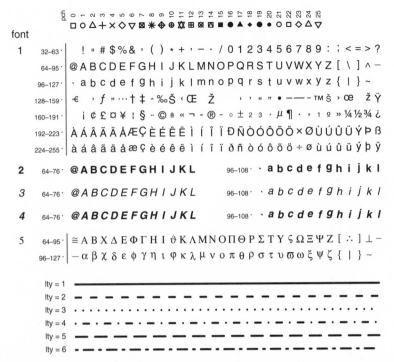

Figure 12.1: Details of symbols and characters that can be specified using the `pch` parameter. For characters in the range 32–255, different font types will give symbols that vary to differing degrees. We note, at the bottom of the figure, available line types.

Note also the functions `rect()` (for plotting rectangles) and `polygon()`. The `polygon()` function can, for example, be used to color areas under curves. Here is a simple example of its use:

```
plot(0:3, 0:3, type="n")
polygon(c(1,1.5,2), c(1,2,1), col="red", border="black")
  # Red triangle
```

Colors

Type `colors()` to get details of available named colors. Many are different shades of the same basic color. Thus the red shades are "red", "red1", "red2", "red3" and "red4". Figure 12.2 shows five shades of each of 52 different colors.

In addition, there are several functions – `rainbow()`, `heat.colors()`, `topo.colors()` and `cm.colors()`, that give a set of graduated colors. These functions all have an argument, set to 17 in Figure 12.2, that controls the number of gradations. The function `rainbow()` has other parameters also; for details see the help page. The function `rgb()` generates colors for given red, green and blue intensities.

The following gives a severely simplified version of Figure 12.2:

```
plot(c(0,11), c(0, 11), type="n", xlab="", ylab="", axes=FALSE)
points(1:10, rep(10, 10), col=rainbow(10), pch=15, cex=4)
```

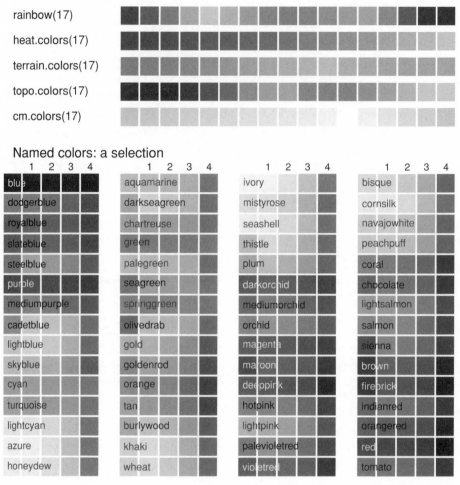

Figure 12.2: A selection of color palettes and named colors that are available in R. For the named colors, we have restricted attention to an incomplete selection of those that come in one of five shades. Thus, in addition to "red", there are "red1", "red2", "red3", and "red4". For color version, see plate section between pages 200 and 201.

```
points(1:10, rep(9, 10), col=heat.colors(10), pch=15, cex=4)
points(1:10, rep(8, 10), col=topo.colors(10), pch=15, cex=4)
points(1:10, rep(7, 10), col=cm.colors(10), pch=15, cex=4)
colnam <- c("gold","khaki","orange","tan","brown",
            "firebrick","indianred","orangered","red",
            "tomato")
colmat <- outer(c("","1","2","3","4"), colnam,
                function(x,y)paste(y,x,sep=""))
points(rep(1:length(colnam), rep(5,length(colnam))),
       6-rep(1:5, length(colnam)), col=as.vector(colmat),
       pch=15, cex=4)
```

The function show.colors() in our *DAAG* package displays the built-in colors. The default (type="singles") is to display colors for which there is a single shade.

Other options are `type="shades"` (display colors that have multiple shades) and `type="gray"` (display shades of gray).

Colors in lattice

By default, seven colors are available for plotting points. Assuming that the *lattice* package is installed, the following will extract the colors:

```
library(lattice)
colr <- trellis.par.get()$superpose.symbol$col
```

Contour and filled contour plots

The relevant functions are `contour()`, `filled.contour()` and `image()`. These functions all take parameters x, y and z, where x and y specify, respectively, *x*- and *y*-coordinates, and z is a matrix of contour level values, with one row corresponding to each element of x, and one column corresponding to each element of y. Alternatively, the argument *x* may be specified as a list, with elements a vector x$x, a vector x$y and a matrix x$z.

The latter two types of plot use a color scale to represent the values of z. Values of x and y must be in increasing order of x and of y within x. For examples of the use of these functions, type in `demo(image)`. The results are often visually appealing.

12.2 Functions – Some Further Details

We begin this section with brief comments on some of the more important built-in functions. Comments then follow on how users may write their own functions. A further possibility is to extend or adapt built-in functions. One of R's major merits is that its resources are at the user's command, to use or modify or enhance at one's discretion. Take care to use that freedom well!

12.2.1 Common Useful Functions

We note some of the more commonly used functions:

```
print()         # Prints a single R object
cat()           # Prints multiple objects, one after the other
class()         # Returns the data class of an R object
length()        # Number of elements in a vector or of a list
mean()          # Mean
median()        # Median
range()         # Range
unique()        # Gives the vector of distinct values
diff()          # Returns the vector of first differences
                # N.B. diff(x) has one less element than x
sort()          # Sort elements into order, but omitting NAs
order()         # x[order(x)] orders elements of x, with NAs last
```

```
cumsum()        # Cumulative sum
cumprod()       # Cumulative product
rev()           # reverse the order of vector elements
table()         # Form a table of counts
sd()            # Calculate the standard deviation
var()           # Calculate the variance
```

Different functions handle missing values in different ways. Several of the above functions allow the parameter setting na.rm=TRUE, i.e., missing values are removed before evaluating the function. The default is na.rm=FALSE. Recall that, by default, table() ignores missing values. The following illustrates the setting na.rm=TRUE:

```
> range(c(1,NA,0,5,6))
[1] NA NA
> range(c(1,NA,0,5,6), na.rm=TRUE)
[1] 0 6
```

The args() *function*

This function gives details of named function parameters, and of any default settings. For example:

```
> args(read.table)      # Version 1.2.3 of R
function (file, header = FALSE, sep = "",
    quote = "\"'", dec = ".", row.names, col.names,
    as.is = FALSE, na.strings = "NA", skip = 0,
    check.names = TRUE, fill = !blank.lines.skip,
    strip.white = FALSE, blank.lines.skip = TRUE)
NULL
```

This function may, when a check is needed on parameter names, be a good alternative to looking up the help page. Note however that in some very common functions such as plot(), many or most of the parameter names do not appear in the argument list. The R system has a mechanism, which we will discuss in Subsection 12.2.2, that makes it possible to pass arguments that are additional to those in the function's named arguments.

Making tables

The function table() makes a table of counts. Specify one vector of values (often a factor) for each table margin that is required. Here are examples:

```
> table(rainforest$species)    # from our DAAG package

Acacia mabellae   C. fraseri   Acmena smithii   B. myrtifolia
             16           12               26              11
```

Specifying *n* vector arguments to table() leads to an *n*-way table. The first argument defines rows, though it is printed horizontally if there is just one column. The second argument defines columns. The table slices (rows by columns) that correspond to different values of the third argument appear in succession down the page, and so on. Here is a two-way table:

```
> library(MASS)
> data(cabbages)
> names(cabbages)
[1] "Cult"    "Date"    "HeadWt" "VitC"
> table(cabbages$Cult, cabbages$Date)

      d16 d20 d21
  c39  10  10  10
  c52  10  10  10
```

Tables are readily converted to a data frame form of representation, with one column for each classifying factor, and one column that holds the counts. For example

```
> cabtab <- table(cabbages$Cult, cabbages$Date)
> as.data.frame(cabtab)
  Var1 Var2 Freq
1  c39  d16   10
2  c52  d16   10
3  c39  d20   10
4  c52  d20   10
5  c39  d21   10
6  c52  d21   10
```

Selection and matching

A highly useful operator is %in%, used for testing set membership. For example:

```
> x <- rep(1:5, rep(3,5))
> x
 [1] 1 1 1 2 2 2 3 3 3 4 4 4 5 5 5
> x[x %in% c(2,4)]
[1] 2 2 2 4 4 4
```

We have picked out those elements of x that are either 2 or 4. To find which elements of x are 2s, which 4s, and which are neither, use match(), thus:

```
> match(x, c(2,4), nomatch=0)
 [1] 0 0 0 1 1 1 0 0 0 2 2 2 0 0 0
```

Specifying nomatch=0 is often preferable to the default, which is NA.

String functions

We note several functions that are important for working with text strings. Type in help.search("string") for a more complete list.

```
> substring("abracadabra",3, 8)  # Select characters 3 to 8
                                 # inclusive
[1] "racada"
> nchar("abracadabra")           # Count the number of characters
[1] 11
> strsplit("abracadabra", "r")   # Split the string wherever "r"
                                 # appears
[[1]]
[1] "ab"      "acadab" "a"
```

All these functions can be applied to a character vector, which is really a vector of text strings. Use is.character() to test whether a vector is a character vector. This can be useful when unsure whether a vector has been coerced to a factor.

The aggregate() *function*

This is convenient when we wish to obtain the mean or other function for each combination of factor levels, getting as output a data frame with one row. We demonstrate the use of this function with the data set cabbages from the *MASS* package. First, we examine the first few rows of the data frame.

```
> data(cabbages)
> cabbages[1:4, ]
  Cult Date HeadWt VitC
1  c39  d16    2.5   51
2  c39  d16    2.2   55
3  c39  d16    3.1   45
4  c39  d16    4.3   42
```

We now calculate means for each combination of levels of Cult and Date.

```
> attach(cabbages)
> cabbages.df <-  aggregate(HeadWt, list(Cult, Date), FUN = mean)
> names(cabbages.df) <- c("Cult", "Date", "Mean")
> cabbages.df
  Cult Date Mean
1  c39  d16 3.18
2  c52  d16 2.26
3  c39  d20 2.80
4  c52  d20 3.11
5  c39  d21 2.74
6  c52  d21 1.47
```

The apply *family of functions*

The functions that are in mind are sapply(), lapply(), apply(), and tapply(). The functions sapply() and lapply() operate on vectors, on lists and on data frames.

We encountered sapply() (s = simplify) in Chapter 1. We will discuss it further, along with the closely related function lapply() (l = list), in the section on lists. We will discuss `apply()` when we come to consider matrices and arrays. Note that `apply()` can also be used to apply a function to rows or columns of a data frame, provided of course that the operations that the function performs are legal for elements in the data frame.

The `tapply()` *function*

The arguments are a variable, a list of factors, and a function that operates on a vector to return a single value. For each combination of factor levels, the function is applied to corresponding values of the variable. The output is an array with as many dimensions as there are factors. Where there are no data values for a particular combination of factor levels, NA is returned. Here is an example.

```
> cabbages.av <- tapply(HeadWt, list(Cult, Date), mean)
> cabbages.av
      d16  d20  d21
c39 3.18 2.80 2.74
c52 2.26 3.11 1.47
> detach(cabbages)
```

The information is the same as from our earlier use of `aggregate()`, but is now presented as a matrix in which the classifying factors correspond to the dimensions.

12.2.2 User-written R *functions*

Here is a function that prints out the mean and standard deviation of a set of numbers:

```
mean.and.sd <- function(x = rnorm(10)){
    av <- mean(x)
    sd <- sqrt(var(x))
    c(mean=av, SD=sd)
    }
```

Notice that we have given the function the default argument `x = rnorm(10)`. This makes it possible to run the function to see how it works, without specifying any data; thus:

```
> mean.and.sd()
    mean          SD
0.6576272 0.8595572
```

Here then is the result of running the function with data that appeared in Table 3.2:

```
> mean.and.sd(x = c(19,8,4,1,6,10,6,-3,6))
    mean          SD
6.333333 6.103278
```

Note that the variables `av` and `sd` are local to the function. They cannot be accessed outside of the internal function environment.

We invite the reader, as an exercise, to add code for calculation of a 95% confidence interval for the mean. Recall that the multiplier for the standard error is `qt(0.975, nu)`, where `nu` is the number of degrees of freedom for the standard deviation estimate.

Rules for creating functions

A function is created using an assignment. On the right hand side, the parameters appear within round brackets. It is often useful to give a default. In the example above the default was `x = rnorm(10)`. This allows users to run the function without specifying a parameter, just to see what it does. The function body follows the ")" that closes the argument list. Except where the function body consists of just one statement, this is enclosed between curly braces (`{ }`). The return value, which must be a single object, is given by the final statement of the function body. Alternatively or additionally, there may be an explicit `return()` statement. In the example above, the return value was the vector consisting of the two named elements `mean` and `sd`.

In the above, the return value was a numeric vector. For returning several objects that are of different types, join them together into a list. Thus, modeling functions return their output as a list that has the technical name *model object*. Thus, we have `lm` objects, `aov` objects, and so on. The list components may be an immediate part of the output or included because they are useful for further calculations.

Issues for the writing and use of functions

Once code is in place for a computation that will be carried out repeatedly, it makes sense to place the code in a function that can be used to automate the computation. This should happen sooner rather than later. Such functions ease the burden of documenting the computations, and of ensuring that there is an audit trail. What precise sequence of changes was applied to the initial data frame, as created from data supplied by a client, in order to create the data frame that was used for the analysis?

Functions are usually the preferred way to parcel code that is to be taken over and used by another person. The R packages do this, for a collection of functions that are built around a common theme, in the carefully structured and documented way that is desirable for code that will be offered to the total R community.

The number of functions may soon become large. Choose their names carefully, so that they are meaningful. Choose meaningful names for arguments, even if this means that they are longer than is preferred. Remember that they can be abbreviated in actual use. Settings that may need to change in later use of the function should appear as parameter defaults. Use lists, where this seems appropriate, to group together parameters that belong together conceptually.

Where convenient, provide a demonstration mode for functions. Such a mode will print out summary information on the data and/or on the results of manipulations prior to analysis, with appropriate labeling. The code needed to implement this feature has the side effect of showing by example what the function does, and may be useful for debugging. For example, the argument x of the function `mean.and.sd()` had the default value `x = rnorm(10)`. The default could equally well have been `x = rnorm(20)` or `x = runif(10)`, or even `x = 1:10`.

Break functions up into a small number of sub-functions or "primitives". Re-use existing functions wherever possible. Write any new "primitives" so that they can be re-used. This helps ensure that functions contain well-tested and well-understood components. The R web pages give details on how to join the r-help electronic mail list. Contributors to this list from time to time offer useful functions for routine tasks. There may be functions in our *DAAG* package that readers will wish to take over and use for their own purposes.

If at all possible, give parameters sensible defaults. Often a good strategy is to use as defaults parameters that are suitable for a demonstration run of the function. NULL is a useful default where the parameter mostly is not required, but where the parameter if it appears may be any one of several types of data structure. Within the function body, an if(!is.null(ParameterOfInterest)) form of construction then determines whether it is necessary to investigate that parameter further.

Structure code to avoid multiple entry of information. As far as possible, make code self-documenting. Use meaningful names for objects. The R system allows the use of names for elements of vectors and lists, and for rows and columns of arrays and data frames. Consider the use of names rather than numbers when extracting individual elements, columns, etc. Thus, dead.tot[, "dead"] is more meaningful and safer than dead.tot[,2].

Where data and labeling must be pulled together from a number of different objects and files, and especially where it may be necessary to retrace steps some months later, take the same care over organizing and naming data as over structuring code.

Inline functions

A function can be defined at the point where it is used. It does not then require a name. The following are equivalent:

```
sapply(elastic1, mean)      # elastic1 is from DAAG
sapply(elastic1, function(x)mean(x))
```

The second line of code sets up an unnamed function that calculates mean. In this example, there is no point in setting up such a function. The syntax is however important; we now have the ability, without setting up a named function, to manipulate the columns of the data frame in arbitrary ways. For example:

```
sapply(elastic1, function(x)sum(log(x)))
```

The . . . argument

Functions may have an argument, written . . ., that allows the passing of parameters that are additional to those that are in the list of named arguments. Parameters that are passed in this way can be accessed by extracting the appropriate list element from list(...). The list elements have the names, if any, that they are given when the function is called.

The function rm() uses this mechanism to target an arbitrary set of objects for removal, as in the following code fragment:

```
x <- c(1,5,7); z <- c(4,9,10,NA); u <- 1:10
# Now do calculations that use x, z and u
rm(x, z, u)
```

The . . . argument can be useful when the number of possible parameters is large, as for plot() and related functions. There are a large number of parameter settings that can be passed to plot(); see help(par). Those that are not formal parameters are passed, if

required, via the . . . list. In the case of plot() the arguments must be given names that
will be recognized within plot(), or within functions to which plot() passes the . . .
parameter list.

12.2.3 *Functions for working with dates*

The *date* package has basic abilities for working with dates. These functions are suitable
only for the simplest applications. For a more general and flexible set of functions for
working with dates and times, see the help for as.POSIXlt(), strptime() and related
functions. See help(DateTimeClasses) for comment on the standards that these more
general and flexible functions implement.

The function as.date() will take a vector of character strings that has an appropriate
format, and convert it into a dates object. By default, dates are stored using January 1 1960
as origin. This becomes apparent when as.integer() is used to convert a date into an
integer value. Here are examples:

```
> library(date)   # package must be installed
> as.date("1/1/60", order="dmy")
[1] 1Jan60
> as.date("1/12/60","dmy")
[1] 1Dec60
> as.date("1/12/60","dmy")-as.date("1/1/60","dmy")
[1] 335
> as.date("31/12/60","dmy")
[1] 31Dec60
> as.date("31/12/60","dmy")-as.date("1/1/60","dmy")
[1] 365
> as.integer(as.date("1/1/60","dmy"))
[1] 0

> as.integer(as.date("1/1/2000","dmy"))
[1] 14610
> as.integer(as.date("29/2/2000","dmy"))
[1] 14669
> as.integer(as.date("1/3/2000","dmy"))
[1] 14670
```

Among the legal formats are 8-31-2000 (or 31-8-2000 if we specify order="dmy"),
8/31/2000 (or 31/8/2000), and August 31 2000. Observe that we can subtract two dates,
thus obtaining the time between them in days. There are several functions for printing out
dates in various different formats. One of these is date.ddmmmyy(); the mnemonic is
that mmm denotes a three-letter abbreviation for the month.

12.3 Data Input and Output

Definitive information is in the help information for the relevant functions or packages,
and in the R Data Import/Export manual that is part of the official R documentation. New

features will appear from time to time. The discussion that follows should be supplemented with examination of up to date documentation.

We will begin by showing how we can create a file that we can read in, using the function `cat()` for this purpose, thus:

```
cat("First of 3 lines", "2 3 5 7", "11 13 17",
    file="ex1.txt",sep="\n") # \n=newline
```

The entries in the file will look like this:

```
First of 3 lines
2 3 5 7
11 13 17
```

12.3.1 Input

The function `read.table()` and its variants

The function `read.table()` is straightforward for reading in rectangular arrays of data that are entirely numeric. Note however that:

- The function `read.table()`, and the variants that are described on its help page, examine each column of data to determine whether it consists entirely of legal numeric values. Any column that does not consist entirely of numeric data is treated as a column of mode character, and by default is stored as a factor. Such a factor has as many different levels as there are unique text strings.[1] It is then immediately available for use as a factor in a model or graphics formula.
- To force the storage of columns of text strings as character, use the parameter setting `as.is = TRUE`. This ensures that columns that contain character strings are stored with mode character. If such a column is later required for use as a factor in a function that does not do the conversion automatically, explicit conversion is necessary.
- Columns that are intended to be numeric data may, because of small mistakes in data entry, be treated as text and (unless `as.is` was set to prevent this) stored as a factor. For example there may be an O (the letter "O") somewhere where there should be a 0 (zero), or the letter "l" where there should be a one (1).
- An unexpected choice of missing value symbol, e.g., an asterisk (*) or period (.), will likewise cause the whole column of data to be treated as text and stored as a factor. Any missing value symbol, other than the default (NA), must be explicitly indicated. With text files from SAS, it will probably be necessary to set `na.strings=c(".")`. There may be multiple missing value indicators, e.g., `na.strings=c("NA",".","*","")`. The `""` will ensure that empty fields are entered as NAs.
- There are variants of `read.table()` that differ from `read.table()` in having defaults that may be more appropriate for data that have been output from spreadsheets. By default, the function `read.delim()` uses tab as the separator, while `read.csv()` uses comma. In addition, both functions have the parameter setting `fill=TRUE`. This

---

[1] Storage of columns of character strings as factors is efficient when a small number of distinct strings are each repeated a large number of times.

ensures that when lines contain unequal numbers of fields, blank fields are implicitly
added as necessary.

These various alternatives to `read.table()` build in alternative choices of defaults.
As with any other function in R, users are free to create their own variants that are
tailored to their own particular demands.

- For other variants of `read.table()`, including variants that use comma (`","`) as
  the decimal point character, see the help page for `read.table()`.

An error message that different rows have different numbers of fields is usually an
indication that one or more parameter settings should be changed. Two of the possibilities,
apart from problems with the choice of separator, are:

- By default, `read.table()` treats double quotes (`"`) and the vertical quote (`'`) as
  character string delimiters. Where a file has text strings that include the vertical single
  quote, it will be necessary to set `quote="\""` (set string delimiter to `"`) or `quote=""`.
  With this last setting, character strings will consist of all characters that appear between
  the successive separators of a character field.
- The comment character, by default #, may be included as part of a field.

The function `count.fields()` is useful in identifying rows where the number of
fields differs from the standard. The style of use is

```
n <- count.fields("ex1.txt")    # Returns number of fields,
table(n)                        # for each row; tabulate
```

Typically, there will be one or two aberrant rows that have fewer fields than other rows. For
example, suppose that the aberrant rows have just 10 fields. To identify the aberrant rows,
enter

```
(1:length(n))[n == 10]
```

Use of an editor to inspect those rows will usually identify the source of the problem.

One line at a time

The function `readLines()` reads one line at a time from a file. For example:

```
## First create a file "ex2.txt" that has four lines
cat("First of 4 lines", "2 3 5 7", "", "11 13 17",
    file="ex2.txt", sep="\n")
## Now use readLines to read the three lines back into zzz
zzz <- readLines("ex2.txt", n=-1)
# unlink("ex2.txt") # Use when needed to tidy up, i.e., delete the
                    # file
```

As we will require this file below, we will not for the moment delete it. Now examine the
contents of `zzz`:

```
> zzz
[1] "First of 4 lines" "2 3 5 7"          ""          "11 13 17"
```

Efficient input of large rectangular files

The function `scan()` provides an efficient way to read a file of data into a vector. By default, `scan()` assumes that data are numeric. The parameter setting `what=" "` allows input of fields into a vector of character strings, i.e., there is no type conversion. The function `scan()` reads data into a list that has one vector for each column of data. Compare

```
> zzz <- scan("ex2.txt", skip = 1, quiet= TRUE)
> zzz
[1]  2  3  5  7 11 13 17
```

with

```
> zzz <- scan("ex2.txt", skip = 1, quiet= TRUE, what="")
> zzz
[1] "2"  "3"  "5"  "7"  "11" "13" "17"
unlink("ex2.txt")   # We have no further use for this file.
```

More generally, the parameter `what` may be a list, with one element for each column of data that are to be input. The type of the list element (character, numeric or logical) identifies the type of successive fields in each row. For example:

```
> cat("First of 4 lines", "a 2 3", "", "b 11 13", file="ex3.txt",
+        sep="\n")
> zz <- scan("ex3.txt", skip = 1, quiet= TRUE,
+            what=c("a",rep(1,3)))
> zz
[1] "a"  "2"  "3"  "b"  "11" "13"
> zz <- scan("ex3.txt", skip = 1, quiet= TRUE,
+            what=as.list(c("a",1,1)))
> zz
[[1]]
[1] "a" "b"

[[2]]
[1] "2"  "11"

[[3]]
[1] "3"  "13"
```

Importing data from other statistical systems

The package `foreign` has functions that are able to handle data in various proprietary formats. In several instances, the data are input as a list. At the time of writing, files in some formats from S-PLUS, Minitab, SPSS, SAS and Stata can be read. There is a function for writing files in the Stata `.dta` format. Type in `help.start()`, look under **Packages** and then under **foreign**.

The function `data.restore()` can read objects that have been created by the function `data.dump()` from various versions of S-PLUS. The function `read.S()` is able to read binary files from a similar range of versions of S-PLUS. Check the help files for up to date information.

Database connections

Such connections allow direct access from the R command line to information stored in a database. This is an area of active R development. Already, there are extensive abilities. See in particular the documentation for the RMySQL package.

12.3.2 Data output

Output of data frames

The function write.table() writes out data frames; thus:

```
write.table(elastic1, file="bandsframe.txt")
```

The file bandsframe.txt then contains the following:

```
"stretch" "distance"
"1" 46 148
"2" 54 182
"3" 48 173
"4" 50 166
"5" 44 109
"6" 42 141
"7" 52 166
```

Note that the row and column labels are printed to the file. Specify row.names=FALSE to suppress row names, and/or col.names=FALSE to suppress column names.

For completeness, note also the function write(), used primarily for writing matrices or for writing vectors. See help(write).

Redirection of screen output to a file

The function sink() takes as argument the name of a file. Screen output is then directed to that file. To direct output back again to the screen, call sink() without specifying an argument. For example:

```
> elastic1           # Below, this will be written to the file
      stretch distance
1       46      148
2       54      182
3       48      173
4       50      166
5       44      109
6       42      141
7       52      166
> sink("bands2.txt")
> elastic1  # NB: No output on screen
> sink()
```

Output to a file using `cat()`

We saw earlier that output from `cat()` could be directed to a file, rather than to the screen. Output objects must at present be scalars or vectors or matrices. Matrices are output as single vectors, with elements in columnwise order. The user can place limited format controls (spaces, tabs and newlines) between the names of output objects.

12.4 Factors – Additional Comments

Factors are an essential adjunct to the use of model formulae and graphics formulae. They find wide use in the statistical modeling functions in the various R packages for applications that extend far beyond analysis of variance and linear regression modeling. Recall that factors are stored as integer vectors, with the integer values interpreted according to the information in the table of levels that holds all the distinct text strings.

Factors have an *attribute* `levels` that holds the level names. They also the potential to cause surprises, so be careful! Note that:

- When a vector of character strings is included as a column of a data frame, R by default turns the vector into a factor, in which the distinct text strings are the level names. To avoid this, enclose the vector of character strings in the wrapper function `I()` when including it in the data frame.
- There are some contexts in which factors become numeric vectors. To be sure of obtaining the vector of text strings, specify e.g., `as.character(country)`.
- To extract the codes $1, 2, \ldots$, specify `as.numeric(country)`.
- For a factor whose levels are character string representations of numeric values (e.g. "10","2","3.1"), use, e.g., `as.numeric(as.character(fac))` to extract the numeric values.

Avoid the (soon to be defunct) function `codes()`. This returns a recoding in which the levels are taken in text string order. Thus "10" is likely to precede "2".

Note three common contexts where factors are taken as vectors of integer codes.

- If the `labels` argument to `text()` is a factor, then it is the codes that appear in the plot.
- When the index variable in a `for` loop takes factor values, the values are the integer codes. For example:

```
> fac <- factor(c("c", "b", "a"))
> for (i in fac) print(i)
[1] 3
[1] 2
[1] 1
```

- The result from `cbind()`, for any factor arguments, depends on whether at least one of the other arguments is a data frame. If none of the arguments is a data frame, factors become integer vectors. See Subsection 12.6.3 for details.

Ordered factors

Actually, it is the levels that are ordered. To create an ordered factor, or to turn a factor into an ordered factor, use the function `ordered()`. The levels of an ordered factor are assumed to specify positions on an ordinal scale. Try

```
> stress.level <- rep(c("low","medium","high"), 2)
> ordf.stress <- ordered(stress.level, levels=c("low", "medium",
  "high"))
> ordf.stress
[1] low     medium high    low     medium high
Levels:  low < medium < high
> ordf.stress < "medium"
[1]   TRUE FALSE FALSE   TRUE FALSE FALSE
> ordf.stress >= "medium"
[1] FALSE  TRUE   TRUE FALSE   TRUE   TRUE
```

Ordered factors have (*inherit*) the attributes of factors, and have a further ordering attribute. All factors, ordered or not, have an order for their levels! The special feature of ordered factors is that there is an ordering relation (<) between factor levels.

Factor contrasts

Given model terms that include one or more factors, factor contrasts make it possible to construct a model matrix. The choice of factor contrasts determines the way that the model is specified, and hence the interpretation of parameter estimates associated with factor levels.

As described in Chapter 7, the R system allows a choice between three different types of factor contrasts: *treatment* contrasts, *sum* contrasts (often known as *anova* contrasts), and *helmert* contrasts, with *treatment* as the default. Users of R may sometimes wish to use sum contrasts in place of treatment contrasts; specify `options(contrasts=c("contr.sum", "contr.poly"))`. It is also possible to set contrasts separately for each factor. See `help(C)`.

The coding for any particular choice of contrasts can be inspected directly. Note the following:

```
> contr.treatment(3)
  2 3
1 0 0
2 1 0
3 0 1
> contr.sum(3)
  [,1] [,2]
1   1    0
2   0    1
3  -1   -1
> cities <- factor(c("Melbourne","Sydney","Adelaide"))
> contr.treatment(cities)
  3 1
```

```
2 0 0
3 1 0
1 0 1
> contr.sum(cities)
  [,1] [,2]
2   1   0
3   0   1
1  -1  -1
```

The default is that Adelaide is level 1, Melbourne is level 2, and Sydney is level 3.

Contrasts – implications for model fitting

Fitted values, residuals, anova tables (with an exception that will be noted below) and model comparisons are not affected by the parameterization.

For a factor that has two levels, helmert contrasts lead to a parameter estimate that is just half that for treatment contrasts. For factors with more than two levels, the use of helmert contrasts leads in general to parameter estimates that are hard to interpret.

Tests for main effects in the presence of interactions

The function anova.lme() in the *nlme* package allows the parameter setting type = "m". This gives marginal tests of the effect of dropping each term in the model in turn, while retaining other terms. This includes tests on the effect of dropping each factor main effect, while retaining all interaction terms – tests that in most circumstances do not make much sense. See Venables (1998) for commentary.

Note that with anova = "m", the analysis of variance table that is presented will, for examination of main effects in the presence of corresponding interactions, change depending on the contrasts that are used. When the *sum* contrasts (options(contrasts=c("contr.sum", "contr.poly"))) are used, the anova table gives information that is relevant to an equivalent, in the multi-level modeling context, of so-called Type III tests.

12.5 Missing Values

The missing value symbol is NA. Any arithmetic operation or relation that involves NA generates NA. This applies also to the relations <, <=, >, >=, ==, !=. The first four compare magnitudes, == tests for equality, and != tests for inequality. Users who do not give adequate consideration to the handling of NAs in their data may be surprised by the results. Specifically, note that x==NA generates NA, whatever the values in x. Be sure to use is.na(x) to test which values of x are NA. The construct x==NA gives a vector in which all elements are NAs, and thus gives no information about the elements of x. For example:

```
> x <- c(1, 6, 2, NA)
> is.na(x)    # TRUE for when NA appears, and otherwise
```

```
FALSE [1] FALSE FALSE FALSE   TRUE
> x == NA                # All elements are set to NA
[1] NA NA NA NA
> NA == NA
[1] NA
```

Missing values in subscripts

Here is behavior that may seem surprising:

```
> x <- c(1, 6, 2, NA, 10)
> x > 2
[1] FALSE  TRUE FALSE    NA  TRUE
> x[x > 2]
[1] 6  NA 10  # NB.  This generates a vector of length 3
> x[x > 2] <- c(101, 102)
Warning message:
number of items to replace is not a multiple of
replacement length
> x
[1]   1 101   2  NA 101
```

To replace elements that are greater than 2, specify

```
> x <- c(1, 6, 2, NA, 10)
> x[!is.na(x) & x > 2] <- c(101, 102)
> x
[1]   1 101   2  NA 102
```

An assignment that in R does give the result that the naïve user might expect, is

```
y <- c(NA, 2, 4, 1, 7)
y[x > 2] <- x[x > 2]  # For S-PLUS behavior, see page 341
```

This replaces elements of y with corresponding elements of x wherever x > 2, ignoring NAs.

An alternative, that explicitly identifies the elements that are substituted, is

```
y[!is.na(x) & x > 2] <- x[!is.na(x) & x > 2]
```

Use of !is.na(x) limits the elements that are identified by the subscript expression, on both sides, to those that are not NAs.

Counting and identifying NA*s*

The following gives information on the number of NAs in subgroups of the data:

```
> table(rainforest$species, !is.na(rainforest$branch))

                  FALSE TRUE
  Acacia mabellae     6   10
```

```
C. fraseri              0   12
Acmena smithii         15   11
B. myrtifolia           1   10
```

Thus for *Acacia mabellae* there are 6 NAs for the variable branch (i.e., number of branches over 2 cm in diameter), out of a total of 16 data values.

The function `complete.cases()` takes as arguments any sequence of vectors, data frames and matrices that all have the same number of rows. It returns a vector that has the value TRUE whenever a row is complete, i.e., has no missing values across any of the objects, and otherwise has the value FALSE.

The expression `any(is.na(x))` returns TRUE if the vector has any NAs, and is otherwise FALSE.

The handling of NAs in R functions

NAs in tables

By default, the function `table()` ignores NAs. The action needed to get NAs tabulated under a separate NA category depends, annoyingly, on whether or not the vector is a factor. If the vector is not a factor, specify `exclude=NULL`. If the vector is a factor then it is necessary, in the call to `table()`, to replace it with a new factor that has "NA" as an explicit level. Specify `x <- factor(x, exclude=NULL)`.

```
> x <- c(1, 5, NA, 8)
> x <- factor(x)
> x
[1] 1   5   NA 8
Levels:  1 5 8
> factor(x, exclude=NULL)
[1] 1   5   NA 8
Levels:  1 5 8 NA
```

NAs in modeling functions

Many of the modeling functions have an argument `na.action`. We can inspect the global setting; thus:

```
> options()$na.action   # Version 1.7.0, following startup
[1] "na.omit"
```

Individual functions may have defaults that are different from the global setting. See `help(na.action)`, and help pages for individual modeling functions, for further details.

Sorting and ordering

By default, `sort()` omits any NAs. The function `order()` places NAs last. Hence:

```
> x <- c(1, 20,  2, NA, 22)
> order(x)
[1] 1 3 2 5 4
```

```
> x[order(x)]
[1]  1  2 20 22 NA
> sort(x)
[1]  1  2 20 22
```

12.6 Lists and Data Frames

Lists make it possible to collect an arbitrary set of R objects together under a single name. We might, e.g., collect together vectors of several different modes and lengths, scalars, matrices or more general arrays, functions, etc. Lists can be, and often are, a rag-tag collection of different objects. We will use for illustration the list object that R creates as output from an lm calculation.

Keep in mind that if zz is a list of length n, then zz[1], zz[2],..., zz[n] are all lists. So, for example, is zz[1:3]. A list is a wrapper into which multiple objects slot in a linear sequence. Think of a bookshelf, where the objects may be books, magazines, boxes of computer disks, trinkets, ... Subsetting takes a selection of the objects, and moves them to another "bookshelf". When just a single object is taken, it goes to another shelf where it is the only object. The individual objects, not now held on any "bookshelf", are zz[[1]], zz[[2]], ..., zz[[n]].

Functions such as c(), length(), and rev() (take elements in the reverse order) can be applied to any vector, including a list. To interchange the first two elements of the list zz, write zz[c(2, 1, 3:length(zz))].

12.6.1 Data frames as lists

Internally, data frames have the structure of a special type of list. In data frames all list elements are lists that consist of vectors of the same length. Data frames are definitely not a rag bag of different objects!

Data frames whose columns are numeric are sufficiently like matrices that it may be an unpleasant surprise to encounter contexts where they do not behave like matrices. If dframe is a data frame, then dframe[,1] is a vector. By contrast dframe[1,] is a list. To turn the data frame into a vector, specify unlist(dframe[,1]). If the columns of dframe are of different modes, there must obviously be some compromise in the mode that is assigned to the elements of the vector. The result may be a vector of text strings.

12.6.2 Reshaping data frames; reshape()

In the data frame immer that is in the *MASS* package, data for the years 1931 and 1932 are in separate columns. Plots that use functions in the *lattice* package may require data for the two years to be in the same column, i.e., we will need to reshape this data frame from its *wide* format into a *long* format. The parameter names for the reshape() function are chosen with this repeated measures context in mind, although its application is more general.

We will demonstrate reshaping from wide to long format, for the first six rows in the immer data frame.

```
> library(MASS)
> data(immer)
```

```
> immer8 <- immer[1:8, ]
> immer8
  Loc Var     Y1     Y2
1  UF   M  81.0   80.7
2  UF   S 105.4   82.3
3  UF   V 119.7   80.4
4  UF   T 109.7   87.2
5  UF   P  98.3   84.2
6   W   M 146.6  100.4
7   W   S 142.0  115.5
8   W   V 150.7  112.2
> immer8.long <- reshape(immer8, varying = list(c("Y1", "Y2")),
+                             direction = "long")
```

In addition to specifying a reshaping to a long format (`direction = "long"`), we need the parameter `varying`. This specifies a list that holds a vector that gives the names of the columns that are stacked one above the other in the long format.

A slightly more complicated form of command gives a result that is better labeled.

```
> immer8.long <- reshape(immer8, varying = list(c("Y1", "Y2")),
+                             timevar = "Year", times = c(1931, 1932),
+                             v.names = "Yield",  direction = "long")
> immer8.long
        Loc Var Year Yield id
1.1931   UF   M 1931  81.0  1
2.1931   UF   S 1931 105.4  2
3.1931   UF   V 1931 119.7  3
4.1931   UF   T 1931 109.7  4
5.1931   UF   P 1931  98.3  5
6.1931    W   M 1931 146.6  6
7.1931    W   S 1931 142.0  7
8.1931    W   V 1931 150.7  8
1.1932   UF   M 1932  80.7  1
2.1932   UF   S 1932  82.3  2
3.1932   UF   V 1932  80.4  3
4.1932   UF   T 1932  87.2  4
5.1932   UF   P 1932  84.2  5
6.1932    W   M 1932 100.4  6
7.1932    W   S 1932 115.5  7
8.1932    W   V 1932 112.2  8
```

It may, in addition, be useful to specify the parameter `ids` that gives values that identify the rows ("subjects") in the *wide* format.

Examples of reshaping from the long to wide format are included in the help page for `reshape()`.

12.6.3 *Joining data frames and vectors –* cbind()

Use cbind() to join, side by side, two or more objects that may be any mix of data frames and vectors. (If all arguments are matrix or vector, then the result is a matrix.) Alternatively, we can use the function data.frame().

In the use of cbind(), note that if all arguments are matrix or vector, then any factors will become integer vectors. If one (or more) argument(s) is a data frame, so that the result is a data frame, factors will remain as factors.

Use rbind() to add one or more rows to a data frame, and/or to stack data frames. When data frames are stacked one above the other, the names and types of the columns must agree.

12.6.4 *Conversion of tables and arrays into data frames*

We may require a data frame form of representation in which the data frame has one classifying column corresponding to each dimension of the table or array, plus a column that holds the table or array entries. Use as.data.frame.table() to handle the conversion.

```
> data(UCBAdmissions)        ## UCBAdmissions is a 3-way table
> dimnames(UCBAdmissions)
$Admit
[1] "Admitted" "Rejected"

$Gender
[1] "Male"     "Female"

$Dept
[1] "A" "B" "C" "D" "E" "F"

> UCB.df <- as.data.frame.table(UCBAdmissions)
> UCB.df[1:2, ]
    Admit Gender Dept Freq
1 Admitted   Male    A  512
2 Rejected   Male    A  313
```

If the argument is an array, it will first be coerced into a table with the same number of dimensions.

12.6.5* *Merging data frames –* merge()

The *DAAG* package has the data frame Cars93.summary, which has as its row names the six different car types in the data frame Cars93 from the *MASS* package. The column abbrev holds one or two character abbreviations for the car types. We show how to merge the information on abbreviations into the data frame Cars93; thus:

```
new.Cars93 <- merge(x=Cars93, y=Cars93.summary[,"abbrev", drop=F],
                    by.x="Type", by.y="row.names")
```

The arguments by.x and by.y specify the keys, the first from the data frame that is specified as the x- and the second from the data frame that is specified as the y-parameter. The new column in the data frame new.Cars93 has the name abbrev.

If there had been rows with missing values of Type, these would have been omitted from the new data frame. We can avoid this by ensuring that Type has NA as one of its levels, in both data frames.

12.6.6 The function sapply() and related functions

Because a data frame has the structure of a list of columns, the functions sapply() and lapply() can be used to apply a function to each of its columns in turn. With sapply() the result is simplified as far as possible, for example into a vector or matrix. With lapply(), the result is a list.

The function apply() can be used to apply a function to rows or columns of a matrix. The data frame is first constrained to be a matrix, so that elements are constrained to be all character, or all numeric, or all logical. This places constraints on the functions that can be applied.

Additional examples for sapply()

Recall that the function sapply() applies to all elements in a vector (usually a data frame or list) the function that is given as its second argument. The arguments of sapply() are the name of the data frame or list (or other vector), and the function that is to be applied. Here are examples, using the data set rainforest from our DAAG package.

```
> sapply(rainforest, is.factor)
    dbh    wood    bark    root  rootsk  branch species
  FALSE   FALSE   FALSE   FALSE   FALSE   FALSE    TRUE
```

These data relate to Ash and Helman (1990). The function is.factor() has been applied to all columns of the data frame rainforest:

```
> sapply(rainforest[,-7], range)     # The final column (7) is a
                                     # factor
     dbh wood bark root rootsk branch
[1,]   4   NA   NA   NA     NA     NA
[2,]  56   NA   NA   NA     NA     NA
```

It is more useful to call range() with the parameter setting na.rm=TRUE. For this, specify na.rm=TRUE as a third argument to the function sapply(). This argument is then automatically passed to the function that is specified in the second argument position when sapply() is called. For example:

```
> sapply(rainforest[,-7], range, na.rm=TRUE)
     dbh wood bark root rootsk branch
[1,]   4    3    8    2    0.3     40
[2,]  56 1530  105  135   24.0    120
```

Note that:

- Any rectangular structure that results from the use of `sapply()` will be a matrix, not a data frame. In order to use `sapply` to carry out further manipulations on its columns, it must first be turned into a data frame.
- If `sapply()` is used with a matrix as its argument, the function is applied to all elements of the matrix – which is not usually the result that is wanted. Be sure, when the intention is to use `sapply()` as described above, that the argument is a data frame.
- The function `sapply()` is one of several functions that have "apply" as part of their name. In the present context, note especially `lapply()`, which works just like `sapply()`, except that the result is a list. It does not try to "simplify" its output into some form of non-list object, such as a vector or matrix.

12.6.7 Splitting vectors and data frames into lists – `split()`

As an example, we work with the data frame `cabbages` from the *MASS* package. Then,

```
data(cabbages)
split(cabbages$HeadWt, cabbages$Date)
```

returns a list with three elements. The list elements have names d16, d20 and d21 and consist, respectively, of the values of `HeadWt` for which `Date` takes the respective values d16, d20 and d21.

One application is to getting side by side boxplots. The function `boxplot()` takes as its first element a list in which the first list element is the vector of values for the first boxplot, the second list element is the vector of values for the second boxplot, and so on. For example:

```
boxplot(split(cabbages$HeadWt, cabbages$Date))
```

The argument to `split()` may alternatively be a data frame, which is split into a list of data frames. For example:

```
split(cabbages[,-1], cabbages$Date)       # Split remaining columns
                                          # by levels of Date
```

12.7* Matrices and Arrays

Matrices are likely to be important for implementation of new regression and multivariate methods. All elements of a matrix have the same mode, i.e., all numeric, or all character. A matrix is a more restricted structure than a data frame. Numeric matrices allow a variety of mathematical operations, including matrix multiplication, that are not available for data frames. See `help(matmult)`.

Where there is a choice between the use of matrix arithmetic to carry out arithmetic or matrix manipulations on a large numeric matrix, and the equivalent operations on a data frame, it is typically much faster and more efficient to use the matrix operations. Even for such simple operations as `x <- x+2` or `x <- log(x)`, the time can be substantially reduced when x is a matrix rather than a data frame. Use `as.matrix()` to handle any conversion that may be necessary.

Additionally, matrix generalizes to array, which may have more than two dimensions. Names may be assigned to the rows and columns of a matrix, or more generally to the different dimensions of an array. We give details below.

Matrices are stored columnwise. Thus consider

```
> xx <- matrix(1:6, ncol=3)   # Equivalently, enter
                              # matrix(1:6, nrow=2)
> xx
     [,1] [,2] [,3]
[1,]   1    3    5
[2,]   2    4    6
```

If xx is any matrix, the assignment

```
x <- as.vector(xx)
```

places columns of xx, in order, into the vector x. In the example above, we get back the elements $1, 2, \ldots, 6$.

Matrices have the attribute "dimension". Thus:

```
> dim(xx)
[1] 2 3
```

In fact a matrix is a vector (numeric or character) whose dimension attribute has length 2.

For use in demonstrating the extraction of columns or rows or submatrices, set

```
> x34 <- matrix(1:12, ncol=4)
> x34
     [,1] [,2] [,3] [,4]
[1,]   1    4    7   10
[2,]   2    5    8   11
[3,]   3    6    9   12
```

The following should be self-explanatory:

```
> x34[2:3, c(1,4)]  # Extract rows 2 & 3 and columns 1 & 4
     [,1] [,2]
[1,]   2   11
[2,]   3   12
> x34[2, ]          # Extract the second row
[1]  2  5  8 11

> x34[-2, ]         # Extract all rows except the second
     [,1] [,2] [,3] [,4]
[1,]   1    4    7   10
[2,]   3    6    9   12

> x34[-2,-3]        # Omit row 2 and column 3
     [,1] [,2] [,3]
```

```
[1,]    1    4    10
[2,]    3    6    12
```

We can use the `dimnames()` function to assign and/or extract matrix row and column names. The `dimnames()` function gives a list, in which the first list element is the vector of row names, and the second list element is the vector of column names. This generalizes in the obvious way for use with arrays, which we will discuss shortly.

The function `t()` takes a matrix as its argument, which it transposes. It may also be used, with the side effect that all elements will be changed to character when columns are of different types, for data frames.

12.7.1 Outer products

There are a variety of contexts where we need to generate a matrix of quantities of the form

$$x_{ij} = a_i b_j$$

or, more generally, to generate an array whose (i, j)th element is a function of a_i and b_j. A simple example is

```
# Multiplication table
> a <- 1:5
> b <- 1:10
> ab <- outer(a, b)
> ab
     [,1] [,2] [,3] [,4] [,5] [,6] [,7] [,8] [,9] [,10]
[1,]    1    2    3    4    5    6    7    8    9    10
[2,]    2    4    6    8   10   12   14   16   18    20
[3,]    3    6    9   12   15   18   21   24   27    30
[4,]    4    8   12   16   20   24   28   32   36    40
[5,]    5   10   15   20   25   30   35   40   45    50
```

We can use an inline function to generate the elements of the array. As an example, we will generate a matrix of color names. The parameter `col` in R plotting functions may be supplied either with a number, or with the name of a color. Many of these colors come in five different shades; thus in addition to "blue" there are "blue1", "blue2", "blue3" and "blue4". The following generates all five shades of the colors "blue", "green", "yellow", "orange", "pink" and "red", storing the result in a matrix.

```
> colormat <- outer(c("blue","green","yellow","orange",
+                     "pink","red"),  c("",paste(1:4)),
+                     function(x,y)paste(x,y, sep=""))
> t(colormat)     # Display the transposed matrix
      [,1]      [,2]       [,3]        [,4]        [,5]       [,6]
[1,] "blue"    "green"    "yellow"    "orange"    "pink"     "red"
[2,] "blue1"   "green1"   "yellow1"   "orange1"   "pink1"    "red1"
[3,] "blue2"   "green2"   "yellow2"   "orange2"   "pink2"    "red2"
[4,] "blue3"   "green3"   "yellow3"   "orange3"   "pink3"    "red3"
[5,] "blue4"   "green4"   "yellow4"   "orange4"   "pink4"    "red4"
```

```
> plot(rep(0:4, rep(6,5)), rep(1:6, 5), col=colormat,
+       pch=15, cex=8)
```

12.7.2 Arrays

An array is a generalization of a matrix (2 dimensions), to allow > 2 dimensions. The dimensions are, in order, rows, columns, ... By way of example, we start with a numeric vector of length 24. So that we can easily keep track of the elements, we make them 1, 2, ..., 24. Thus, given

```
x <- 1:24
```

we can set

```
dim(x) <- c(4, 6)
```

This turns x into the 4 × 6 matrix

```
> x
     [,1] [,2] [,3] [,4] [,5] [,6]
[1,]    1    5    9   13   17   21
[2,]    2    6   10   14   18   22
[3,]    3    7   11   15   19   23
[4,]    4    8   12   16   20   24
```

Now try

```
> dim(x) <- c(3, 4, 2)
> x

, , 1
     [,1] [,2] [,3] [,4]
[1,]    1    4    7   10
[2,]    2    5    8   11
[3,]    3    6    9   12

, , 2
     [,1] [,2] [,3] [,4]
[1,]   13   16   19   22
[2,]   14   17   20   23
[3,]   15   18   21   24
```

The apply() *function*

The function apply() can be used with arrays or data frames. With data frames it can often be used to give the same end result as sapply(). Its first argument is an array or data frame. The second argument specifies the dimension – 1 for applying the function to each row in turn, and 2 when the function is to be applied to each column in turn (if the first argument is an array of more than two dimensions, then a number greater than 2 is possible). Here is an example:

```
> data(airquality)            # base library
> apply(airquality, 2, mean) # All elements must be numeric!
  Ozone Solar.R    Wind    Temp    Month      Day
     NA      NA    9.96   77.88     6.99    15.80
> apply(airquality, 2, mean, na.rm=TRUE)
  Ozone Solar.R    Wind    Temp    Month      Day
  42.13  185.93    9.96   77.88     6.99    15.80
```

The use of `apply(airquality, 1, mean)` will give means for each row. These are not, for these data, useful information!

12.8 Classes and Methods

Functions such as `print()`, `plot()` and `summary()` are generic functions. Their action varies according to the *class* of the object that is given as argument. Thus consider `print()`. For a factor `print.factor()` is used, for a data frame `print.data.frame()` is used, and so on. Ordered factors "inherit" the print method for factors. For objects (such as numeric vectors) that do not otherwise have a print method, `print.default()` handles the printing.

Generic functions do not call the specific method, such as `print.factor()`, directly. Instead they call the `UseMethod()` function, which then calls the relevant method for that class of object, e.g., the factor method (such as `print.factor()`) for a factor object. For example, here is the function `print()`.

```
> print
function (x,...)
UseMethod("print")
```

The function `UseMethod()` notes the class of the object, now identified as x, and calls the print function for that class.

Use the function `class()` to determine the class of an object.

12.8.1 Printing and summarizing model objects

Just as for any other R object, typing the name of a model object on the command line invokes the print function, if any, for that class of object. Thus typing `elastic.lm`, where `elastic.lm` is an `lm` object, has the same effect as `print.lm(elastic.lm)` or `print(elastic.lm)`.

Print functions for model objects, e.g., `print.lm()` for printing the model object `elastic.lm`, process output into a form that is, broadly, suitable for immediate inspection. Additional or different information may be available by directly accessing the list elements of the model object. For most classes of object there is, in addition, a `summary()` function that gives a different and often more detailed summary. For example

```
elastic.sum <- summary(elastic.lm)
```

stores summary information for `elastic.lm` in the `elastic.sum` lm summary object. Typing `elastic.sum` (or `summary(elastic.lm)`) on the command line invokes the function `print.summary.lm()`, thus printing the summary information.

Other generic functions that commonly find use with model objects are `coef()` (alias `coefficients()`), `residuals()` (alias `resid()`), `fitted()`, `predict()` and `anova()`.

12.8.2 Extracting information from model objects

Consider for example the linear model (`lm`) object `elastic.lm` that we create by specifying

```
elastic.lm <- lm(distance ~ stretch, data=elasticband)
```

The object `elastic.lm` is a list that holds a variety of quite different objects. To get the names of the list elements, type

```
> names(elastic.lm)
 [1] "coefficients"  "residuals"   "effects"    "rank"
 [5] "fitted.values" "assign"      "qr"         "df.residual"
 [9] "xlevels"       "call"        "terms"      "model"
```

Here are three different and equivalent ways to examine the contents of the first list element:

```
> elastic.lm$coefficients
(Intercept)       stretch
 -63.571429      4.553571
> elastic.lm[["coefficients"]]
(Intercept)       stretch
 -63.571429      4.553571
> elastic.lm[[1]]
(Intercept)       stretch
 -63.571429      4.553571
```

Note that we can also ask for `elastic.lm["coefficients"]` or equivalently `elastic.lm[1]`. These are subtly different from `elastic.lm$coefficients` and its equivalents. Either of these gives the list whose only element is the vector `elastic.lm$coefficients`. This is reflected in the result that is printed. The information is preceded by `$coefficients`, meaning "list element with name coefficients".

```
> elastic.lm[1]
$coefficients
(Intercept)       stretch
 -63.571429      4.553571
```

The reason is that a list is a vector object. Lists have a recursive structure, so that subscripting a list yields a list. Use of `elastic.lm[1]` extracts the list element whose contents are `elastic.lm[[1]]`.

The second list element is a vector of length 7:

```
> options(digits=3)
> elastic.lm$residuals
      1        2        3        4        5        6        7
  2.107   -0.321   18.000    1.893  -27.786   13.321   -7.214
```

The tenth list element documents the function call:

```
> elastic.lm$call
lm(formula = distance ~ stretch, data = elasticband)
> mode(elastic.lm$call)
[1] "call"
```

12.9 Databases and Environments

When R starts up, it has a set of names of directories ("databases") where it looks, in order, for objects. The workspace comes first on this *search path* or *search list*. Thus, `ls (pos=1)` is equivalent to `ls ()`. Use `ls (pos=2)` to print the names of the objects in the database that is second on the search path, and so on.

For version 1.7.0, immediately after startup, the search path is:

```
> search()
[1] ".GlobalEnv"      "package:methods" "package:ctest" "package:mva"
[5] "package:modreg" "package:nls"      "package:ts"     "Autoloads"
[9] "package:base"
```

Notice that *methods*, *ctest* and several other packages are automatically loaded, in addition to the base package. The `Autoloads` database need not concern us here. See `help (autoloads)` for details.

The functions `library()` and `attach()` extend the search path. Consider first the use of `library()` to load a package. For example:

```
> library(ts)          # Time series package, included with most
                        # distributions
> search()
[1] ".GlobalEnv"   "package:ts" "Autoloads"   "package:base"
```

For use of `attach()` with data frames or other list objects, recall the syntax:

```
> attach(primates)     # NB: No quotes
> detach(primates)     # NB: In R quotes are optional
```


Try the following:

```
> data(cars)
> search()
[1] ".GlobalEnv" "package:ctest" "Autoloads"   "package:base"
> attach(cars)
> search()
[1] ".GlobalEnv"     "cars"          "package:ctest" "Autoloads"
[5] "package:base"
> ls(pos=2)   # NB: The data frame cars has columns dist and speed
```

```
[1] "dist"  "speed"
> detach(cars)
```

Another use of `attach()`, to give access to objects that have been saved in an image file, will be described in the next subsection.

12.9.1 Workspace management

It is undesirable to keep adding new objects to the workspace, while leaving existing objects in place. The workspace becomes cluttered, and data and other objects that are in use must be identified from among the clutter. Large data objects, such as are common in expression array work, take up what may for some tasks be a scarce memory resource. There are two complementary strategies:

- Objects that cannot easily be reconstructed or copied from elsewhere, but are not for the time being required, are conveniently saved to an image file, using `save()`.
- Use a separate working directory for each major project.

To save one or more objects, here `possum` and `possumsites` in save file format, enter

```
save(possum, possumsites, file="possum.RData")
```

To recover the objects later, enter

```
load("possum.RData")
```

The default action of `save.image()`, i.e., save all objects in the workspace to the default **.Rdata** file, is equivalent to `save(list=ls(), file=".Rdata")`

The command

```
attach("possum.RData")
```

gives access to objects that have been saved in the file **possum.RData**. Technically, the file then becomes a database. By default, it appears at the second position on the search path, i.e. immediately following `.Globalenv`, which is the the workspace. Workspace objects, even if they happen to have the same names as objects in the file that has been attached, are unchanged. Any workspace image file, if necessary with path included to indicate the directory in which it resides, can be attached in this same way.

It is straightforward, within an R session, to move from one workspace to another. Use `save.image()` to save the contents of the current workspace in the current working directory, `rm(list=ls())` to clear the workspace, `setwd("newdir")` to set the new working directory to, here, `newdir`, and `load(".Rdata")` to load the default workspace that is in the new working directory. These operations can all be carried out from the menu, where available. The effect is to change `.Globalenv`.

12.9.2 Function environments, and lazy evaluation

When a function is defined, this sets up an evaluation environment for that function. Where objects have not been defined within the function itself or passed as parameters, R looks in the environment in which the function was itself defined.

The consequences for lazy evaluation, which we now discuss, are mildly subtle.

Lazy evaluation

Expressions may be specified as arguments to R functions. The expression is evaluated only when it is encountered in the R function. For example

```
> lazyfoo <- function(x=4, y = x^2)y
> lazyfoo()
[1] 16
> lazyfoo(x=3)
[1] 9
```

This is unsurprising. Expressions that appear in default parameter settings are evaluated within the function environment.

By contrast, however, expressions that are specified when the function is called are evaluated in the parent environment, i.e., in the environment from which the function was called. Hence the following behavior:

```
> lazyfoo(y=x^3)    # We specify a value for y in the function call
Error in lazyfoo(y = x^3) : Object "x" not found
> x <- 9
> lazyfoo(y=x^3)
[1] 729
```

For the call `lazyfoo()`, the parameter setting was the default, and the function looked for an x in the environment interior to the function. For the call `lazyfoo(y=x^3)`, the function looked for an x in the environment from which the function was called.

Example – a function that identifies objects added during a session

This illustrates points that were made in the discussion above.

At the beginning of a new session, we might store the names of the objects in the workspace in the vector dsetnames, thus:

```
dsetnames <- objects()
```

Now suppose that we have a function additions(), defined thus:

```
additions <- function(objnames = dsetnames){
    newnames <- objects(envir=sys.frame(0))
    existing <- newnames %in% objnames
    newnames[!existing]
    }
```

The function call sys.frame(0) returns the name of the workspace. At some later point in the session, we can enter

```
additions(dsetnames)
```

to obtain the names of objects that have been added since the start of the session.

Use of newnames <- objects() in the above function, i.e., leaving parameter settings at their defaults, would have returned the names of objects in the function environment.

12.10 Manipulation of Language Constructs

We will demonstrate manipulations involving formulae and expressions.

Model and graphics formulae

On some occasions, it can be helpful to construct model or graphics formulae from text strings. For example, here is a function that takes two named columns from the data frame `mtcars`, plotting them one against another:

```
plot.mtcars <- function(xvar="disp", yvar="hp"){
    attach(mtcars)
    mt.txt <- paste(yvar, "~", xvar)
    plot(formula(mt.txt))
    }
```

We can now, when we call the function, set `xvar` and `yvar` to be any columns we choose:

```
> data(mtcars)            # base package
> names(mtcars)
 [1] "mpg"  "cyl"  "disp" "hp"  "drat" "wt"  "qsec" "vs"
     "am"   "gear"
[11] "carb"
> plot.mtcars(xvar="disp", yvar="mpg")
```

Expressions

An expression is anything that can be evaluated. Thus `x^2` is an expression, `y <- x^2` is an expression that returns the value that is assigned to `y`, and `y == x^2` is an expression. Recall that `y <- x^2` assigns the value of `x^2` to `y`, while `y == x^2` tests whether `y` equals `x^2`.

The following can be convenient when there is a complicated expression that we want to evaluate repeatedly from the command line, changing one or more of the parameters each time.

```
> local(a+b*x+c*x^2, envir=list(x=1:4, a=3, b=5, c=1))
[1]   9  17  27  39
```

Formatting and plotting of text and equations

An extension of the syntax for expressions makes it possible to format and plot mathematical symbols and formulae, either on their own or as part of text strings. Wherever a plot command, such as `text()` or `mtext()` or `title()`, allows a text string, an expression can be substituted. Text and/or text strings can form part of the "expression", and can appear where a mathematical expression would require a variable. The output is formatted according to LaTeX-like rules.

The following, which follows normal text with italic text, demonstrates the mixing of different text styles in a single line of text:

```
library(DAAG); data(rainforest)
plot(wood ~ dbh, data = rainforest,
```

```
        subset = species=="Acmena smithii")
title(main = expression("Graph is for" * phantom(0) *
                        italic("Acmena smithii")))
# phantom(0) inserts a space that is the width of a zero
```

The asterisk juxtaposes the two text strings, in just the same way as `expression(u*v)` will be written uv.

The following demonstrates the inclusion of a mathematical expression in an annotation for a graph. It uses both `expression()` and `substitute()`. This latter function is a generalization of `expression()` that allows the substitution of selected symbols in the expression:

```
## First plot the "angular" transformation
curve(asin(sqrt(x)), from = 0, to = 1.0, n=101,
      xlab = expression(italic(p)),
      ylab = expression(asin(sqrt(italic(p)))))
# Place a title above the graph
mtext(side = 3, line = 1,
      substitute(tx * italic(y) == asin(sqrt(italic(p))),
      list(tx = "Plot of the angular function ")))
# Use "==" where "=" will appear in the plotted text
```

The function `substitute()` may, additionally, be used to extract the names of the actual function parameters when they are passed at run time. This can be useful for graphical annotation. The following demonstrates the syntax:

```
> testfun <- function(x, y){
+   xpar <- deparse(substitute(x))
+   xpar}
> testfun(x = AnyValidName)
[1] "AnyValidName"
```

See `help(plotmath)` for further details.

12.11 Further Reading

The definitive document is the relevant up to date version of the R Language Definition document (R Core Development Team). Venables and Ripley (2000) may be consulted as a useful commentary on that document.

For citing R in a publication, use Ihaka and Gentleman (1996).

References for further reading

Ihaka, R. and Gentleman, R. 1996. R: a language for data analysis and graphics. *Journal of Computational and Graphical Statistics* 5: 299–314.

R Core Development Team, updated regularly. R *Language Definition*. Available from CRAN sites.

Venables, W.N. and Ripley, B.D. 2000. S *Programming*. Springer-Verlag.

12.12 Exercises

1. Compare the different outputs from `help.search("print")`, `apropos(print)` and `methods(print)`. Look up the help for each of these three functions, and use what you find to explain the different outputs.

2. Identify as many R functions as possible that are specifically designed for manipulations with text strings.

3. Test whether `strsplit()` is vectorized, i.e., does it accept a vector of character strings as input, then operating in parallel on all elements of the vector?

4. For the data frame `Cars93`, get the information provided by `summary()` for each level of Type. [Use `split()`.]

5. Determine the number of cars, in the data frame `Cars93`, for each Origin and Type.

6. In the data frame `Insurance` (MASS package):

   (a) determine the number of rows of information for each age category (`Age`) and car type (`Group`);
   (b) determine the total number of claims for each age category and car type;

7. Enter the following, and explain the steps that are performed in the process of obtaining the result:

```
library(DAAG); data(science)
attach(science)
sapply(split(school, PrivPub),
       function(x)length(unique(x)))
```

8. Save the objects in your workspace, into an image **(.RData)** file with the name **archive.RData**. Then remove all objects from the workspace. Demonstrate how, without loading the image file, it is possible to list the objects that were included in **archive.RData** and to recover a deleted object that is again required.

9. Determine the number of days, according to R, between the following dates:

   (a) January 1 in the year 1700, and January 1 in the year 1800;
   (b) January 1 in the year 1998, and January 1 in the year 2007.

10.* The following code concatenates (x, y) data values that are random noise to data pairs that contain a "signal", randomly permutes the pairs of data values, and finally attempts to reconstruct the signal:

```
### Thanks to Markus Hegland (ANU),
    who wrote the initial version
##1 Generate the data
  cat("generate data \n")
  n <- 800      # length of noise vector
  m <- 100      # length of signal vector
  xsignal <- runif(m)
  sig <- 0.01
  enoise <- rnorm(m)*sig
  ysignal <- xsignal**2+enoise
```

```
   maxys <- max(ysignal)
   minys <- min(ysignal)
   x <- c(runif(n), xsignal)
   y <- c(runif(n)*(maxys-minys)+minys, ysignal)
   # random permutation of the data vectors
   iperm <- sample(seq(x))
   x <- x[iperm]
   y <- y[iperm]
   # normalize the data, i.e., scale x & y values to
   # lie between 0 & 1
   xn <- (x - min(x))/(max(x) - min(x))
   yn <- (y - min(y))/(max(y) - min(y))
##1 End

##2 determine number of neighbors within
# a distance  <= h = 1/sqrt(length(xn))
   nx <- length(xn)
   # determine distance matrix
   d <- sqrt( (matrix(xn, nx, nx) - t(matrix(xn, nx, nx)) )**2 +
              (matrix(yn, nx, nx) - t(matrix(yn, nx, nx)) )**2 )
   h <- 1/sqrt(nx)
   nnear <- apply(d <= h, 1, sum)
##2 End

##3 Plot data, with reconstructed signal overlaid.
   cat("produce plots \n")
   # identify the points which have many such neighbors
   ns <- 8
   plot(x,y)
   points(x[nnear > ns], y[nnear > ns], col="red", pch=16)
##3 End
```

(a) Run the code and observe the graph that results.

(b) Work through the code, and write notes on what each line does.
      [The key idea is that points that are part of the signal will, on average, have more near neighbors than points that are noise.]

(c) Split the code into three functions, bracketed respectively between lines that begin ##1, lines that begin ##2, and lines that begin ##3. The first function should take parameters m and n, and return a list xy that holds data that will be used subsequently. The second function should take vectors xn and yn as parameters, and return values of nnear, i.e., for each point, it will give the number of other points that lie within a circle with the point as center and with radius h. The third function will take as parameters x, y, nnear and the constant ns such that points with more than ns near neighbors will be identified as part of the signal. Run the first function, and store the output list of data values in xy.

(d) Run the second and third functions with various different settings of h and ns. Comment on the effect of varying h. Comment on the effect of varying ns.

(e) Which part of the calculation is most computationally intensive? Which makes the heaviest demands on computer memory?

(f) Suggest ways in which the calculation might be made more efficient.

11. Try the following, for a range of values of n between, e.g., 2×10^5 and 10^7. (On systems that are unable to cope with such large numbers of values, adjust the range of values of n as necessary.)

```
n <- 10000; system.time(sd(rnorm(n)))
```

The first output number is the user cpu time, while the third output number is the elapsed time. Plot each of these numbers, separately, against n. Comment on the graphs. Is the elapsed time roughly linear with n? Try the computations both for an otherwise empty workspace, and with large data objects (e.g., with 10^7 or more elements) in the workspace. (On a multi-user system, it may be well to check with the system manager before undertaking this exercise.)

Epilogue – Models

Our first several chapters were introductory and elementary in style, though with more attention to the subtleties of practical application than is common in elementary treatments. The last several chapters treated topics that have traditionally been reserved for advanced courses. We have thought it important to give readers a taste of methods for data for which linear models may be inadequate, and/or that do not have independently and identically distributed errors.

Models that are not strictly correct, or even perhaps badly broken, may nevertheless be useful, giving acceptably accurate predictions. The validity of model assumptions remains an important issue. We need to know the limits of the usefulness of our models. Experience from comparing results from a simplistic model with results from a less simplistic model can be a huge help in developing intuition. The only satisfactory way to determine whether assumptions matter may be to compare results from the simpler model with results from a model that takes better account of the data. Chapters 7–11 are, for those who hope to do a good job of the analyses described in Chapters 2–6, essential background! An understanding of the ideas of multi-level and repeated measures models seems particularly important, since they introduced the idea that the noise components of the models have a structure that frequently requires attention.

Whether or not faulty assumptions matter will depend on the circumstances. At least for simpler models, the independently and identically distributed errors assumption typically makes little difference to estimates of model parameters and to fitted values, but can have a large effect on standard errors. Consider our analysis of the `frogs` data in Chapter 8. Models that allow for the likely spatial correlation are beyond the scope of this text. We therefore limited ourselves to the standard form of multiple logistic regression. Because we did not take account of spatial autocorrelation, the standard errors were more than otherwise unreliable, and the best we could do was to make a tentative distinction between coefficients that seemed clearly statistically significant, and those that were not.

What would we gain from modeling the correlation structure? We would have a description that should generalize better to sites in the vicinity of those that were studied, with more believable indications of the accuracy of such a description. For generalizing in time, e.g., to a subsequent year, the benefits are more doubtful. If data from multiple years were available, then for predictive purposes the modeling of the temporal structure should be the priority.

Statistical models in genomics

Human and other chromosomes carry long sequences of four nucleotides that have the codes A, T, C and G, constituting a four-letter alphabet. By studying the coded messages of this four-letter alphabet, one aim is to identify those segments of DNA that are likely to code for genes. Once identified, such candidate gene sequences can then be investigated by more direct laboratory investigation. To date, the most successful approaches have relied on what are known as hidden Markov models (HMMs), and on extensions of such models. What is important is that the model should reach the same conclusions, i.e., non-gene or gene, as the processes that the cell uses.

Although these models are quite different in character from the models that we have discussed in earlier chapters, there are very similar predictive validation issues. Empirical testing, testing the model to new data, is the only way to decide whether the model is working. How should the test set, the data that will be used for testing, be chosen? A model that has a high predictive power for genes that have been found to date may not do well at finding new genes. The crucial test will come from making predictions that can then be tested out in the laboratory. We want test sets such that models that perform well on the test sets also do well when their predictions are checked out in the laboratory.

We can think of an HMM as a gene-making machine that uses probabilistic mechanisms to determine the sequence of nucleotides. They have some of the characteristics of an unusually complicated poker machine. The probabilities assigned to successive outputs depend in defined ways both on hidden states that are internal to the machine and on previous outputs. The aim is to identify states, occupying less than 10% of the human genome, in which the strings are parts of genes. For mathematical details, see Durbin et al. (1998) and Baldi and Brunak (2001). Hidden Markov models had earlier been used in linguistics and investigated for use in financial modeling.

Do these models closely reflect the way that the cell distinguishes between non-coding and coding parts of the gene? This seems unlikely. Are the models useful? Yes, but only if carefully validated.

Other models yet!

Each of Chapters 6 to 11 might be expanded into a book. Our discussions of generalized linear models, of multi-level and repeated measures models, and of time series models, have been especially cursory. There has been no discussion of models for data where the outcome is multivariate. We have not discussed clustering methods. Bayesian approaches, as in the *GLMMGibbs* package, have not featured in our discussion. The combination of the Bayesian framework with heavy use of computer simulation allows a computationally tractable approach in some contexts where the interplay between theoretical development and computer implementation has not yet reached a point that allows the use of more direct methods.

New areas of statistical application often have their own new requirements, which act as a stimulus to the development of new methodology. Bioinformatics – genomics, proteomics and related specialities – are the latest of many areas of application where this has happened. The chapter headings of Ewens and Grant (2001) and Baldi and Brunek (2001) give an indication of the present scope of such applications.

The demand for powerful and flexible tools for the processing of microarray data has been the motivation for the Bioconductor project (see http://biowww.dfci.harvard. edu/ bioconductor/). The first of the R Bioconductor packages appeared in early 2002. Version 1.0 of the Bioconductor bundle, including seven packages, appeared in May 2002. The scope of the packages is of course wider than statistical analysis, extending to annotation, to data management and organization, and to data storage and retrieval. The bundle includes functions that handle web access.

We encourage readers to continue their explorations beyond the limits that we have set ourselves. We have given references that will often be useful starting points. There is much information, and further references, in the R help pages. There are many more R packages to explore, in addition to those that we have mentioned. It is interesting (and daunting) to check through the summary details of packages that are available from the CRAN sites.

References for further reading

Baldi, P. and Brunak, S. 2001. *Bioinformatics. The Machine Learning Approach.* MIT Press.

Durbin, R.S., Eddy, A., Krogh, A. and Mitchison, G. 1998. *Biological Sequence Analysis.* Cambridge University Press.

Ewens, W.J. and Grant, G.R. 2001. *Statistical Methods in Bioinformatics: an Introduction.* Springer-Verlag.

Appendix – S-PLUS Differences

For a relatively complete account of current differences between R and S-PLUS, we refer the reader to Venables and Ripley (2000). Our purpose here is to draw attention to those differences that may be important in the use or simple adaptation of code that appears in this book.

Aside from R, there are two main dialects of the S language – S version 3 and S version 4. Versions of S-PLUS up to and including S-PLUS 3.4 (on Unix), S-PLUS 4.5 and S-PLUS 2000 (on Windows systems) were based on S version 3. Versions of S-PLUS with a version number of 5.0 or greater are based on S version 4. While the differences between versions 3 and 4 of S are important for the writing of libraries and of substantial functions, they are rarely important for the code that appears in this book.

The R language syntax and semantics are closer to version 3 than to version 4 of S. It may in future incorporate changes that will bring it closer to version 4. The differences that we note here affect all versions of S-PLUS.

The Handling of NAs

In both R and S-PLUS, the expression x == 0 will have the value NA whenever x is NA, and similarly for other relational expressions. For example:

```
> x <- c(1, 4, NA, 0, 8)
> x == 0
[1] FALSE FALSE    NA  TRUE FALSE
```

The S-PLUS output has F in place of FALSE, and T in place of TRUE, but is otherwise identical.

There is however an important difference between S-PLUS and R when subscripts have one or more elements that evaluate to NA on both sides of an assignment. Set

```
x <- c(1, 4, NA, 0, 8)
y <- 1:5
z <- 11:15
```

In S-PLUS the following two statements do not have the same effect, while in R they do:

```
y[x == 0] <- z[x == 0]    # Avoid in S-PLUS, unless x
                          # is free of NAs
y[!is.na(x) & x==0] <- z[!is.na(x) & x==0]
```

In S-PLUS elements where the subscript expression evaluates to NA drop out on the left of the assignment, but evaluate to NA on the right of the assignment. Even though y and z were of equal length, the vectors on the two sides no longer match up. In our example there are five elements (including an NA in position 3) to assign, and only four positions to which to assign them. S-PLUS recycles y[x == 0] until all elements of z[x == 0] have been assigned, and prints a warning message. The result is unlikely to be what the user wants.

In R, elements where the subscript expression evaluates to NA drop out on both sides. Thus, in y[x == 0] <- z[x == 0], the third element of y[x == 0] drops out on the left hand side, while the third element of z[x == 0] drops out on the right hand side of the assignment.

The Default Parameterization of Factors

By default S-PLUS uses *helmert* contrasts, whereas R uses *treatment* contrasts. Each different choice of contrasts leads to printed output that has different parameter estimates. For details, refer back to Section 12.4. For a factor that has two levels, the helmert contrasts that S-PLUS has as its default lead to a parameter estimate that is just half that for treatment contrasts.

S-PLUS users are advised to make a habit of changing the default S-PLUS setting, thus:

```
options(contrasts=c("contr.treatment", "contr.poly"))
```

If this statement is put into a function .First(), it will be executed each time that a session is started in the directory where the function appears.

Graphics

The function plot() *and allied functions*

Functions such as plot(), points(), text(), and mtext() allow a broadly similar range of parameter settings in both R and S-PLUS. One difference is that parameters such as col, pch and cex are allowed to be vectors in R; in S-PLUS they are not. Thus in R the command

```
plot(1:5, 1:5, pch=1:5, col=1:5, cex=1:5)
```

generates five different plot symbols, each in a different colour, in five different sizes. In S-PLUS the settings that are used are pch=1, col=1 and cex=1, for all five points. There is a warning message.

S-PLUS does not allow the plotting of expressions. Alan Zavlavsky's *postscriptfonts* package that is available from the *statlib* library (see http://lib.stat.cmu.edu) is a partial substitute. In S-PLUS under windows, graphs can be edited to include mathematical text and annotations. For details, see the User's Guide.

S-PLUS *trellis graphics;* R *lattice graphics*

The R *lattice* graphics package is an implementation, differing in important respects from that in the S-PLUS trellis graphics package, of trellis graphics. Whereas the *trellis* package is complete in itself, the *lattice* package sits on top of R's *grid* package. Thus while the functions in the *lattice* package itself have the same names and a similar syntax as functions in the trellis library, the underlying implementations are different.

This affects the writing of panel functions. In *trellis*, panel functions are written using functions such as `points()`, `lines()` and `text()`. By contrasts R's lattice graphics package has its own set of primitives. Thus `grid.points()` replaces `points()`, and so on.

Packages

In R, and to a lesser extent in S-PLUS, the handling of packages and the deployment of widely useful functions within packages are under continual review. Users should check the documentation for their version of R or S-PLUS.

From time to time we have used the Venables and Ripley *MASS* package. This is included with more recent S-PLUS distributions. As in R, it must be loaded as required.

R's *boot* package is available also for S-PLUS. It must be downloaded from the the web.

In recent versions of S-PLUS the *nlme*, *modreg* and *survival* packages are automatically loaded. The `trellis` package, on which the R `lattice` package is modeled, is also loaded automatically.

The principal components analysis functions in R's *mva* package are similar to S-PLUS's built-in functions with the same names. For discriminant analysis, we used the function `lda()` from the *MASS* package. Just as in R, S-PLUS users can load this as required.

While S-PLUS has built-in functions for time series analysis that are roughly comparable to those in the *ts* package, there are implementation differences.

Packages that are available for R, and that are either built into S-PLUS or available for S-PLUS, include *nlme* (note especially `gls()` and `gnls()`), *pear* (periodic time series), and *fracdiff* (long memory series). There is some overlap between the R *tseries* time series package that has a focus on financial modeling and the S-PLUS *GARCH* package that must be purchased seperately. The R *strucchange* package, for testing for temporal or other sequential structural change in linear regression, is not at the time of writing available for *S-PLUS*.

Data sets

S-PLUS data sets in any of the packages that have been loaded are automatically available, just as is the case for functions that are in those packages. There is no need to copy data sets that may be required into the working directory. S-PLUS has no equivalent to `data()`, because it has no need for such a function.

In R, data sets that are in a package that is on the search path must be loaded individually into the workspace before they can be accessed. One uses the `data()` function for this. Once loaded, they remain in the workspace until the end of the session, unless explicitly removed. If not removed, they will be saved as part of the workspace whenever the workspace is saved.

Differences that relate to Chapter 12

The function `strsplit()`

The `strsplit()` function (Subsection 12.2.1) is not available in S-PLUS. The `unpaste()` function, which however expects a character string for its `sep` parameter where `strsplit()` expects a regular expression for its `split` parameter, is a close equivalent.

Import of data from other systems: Subsection 12.3.1

Under Microsoft Windows, S-PLUS users should use the **Import** menu item, under the **File** menu. Users on other systems can use the function `import.file()`.

The function `date()` *and related functions: Subsection 12.2.3*

See under `help(dates)` for the equivalent S-PLUS functions.

Reshaping data frames: Subsection 12.6.2

S-PLUS does not have the R function `reshape()`. The preferred way to do such operations in Microsoft Windows versions of S-PLUS may be to use the menu items **Stack** and **Unstack** under **Data | Restructure**.

Databases and environments: Section 12.9

The S-PLUS function `detach()` expects as its argument either a number or a text string, e.g. `detach("mtcars")`, which R also allows. The S-PLUS usage contrasts with the usage `attach(mtcars)`, which is common to S-PLUS and R.

   We now make brief comments on issues that arise when functions are called from within other functions. Objects that may potentially be large are typically not passed as parameters – this will lead to the creation of a new copy of the object. Instead steps are taken, or should be taken, to ensure that the initial copy of the object is accessible within any function when one may require access to it.

   Technically, the rules that govern such accessibility have the name *scoping* rules. S-PLUS has different rules from R. However, for elementary use of either language, implementation issues are likely to be more important than the different scoping rules. In general R modeling functions are written so that the behavior is the same whether functions are called from the command line or from within other functions.

   As an example, consider

```
regfun <- function(){
  logdist <- log(hills$dist)
  logtime <- log(hills$time)
  hills.loglm <- lm(logtime ~ logdist)
  hattime <- predict(hills.loglm, se=T)
  invisible(hattime)
```

```
      }
      library(MASS)       # In R, add data(hills)
      hatvals <- regfun()
```

In versions of S-PLUS in which we have tried it, the call to regfun() fails. One of the functions that is invoked as a result of the call to predict() cannot find logtime and logdist. The statements in regfun() do however work perfectly well if typed in from the command line. One solution is to modify regfun() to include, prior to the call to predict(), the assignments assign("logdist", logdist, frame=1) and assign("logtime", logtime, frame=1). Objects that are in frame 1 are available for calculations that result from the function call. A solution that works in this instance, with all more recent versions of S-PLUS, is to modify the function thus:

```
      . . .
      loghills <- data.frame(logtime=log(time), logdist=log(dist))
      hills.loglm <- lm(logtime ~ logdist, data=loghills)
      . . .
```

In some uses of modeling functions, this will not work. One then adds, after creating loghills,

```
              assign("loghills", loghills, frame=1)
```

In S-PLUS there is automatic access to objects in the workspace, and in the environment from which the function is called. In R, objects that are visible in the environment in which a function has been defined are automatically available in any environment, including the environments of other functions, from which the function may be called. In R, type in demo(scoping) to get an example whose result is different in R from that in S-PLUS. See also the subsection on lexical scoping in the R FAQ (Frequently Asked Questions) that can be accessed from the CRAN web sites, or from the **Help** menu item in most R installations.

Colors: Section 12.1

Colors are handled differently in S-PLUS from R. There are no equivalents of R's "mix-your-own colors" rgb() and hsv() functions. In Microsoft Windows versions of S-PLUS, graphical color schemes are conveniently changed from the menu. Go to **Options | Color Scheme**. See help(graphsheet) for details on setting up a graphics window that uses a color scheme that is different from the default.

References

References for data sets and examples are in a separate list that follows this list.

Andersen, B. 1990. *Methodological Errors in Medical Research: an Incomplete Catalogue*. Blackwell Scientific.

Atkinson, A.C. 1986. Comment: aspects of diagnostic regression analysis. *Statistical Science* 1: 397–402.

Atkinson, A.C. 1988. Transformations unmasked. *Technometrics* 30: 311–318.

Baldi, P. and Brunak, S. 2001. *Bioinformatics. The Machine Learning Approach*. MIT Press.

Bates, D.M. and Watts, D.G. 1988. *Nonlinear Regression Analysis and Its Applications*. Wiley.

Bickel, P.J., Hammel, E.A. and O'Connell, J.W. 1975. Sex bias in graduate admissions: data from Berkeley. *Science* 187: 398–403.

Box, G.E.P. and Cox, D.R. 1964. An analysis of transformations (with discussion). *Journal of the Royal Statistical Society* B 26: 211–252.

Breiman, L. 2001. Statistical modeling: the two cultures. *Statistical Science* 16: 199–215.

Brockwell, P. and Davis, R.A. 2002. *Time Series: Theory and Methods*, 2nd edn, Springer-Verlag.

Chalmers, I. and Altman, D.G. 1995. *Systematic Reviews*. BMJ Publishing Group, London.

Chambers, J.M. 1998. *Programming with Data: a Guide to the* S *Language*. Springer-Verlag.

Chambers, J.M. and Hastie, T.J. 1992. *Statistical Models in* S. Wadsworth and Brooks-Cole Advanced Books and Software.

Chanter, D.O. 1981. The use and misuse of regression methods in crop modelling. In *Mathematics and Plant Physiology*, ed. D.A. Rose and D.A. Charles-Edwards. Academic Press.

Chatfield, C. 2002. Confessions of a statistician. *The Statistician* 51: 1–20.

Chatfield, C. 2003. *The Analysis of Time Series: an Introduction*, 6th edn. Chapman and Hall.

Chatfield, C. 2003. *Problem Solving. A Statistician's Guide*, 2nd edn. Chapman and Hall/CRC.

Clarke, D. 1968. *Analytical Archaeology*. Methuen.

Cleveland, W.S. 1981. LOWESS: a program for smoothing scatterplots by robust locally weighted regression. *The American Statistician*, 35: 54.

Cleveland, W.S. 1993. *Visualizing Data*. Hobart Press.

Cleveland, W.S. 1994. *The Elements of Graphing Data*, revised edn. Hobart Press.

Cochran, W.G. and Cox, G.M. 1957. *Experimental Designs*, 2nd edn. Wiley.

Collett, D. 2003. *Modelling Survival Data in Medical Research*, 2nd edn. Chapman and Hall.

Cook, R.D. and Weisberg, S. 1999. *Applied Regression Including Computing and Graphics*. Wiley.

Cox, D.R. 1958. *Planning of Experiments*. Wiley.

Cox, D.R. and Reid, N. 2000. *Theory of the Design of Experiments*. Chapman and Hall.

Cox, D.R. and Wermuth, N. 1996. *Multivariate Dependencies: Models, Analysis and Interpretation*. Chapman and Hall.

Dalgaard, P. 2002. *Introductory Statistics with* R. Springer-Verlag.

Davison, A.C. and Hinkley, D.V. 1997. *Bootstrap Methods and Their Application*. Cambridge University Press.

Dehejia, R.H. and Wahba, S. 1999. Causal effects in non-experimental studies: re-evaluating the evaluation of training programs. *Journal of the American Statistical Association* 94: 1053–1062.

Diggle, P. 1990. *Time Series: a Biostatistical Introduction*. Clarendon Press.

Diggle, P.J., Heagerty, P.J., Liang, K.-Y. and Zeger, S.L. 2002. *Analysis of Longitudinal Data*, 2nd edn. Clarendon Press.

Dobson, A.J. 2001. *An Introduction to Generalized Linear Models*, 2nd edn. Chapman and Hall/CRC.

Dudoit, S., Yang, Y.H., Callow, M.J. and Speed, T.P. 2002. Statistical methods for identifying differentially expressed genes in replicated cDNA microarray experiments. *Statistica Sinica* 12: 111–140.

Durbin, R.S., Eddy, A., Krogh, A. and Mitchison, G. 1998. *Biological Sequence Analysis*. Cambridge University Press.

Edwards, D. 2000. *Introduction to Graphical Modelling*, 2nd edn. Springer-Verlag.

Efron, B. and Tibshirani, R. 1993. *An Introduction to the Bootstrap*. Chapman and Hall.

Eubank, R.L. 1999. *Nonparametric Regression and Spline Smoothing*, 2nd edn. Marcel Dekker.

Ewens, W.J. and Grant, G.R. 2001. *Statistical Methods in Bioinformatics: an Introduction*. Springer-Verlag.

Fan, J. and Gijbels, I. 1996. *Local Polynomial Modelling and Its Applications*. Chapman and Hall.

Finney, D.J. 1978. *Statistical Methods in Bioassay*, 3rd edn. Macmillan.

Fisher, R.A. 1935 (7th edn. 1960). *The Design of Experiments*. Oliver and Boyd.

Fox, J. 2002. *An* R *and* S-PLUS *Companion to Applied Regression*. Sage Books.

Gardner, M.J., Altman, D.G., Jones, D.R. and Machin, D. 1983. Is the statistical assessment of papers submitted to the *British Medical Journal* effective? *British Medical Journal* 286: 1485–1488.

Gaver, D.P., Draper, D.P., Goel, K.P., Greenhouse, J.B., Hedges, L.V., Morris, C.N. and Waternaux, C. 1992. *Combining Information: Statistical Issues and Opportunities for Research*. National Research Council, National Academy Press.

Gelman, A.B., Carlin, J.S., Stern, H.S. and Rubin, D.B. 1995. *Bayesian Data Analysis*. Chapman and Hall/CRC.

Gigerenzer, G. 1998. We need statistical thinking, not statistical rituals. *Behavioural and Brain Sciences* 21: 199–200.

Gigerenzer, G. 2002. *Reckoning with Risk: Learning to Live with Uncertainty*. Penguin Books.

Gigerenzer, G., Swijtink, Z., Porter, T., Daston, L., Beatty, J. and Krüger, L. 1989. *The Empire of Chance*. Cambridge University Press.

Goldstein, H. 1995. *Multilevel Statistical Models*. Arnold. (Available from http://www.arnoldpublishers.com/support/goldstein.htm)

Hall, P. 2001. Biometrika centenary: nonparametrics. *Biometrika* 88: 143–165.

Harlow, L.L., Mulaik, S.A., and Steiger, J.H. (eds.) 1997. *What If There Were No Significance Tests?* Lawrence Erlbaum Associates.

Harrell, F.E. 2001. *Regression Modelling Strategies, with Applications to Linear Models, Logistic Regression and Survival Analysis*. Springer-Verlag.

Hastie, T.J. and Tibshirani, R.J. 1990. *Generalized Additive Models*. Chapman and Hall.

Hastie, T., Tibshirani, R. and Friedman, J. 2001. *The Elements of Statistical Learning. Data Mining, Inference and Prediction*. Springer-Verlag.

Ihaka, R. and Gentleman, R. 1996. R: a language for data analysis and graphics. *Journal of Computational and Graphical Statistics* 5: 299–314.

Johnson, D.H. 1995. Statistical sirens: the allure of nonparametrics. *Ecology* 76: 1998–2000.

Krantz, D.H. 1999. The null hypothesis testing controversy in psychology. *Journal of the American Statistical Association* 44: 1372–1381.

Krzanowski, W.J. 2000. *Principles of Multivariate Analysis. A User's Perspective*, revised edn. Clarendon Press.

Lalonde, R. 1986. Evaluating the economic evaluations of training programs. *American Economic Review* 76: 604–620.

Lim, T.-S. and Loh, W.-Y. 2000. A comparison of prediction accuracy, complexity, and training time of thirty-three old and new classification algorithms. *Machine Learning* 40: 203–228.

Maindonald, J.H. 1984. *Statistical Computation*. Wiley.

Maindonald J.H. 1992. Statistical design, analysis and presentation issues. *New Zealand Journal of Agricultural Research* 35: 121–141.

Maindonald, J.H. 2001. Using R for data analysis and graphics. Available as a pdf file at http://wwwmaths.anu.edu.au/~johnm/r/usingR.pdf

Maindonald, J.H. and Cox, N.R. 1984. Use of statistical evidence in some recent issues of PSIR agricultural journals. *New Zealand Journal of Agricultural Research* 27: 597–610.

Maindonald, J.H., Waddell, B.C. and Petry, R.J. 2001. Apple cultivar effects on codling moth (Lepidoptera: Tortricidae) egg mortality following fumigation with methyl bromide. *Postharvest Biology and Technology* 22: 99–110.

McCullagh, P. and Nelder, J.A. 1989. *Generalized Linear Models*, 2nd edn. Chapman and Hall.

Miller R.G. 1986. *Beyond ANOVA, Basics of Applied Statistics*. Wiley.

Myers, R.H. 1990. *Classical and Modern Regression with Applications*, 2nd edn. Brooks Cole.

Nelder, J.A. 1999. From statistics to statistical science. *Journal of the Royal Statistical Society*, Series D, 48: 257–267.

Nicholls, N. 2000. The insignificance of significance testing. *Bulletin of the American Meteorological Society* 81: 981–986.

Payne, R.W., Lane, P.W., Digby, P.G.N., Harding, S.A., Leech, P.K., Morgan, G.W., Todd, A.D., Thompson, R., Tunnicliffe Wilson, G., Welham, S.J. and White, R.P. 1997. *Genstat 5 Release 3 Reference Manual*. Oxford University Press.

Pinheiro, J.C. and Bates, D.M. 2000. *Mixed Effects Models in S and S-PLUS*. Springer-Verlag.

R Core Development Team. *An Introduction to* R. Available from CRAN sites, updated regularly (for a list of CRAN sites, go to http://cran.r-project.org).

R Core Development Team. R *Language Definition*. Available from CRAN sites, updated regularly.

Ripley, B.D. 1996. *Pattern Recognition and Neural Networks*. Cambridge University Press.

Rosenbaum, P.R. 1999. Choice as an alternative to control in observational studies. *Statistical Science* 14: 259–278, with following discussion, pp. 279–304.

Rosenbaum, P. and Rubin, D. 1983. The central role of the propensity score in observational studies for causal effects. *Biometrika* 70: 41–55.

Schmidt-Nielsen, K. 1984. *Scaling. Why Is Animal Size So Important?* Cambridge University Press.

Sharp, S.J., Thompson, S.G., and Altman, D.G. 1996. The relation between treatment benefit and underlying risk in meta-analysis. *British Medical Journal* 313: 735–738.

Snijders, T.A.B. and Bosker, R.J. 1999. *Multilevel Analysis. An Introduction to Basic and Advanced Multilevel Modelling*. Sage.

Steel, R.G.D., Torrie, J.H. and Dickie, D.A. 1993. *Principles and Procedures of Statistics. A Biometrical Approach*, 3rd edn. McGraw-Hill, New York.

Streiner, D.L. and Norman, G.R. 1995. *Health Measurement Scales. A Practical Guide to their Development and Use*, 2nd edn. Oxford University Press.

Talbot, M. 1984. Yield variability of crop varieties in the U.K. *Journal of the Agricultural Society of Cambridge* 102: 315–321.

Therneau, T.M. and Atkinson, E.J. 1997. *An Introduction to Recursive Partitioning Using the RPART Routines.* Technical Report Series No. 61, Department of Health Science Research, Mayo Clinic, Rochester, MN, 2000. Available from http://www.mayo.edu/hsr/techrpt.html

Therneau, T.M. and Grambsch P.M. 2001. *Modeling Survival Data: Extending the Cox Model.* Springer-Verlag.

Tufte, E.R. 1997. *Visual Explanations.* Graphics Press.

Tukey, J.W. 1991. The philosophy of multiple comparisons. *Statistical Science* 6: 100–116.

Venables, W.N. 1998. Exegeses on linear models. Proceedings of the 1998 international S-PLUS user conference. Available from www.stats.ox.ac.uk/pub/MASS3/Compl.html

Venables, W.N. and Ripley, B.D. 2000. S *Programming.* Springer-Verlag.

Venables, W.N. and Ripley, B.D. 2002. *Modern Applied Statistics with* S, 4th edn. Springer-Verlag. See also 'R' Complements to Modern Applied Statistics with S, available from http://www.stats.ox.ac.uk/pub/MASS4/

Wainer, H. 1997. *Visual Revelations.* Springer-Verlag.

Weisberg, S. 1985. *Applied Linear Regression*, 2nd edn. Wiley.

Welch, B.L. 1949. Further note on Mrs. Aspin's tables and on certain approximations to the tabled function. *Biometrika* 36: 293–296.

Wilkinson, L. and Task Force on Statistical Inference 1999. Statistical methods in psychology journals: guidelines and explanation. *American Psychologist* 54: 594–604.

Williams, E.R., Matheson, A.C. and Harwood, C.E. 2002. *Experimental Design and Analysis for Use in Tree Improvement*, revised edn. CSIRO Information Services.

Williams, G.P. 1983. Improper use of regression equations in the earth sciences. *Geology* 11: 195–197.

References for data sets and examples

Ash, J. and Helman, C. 1990. Floristics and vegetation biomass of a forest catchment, Kioloa, south coastal N.S.W. *Cunninghamia* 2(2): 167–182 (published by Royal Botanic Gardens, Sydney).

Boot, H.M. 1995. How skilled were the Lancashire cotton factory workers in 1833? *Economic History Review* 48: 283–303.

Burns, N.R., Nettlebeck, T., White, M. and Willson, J., 1999. Effects of car window tinting on visual performance: a comparison of elderly and young drivers. *Ergonomics* 42: 428–443.

Callow, M.J., Dudoit, S., Gong, E.L. and Rubin, E.M. 2000. Microarray expression profiling identifies genes with altered expression in HDL-deficient mice. *Genome Research* 10: 2022–2029.

Christie, M. 2000. *The Ozone Layer: a Philosophy of Science Perspective.* Cambridge University Press.

Darwin, Charles. 1877. *The Effects of Cross and Self Fertilisation in the Vegetable Kingdom.* Appleton and Company.

Dehejia, R.H. and Wahba, S. 1999. Causal effects in non-experimental studies: re-evaluating the evaluation of training programs. *Journal of the American Statistical Association* 94: 1053–1062.

Ezzet, F. and Whitehead, J. 1991. A random effects model for ordinal responses from a crossover trial. *Statistics in Medicine* 10: 901–907.

Gihr, M. and Pilleri, G. Anatomy and biometry of Stenella and Delphinus. 1969. In *Investigations on Cetacea*, ed. G. Pilleri. Hirnanatomisches Institute der Universität Bern.

Hales, S., de Wet, N., Maindonald, J. and Woodward, A. 2002. Potential effect of population and climate change global distribution of dengue fever: an emprical model. *The Lancet.* Published online August 6, 2002. http://image.thelancet.com/extras/Olart11175web.pdf

Harker, F.R. and Maindonald J.H. 1994. Ripening of nectarine fruit. *Plant Physiology* 106: 165–171.

Hobson, J.A. 1988. *The Dreaming Brain.* Basic Books.

Hunter, D. 2000. The conservation and demography of the southern corroboree frog (*Pseudophryne corroboree*). M.Sc. thesis, University of Canberra.

King, D.A. 1998. Relationship between crown architecture and branch orientation in rain forest trees. *Annals of Botany* 82: 1–7.

King, D.A. and Maindonald, J.H. 1999. Tree architecture in relation to leaf dimensions and tree stature in temperate and tropical rain forests. *Journal of Ecology* 87: 1012–1024.

Lalonde, R. 1986. Evaluating the economic evaluations of training programs. *American Economic Review* 76: 604–620.

Latter, O.H. 1902. The egg of *Cuculus canorus*. An inquiry into the dimensions of the cuckoo's egg and the relation of the variations to the size of the eggs of the foster-parent, with notes on coloration, &c. *Biometrika* 1: 164–176.

Linacre, E. 1992. *Climate Data and Resources. A Reference and Guide*. Routledge.

Linacre, E. and Geerts, B. 1997. *Climates and Weather Explained*. Routledge.

Lindenmayer, D.B., Viggers, K.L., Cunningham, R.B., and Donnelly, C.F. 1995. Morphological variation among columns of the mountain brushtail possum, *Trichosurus caninus* Ogilby (Phalangeridae: Marsupiala). *Australian Journal of Zoology* 43: 449–458.

Matthews, D.E. and Farewell, V.T. 1996. *Using and Understanding Medical Statistics*. Karger.

McLellan, E.A, Medline, A. and Bird, R.P. 1991. Dose response and proliferative characteristics of aberrant crypt foci: putative preneoplastic lesions in rat colon. *Carcinogenesis* 12: 2093–2098.

McLeod, C.C. 1982. Effect of rates of seeding on barley grown for grain. *New Zealand Journal of Agriculture* 10: 133–136.

Nicholls, N., Lavery, B., Frederiksen, C. and Drosdowsky, W. 1996. Recent apparent changes in relationships between the El Niño – southern oscillation and Australian rainfall and temperature. *Geophysical Research Letters* 23: 3357–3360.

Perrine, F.M., Prayitno, J., Weinman, J.J., Dazzo, F.B. and Rolfe, B. 2001. Rhizobium plasmids are involved in the inhibition or stimulation of rice growth and development. *Australian Journal of Plant Physiology* 28: 923–927.

Roberts, H.V. 1974. *Conversational Statistics*. Hewlett-Packard University Business Series. The Scientific Press.

Shanklin, J. 2001. Ozone at Halley, Rothera and Vernadsky/Faraday. Available from http://www.nerc-bas.ac.uk/public/icd/jds/ozone/

Snelgar, W.P., Manson, P.J. and Martin, P.J. 1992. Influence of time of shading on flowering and yield of kiwifruit vines. *Journal of Horticultural Science* 67: 481–487.

Stewardson, C.L., Hemsley, S., Meyer, M.A., Canfield, P.J. and Maindonald, J.H. 1999. Gross and microscopic visceral anatomy of the male Cape fur seal, *Arctocephalus pusillus pusillus* (Pinnipedia: Otariidae), with reference to organ size and growth. *Journal of Anatomy (Cambridge)* 195: 235–255. (WWF project ZA-348).

Stewart, K.M., Van Toor, R.F. and Crosbie, S.F. 1988. Control of grass grub (Coleoptera: Scarabaeidae) with rollers of different design. *New Zealand Journal of Experimental Agriculture* 16: 141–150.

Stiell, I.G. Wells, G.A., Vandemheen, K., Clement, C., Lesiuk, H., Laupacis, A., McKnight, R.D., Verbeek, R., Brison, R., Cass, D., Eisenhauer, M.A., Greenberg, G.H. and Worthington, J., for the CCC Study Group 2001. The Canadian CT Head Rule for patients with minor head injury. *The Lancet* 357: 1391–1396.

Tippett, L.H.C. 1931. *The Methods of Statistics*. Williams and Norgate.

Thall, P.F. and Vail, S.C. 1990. Some covariance models for longitudinal count data. *Biometrics* 46: 657–671.

Wainright P., Pelkman, C. and Wahlsten, D. 1989. The quantitative relationship between nutritional effects on preweaning growth and behavioral development in mice. *Developmental Psychobiology* 22: 183–193.

Web sites: sources of software, data and accompanying documentation

Comprehensive R Archive Network: http://cran.r-project.org/

Data set `bomsoi`: http://www.bom.gov.au/climate/change/rain02.txt and
   http://www.bom.gov.au/climate/current/soihtm1.shtml

Data set `nsw74psid1`: http://www.columbia.edu/~rd247/nswdata.html

Data set `ozone`: http://www.antarctica.ac.uk/met/jds/ozone/

GGobi: http://www.ggobi.org

Machine Learning Repository: http://www.ics.uci.edu/~mlearn/MLRepository.html

MONICA project: http://www.ktl.fi/monicaindex.html

S-PLUS: http://www.insightful.com/products/default.asp

XGobi: http://www.research.att.com/areas/stat/xgobi/

Acknowledgements for use of data

We thank the following for permission to reproduce graphs or tables that have appeared in published material: *Journal of Ecology*, for permission to reproduce Figure 11.3B, which is a redrawn version of the fourth panel in Figure 3 in King and Maindonald (1999); *Plant Physiology*, in relation to the right panel of Figure 2.6, which is a redrawn version of Figure 3 in Harker and Maindonald (1994), copyrighted by the American Society of Plant Biologists; SIR Publishing (The Royal Society of New Zealand), for permission to reproduce data in Subsection 5.1.2 (from Stewart et al. 1988), Table 7.4 (from Stewart et al. 1988), Table 9.1 (from Maindonald 1992), and Figure 9.3 which is similar to Figure 1 in Maindonald (1992); CSIRO Publishing, for permission to reproduce (in Section 4.5) data that appear in Table 4 in Perrine et al. (2001); *Australian Journal of Zoology*, for permission to reproduce a graph that is similar to a part of Figure 2 in Lindenmayer et al. (1995).

We acknowledge the help of the following individuals and organizations in making available their data.

Darren Kriticos (CSIRO Division of Entomology): data frame `houseprices`. J.D. Shanklin (British Antarctic Survey): `ozone`. W.S. Snelgar (Horticulture and Food Research Institute of NZ Ltd): `kiwishade`. Francine Adams and supervisors Rosemary Martin and Murali Nayadu (all from ANU): `science`. Jasmyn Lynch (ANU), for the rare and endangered plant species data in Subsection 4.4.1. D.B. Lindenmayer and coworkers (ANU): `possum`. D.A. King (formerly ANU): `leafshape` and `leafshape17`. R.F. Harker (Horticulture and Food Research Institute of NZ Ltd): `fruitohms`. E. Linacre (ANU): `dewpoint`. N.R. Burns (University of Adelaide): `tinting`. M. Boot (ANU): `wages1833`. J. Erickson (University of Chicago) and A.H. Welsh (University of Southampton): `anesthetic`. D. Hunter (University of Canberra): `frogs`. Sharyn Wragg (formerly ANU): `moths`. Claudia Haarman (ANU), for the flow inhibition data of Exercise 1 in Section 8.9. Melissa Manning (ANU): `socsupport`. J. Ash (ANU): `rainforest`. Katharina Siebke and Susan von Cammerer (ANU): `leaftemp`. Ranjana Bird (University of Manitoba), for the aberrant crypt foci data in Section 3.7; Australian Bureau of Meteorology: `bomsoi`. Note the abbreviation ANU for Australian National University.

Index of R Symbols and Functions

Index of Terms

Index of Names